Universitext

Universitext

Editors: F.W. Gehring, P.R. Halmos, C.C. Moore

Booss/Bleecker: Topology and Analysis
Chern: Complex Manifolds Without Potential Theory
Chorin/Marsden: A Mathematical Introduction to Fluid Mechanics
Cohn: A Classical Invitation to Algebraic Numbers and Class Fields
Curtis: Matrix Groups, 2nd ed.
van Dalen: Logic and Structure
Devlin: Fundamentals of Contemporary Set Theory
Edwards: A Formal Background to Mathematics I a/b
Edwards: A Formal Background to Higher Mathematics II a/b
Endler: Valuation Theory
Frauenthal: Mathematical Modeling in Epidemiology
Gardiner: A First Course in Group Theory
Godbillon: Dynamical Systems on Surfaces
Greub: Multilinear Algebra
Hermes: Introduction to Mathematical Logic
Kelly/Matthews: The Non-Euclidian, The Hyperbolic Plane
Kostrikin: Introduction to Algebra
Luecking/Rubel: Complex Analysis: A Functional Analysis Approach
Lu: Singularity Theory and an Introduction to Catastrophe Theory
Marcus: Number Fields
Meyer: Essential Mathematics for Applied Fields
Moise: Introductory Problem Course in Analysis and Topology
Øksendal: Stochastic Differential Equations
Rees: Notes on Geometry
Reisel: Elementary Theory of Metric Spaces
Rey: Introduction to Robust and Quasi-Robust Statistical Methods
Rickart: Natural Function Algebras
Schreiber: Differential Forms
Smoryński: Self-Reference and Modal Logic
Stanisić: The Mathematical Theory of Turbulence
Stroock: An Introduction to the Theory of Large Deviations
Tolle: Optimization Methods

C. Smoryński

Self-Reference and Modal Logic

Springer-Verlag
New York Berlin Heidelberg Tokyo

C. Smoryński
Department of Mathematics
and Computer Science
San Jose State University
San Jose, CA 95192
U.S.A.

AMS Subject Classification: 03-02, 03B45

Library of Congress Cataloging-in-Publication Data
Smoryński, C.
Self-reference and modal logic.
(Universitext)
Bibliography: p.
1. Modality (Logic) I. Title.
QA9.46.S6 1985 511.3 85-17219

Printed and bound by Halliday Lithograph, West Hanover, Massachusetts.
Printed in the United States of America.

9 8 7 6 5 4 3 2 1

ISBN 0-387-96209-3 Springer-Verlag New York Berlin Heidelberg Tokyo
ISBN 3-540-96209-3 Springer-Verlag Berlin Heidelberg New York Tokyo

TO

GOOD OLD WATSON ...

THE ONE FIXED POINT IN A CHANGING AGE

Foreword

In the fall of 1980 I had the great pleasure of visiting the University of Warsaw at the invitation of Cecylia Rauszer and lecturing on the modal logic of provability. When one lectures in English in a country in which English is not the native language, one must, of course, try to remember not to get nervous and speak too rapidly, as one's hosts can often be too polite to interrupt. In the end, the best strategy is to write up notes to be passed around. These are not those.

Between the time of submission of my original lecture notes for the series Lecture Notes in Mathematics and Roberto Minio's suggestion of Universitext, I learned a little more about the subject from my friend Albert Visser. I then announced to Springer that I would rewrite the notes completely, but that it wouldn't take long-- after all, the original only took 3 months. Now, several years later, the job is finished; the exposition may be a little more polished in spots, and still lacking in others. A greater self-consciousness is, of course, one source of delay, as is the ease with which a larger task is set aside for a smaller. But, a certain amount of research is also to blame: Chapters 4 and 5 contain material that simply did not exist in 1980, some of which material (particularly my work in Chapter 4) actually grew out of the task of exposition. More should have resulted in this manner, but at some point one must decide to resist the temptation to do research and report on what one knows. The task of further fleshing out the material is thus left to the potential researchers among the readers. I have not cited open problems, but they are there and the attentive reader can not help but find them.

One must not only decide against pursuing new material to include, but also decide which known material not to include. Here, my choice is easily explained: I am interested in exploring self-reference through modal techniques, for the purposes both of understanding self-reference and of applying it. Those directions not completely parallel to these are simply not followed or even hinted at. My apologies extend to those whom I have slighted in this manner. (While I am at it, I should apologise to any whom I have slighted by not giving credit where credit is due. Originally, I intended to include a list of credits, but (i) such seems out of place in a book that purports to be an universi(ty)text, and (ii) I got so sick of typing.)

As for the use of this book as a text: As there generally are no courses on this material, I envision it for occasional graduate seminars and self-study. Chapters 1-3, section 1 of Chapter 4, Chapter 6, and the first two sections of Chapter 7 form the core material; the rest is aimed at the potential specialist and can be sacrificed if time demands. What about Chapter 0? Well, I wanted this book to be as self-contained as possible and have included a bit too much in this Chapter. To get into the modal theory most quickly, I recommend skipping the latter material in Chapter 0 (say, sections 4, 5, and 6) until one finds it necessary to refer back to it when one is in Chapter 3.

Having said a few words about how this book came about and how to read it, I should explain why it came about and why one should read it. For one thing, this series pays royalties and if enough people read it I will recover my typewriter rental expenses... Less personally, I hope this monograph will set the record straight: Gödel's Theorem is not artificial; the use of self-reference has not been obsoleted by recursion theory or combinatorics; and self-reference is not that mysterious. This monograph reports on the beginnings of a coherent theory of self-reference and incompleteness phenomena, a theory I hope will be furthered by some of its readers.

There is already a number of people working on various aspects of self-reference and modal logic. There are not too many to name most of them here, but there are too many for me to name and expect any but those who should be named to read the list, and most of them are mentioned in the text anyway. Hence I need not name them here.

Another list of names-- the acknowledgements-- is more customary and it is not really safe to forego this pleasure: Mathematically, I must thank those who developed material for me to write about. This list I've already promised not to give. I also thank Dick de Jongh for first introducing me to the subject, Albert Visser for the insights he has given me on the subject, Cecylia Rauszer for providing me with the excuse for writing this monograph, Tim Carlson for permission to include some of his results which he has *finally* submitted for publication, Mel Fitting for the inspiration for one of the Exercises, and the editorial staff at Springer-Verlag.

San José
June 1985

Contents

Chapter 0
Introduction

It is Sunday, the 7th of September 1930. The place is Königsberg and the occasion is a small conference on the foundations of mathematics. Arend Heyting, the foremost disciple of L.E.J. Brouwer, has spoken on intuitionism; Rudolf Carnap of the Vienna Circle has expounded on logicism; Johann (formerly Janos and in a few years to be Johnny) von Neumann has explained Hilbert's proof theory-- the so-called formalism; and Hans Hahn has just propounded his own empiricist views of mathematics. The floor is open for general discussion, in the midst of which Heyting announces his satisfaction with the meeting. For him, the relationship between formalism and intuitionism has been clarified: There need be no war between the intuitionist and the formalist. Once the formalist has successfully completed Hilbert's programme and shown "finitely" that the "idealised" mathematics objected to by Brouwer proves no new "meaningful" statements, even the intuitionist will fondly embrace the infinite. To this euphoric revelation, a shy young man cautions, "According to the formalist conception one adjoins to the meaningful statements of mathematics transfinite (pseudo-)statements which in themselves have no meaning but only serve to make the system a well-rounded one just as in geometry one achieves a well-rounded system by the introduction of points at infinity. This conception presupposes that when one adds to the system S of meaningful statements the system T of transfinite statements and axioms and then proves a statement from S via a detour over statements from T then this statement is also correct (inhaltlich richtig) in its content so that through the addition of the transfinite axioms no contentually false statements become provable. One commonly replaces this requirement with that of consistency. I would like to indicate that these two requirements cannot by any means be immediately regarded as equivalent. For, if a meaningful sentence p is provable in a consistent formal system A (say that of classical mathematics), then all that follows from the consistency of A is that *not-p* is not provable within the system A. Nevertheless it remains conceivable that one could recognise *not-p* through some conceptual (intuitionistic) considerations which can*not* be formally represented in A. In this case, despite the consistency of A, a sentence would be provable in A the falsehood of which one could recognise through finite considerations. However, as soon as one construes the concept "meaningful statement" sufficiently narrowly (for example restricted to finite numerical equations) such a thing cannot occur. On the contrary it would be, e.g., entirely possible that one could prove with the transfinite methods of classical mathematics a sentence of the form $\exists x F(x)$ where F is a finite property of natural numbers (e.g. the negation of the Goldbach conjecture has this form) and on the other hand recognise through conceptual considerations that all numbers have the property *not-F*; and what I want to indicate is that this remains possible even if one had verified the consistency of the formal system of classical mathematics. For, one cannot claim with certainty of any formal system that all conceptual considerations are representable in it."

This incisive critique of Hilbert's programme draws only the comment by von Neumann that, "It is not settled that all modes of inference that are intuitionistically permitted can be repeated formally." The young man makes his point more firmly: "One can (under the assumption of the consistency of classical mathematics) even give examples of statements (and even such of the sort of Goldbach's or Fermat's), which are conceptually correct but unprovable in the formal system of classical mathematics. Therefore, if one adjoins the negation of such a statement to the axioms of classical mathematics, then one obtains a consistent system in which a conceptually false sentence is provable." Kurt Reidemeister closes the discussion with a few unrelated remarks.

Kurt Gödel has just made the first public announcement of his celebrated First Incompleteness Theorem and no-one present appears to have understood it. However, one cannot keep a good theorem down: Von Neumann quickly understood the importance of Gödel's result and it is reported he spoke several hours with Gödel at the meeting; and later the following month Hahn, who had been Gödel's thesis advisor, presented Gödel's explicit announcement of both the First and Second Incompleteness Theorems to the Vienna Academy of Sciences. Early the following year, Gödel's paper was published and, just over a year after his initial cautious announcement, on Tuesday, 15 September 1931, Gödel presented his result in a meeting of the Deutsche Mathematiker-Vereinigung in Bad Elster.

Gödel's paper, "Über formal unentscheidbare Sätze der Principia Mathematica und verwandter Systeme I", contained a detailed proof of his First Incompleteness Theorem, a few related results, and an announcement of the Second Incompleteness Theorem, a full proof of which was to appear in a sequel. This second paper never materialised-- partly (I am told) due to Gödel's health and partly (I am told) due to the immediate acceptance of his results: Unlike his earlier cautious and nearly inexplicit announcement, the effect of which was less than dramatic, his paper is a paradigm of clarity (and, incidentally, remains one of the most readable expositions of his First Incompleteness Theorem today).

The task of expositing the details of the proof of the Second Incompleteness Theorem thus fell to the textbook writers; Paul Bernays included such a proof in the second volume of Grundlagen der Mathematik, which he penned under the names of Hilbert and Bernays. This volume, published in 1939, remains the only source for a fully detailed proof of the Second Incompleteness Theorem. (To be fair, I should credit Shoenfield and Monk with sufficiently many details in their texts to render the completion of the proof routine.)

1. THE INCOMPLETENESS THEOREMS

Hilbert's Programme can be described thus: There are two systems, nowadays called formal theories, S and T of mathematics. S consists of the finite, meaningful statements and methods of proof and T the transfinite, idealised such statements and methods. The goal is to show that, for any meaningful assertion ϕ, if $T \vdash \phi$ then $S \vdash \phi$. Moreover, this is to be shown in the system S.

Hilbert's Programme was not *a priori* unreasonable. Through the medium of formalisation, the transfinite character of T can be by-passed. In the formalism, one has mere tokens manipulated in concrete fashion-- just the sort of finite activity one wanted S to reason about. The theory S would analyse the formal simulation of T and prove the crucial conservation result.

Now, Hilbert had made the observation that (as Gödel remarked on 7 September 1930), if one construed "meaningful" narrowly enough, the conservation of T over S with respect to meaningful statements would follow from a proof in S of the weaker property that T was consistent. The example Hilbert used to illustrate this point was the Fermat problem: Suppose

$$T \vdash \forall xyz > \overline{0}\ \forall n > \overline{2}\ (x^n + y^n \neq z^n)$$

and that S proves the consistency of T. Then,

$$S \vdash \forall xyz > \overline{0}\ \forall n > \overline{2}\ (x^n + y^n \neq z^n).$$

To see this, reason in S: If for some x, y, $z > 0$ and some $n > 2$, $x^n + y^n = z^n$, then a mere computation of this fact would verify this and T would be able to prove this, i.e.

$$S \vdash x^n + y^n = z^n \rightarrow \text{"}x^n + y^n = z^n \textit{ is provable in } T\text{"}.$$

But, if T proves there are no such x, y, z, n, then T is inconsistent; thus

$$S \vdash x^n + y^n = z^n \rightarrow \text{"}T \textit{ is inconsistent}\text{"}.$$

Since S proves the consistency of T, the contraposition yields

$$S \vdash x^n + y^n \neq z^n.$$

Thus, for some sentences, Hilbert's consistency programme-- successfully completed-- yields his conservation programme. However, as Gödel announced in Königsberg, this doesn't hold for all sentences.

Gödel destroyed Hilbert's Programme with his First Incompleteness Theorem by which he produced a sentence ϕ satisfying a sufficiently narrow criterion of meaningfulness and which, though readily recognised as true-- hence a theorem of the transfinite system T, was unprovable in S. In short, he produced a direct counter-example to Hilbert's desired conservation result. Moreover, adding new axioms to S would not help-- Gödel's method would yield a new sentence ϕ' and an eminently reasonable new system T' for which the conservation result failed. It followed, of course, that S could not prove the consistency of T and Hilbert's derivative consistency programme was killed with its conservative sire. To further defile the corpses, as it were, Gödel, via his Second Incompleteness Theorem, showed that S could not even prove its own consistency.

How did Gödel do all of this? A modern answer would be: applied word processing. Sitting on my desk is a personal computer with a couple of disk drives. When I shove a particular floppy disk into one of the drives, my computer becomes a friendly typewriter. When I type, "The quick blue fox jumped over the lazy dog," and realise I got the colour wrong, I type a special key and get a display,

Repl: ... .

When I type in "blue" and hit the Return key, the screen displays,

With: ... .

I now type in "brown", hit Return, am asked if each occurrence of "blue" is to be replaced by "brown", and, if I say "yes", my text suddenly reads, "The quick brown fox jumped over the lazy dog." What has happened is this: My computer codes all letters as strings of *0*'s and *1*'s-- as numbers written in binary if you will. Words, sentences, and paragraphs are merely longer such strings. My word processor has a replacement, or substitution, function built into it. This function $subst(\cdot,\cdot,\cdot)$ will replace all (say) occurrences of one string in a second string by a third string. I illustrate this thus: $subst(\ldots\_\_\ldots, \_\_, --) = \ldots--\ldots$. Of course, in the computer it is not the actual strings but their *0-1* numerical codes that are handled. Thus, we could write $subst(\ulcorner\ldots\_\_\ldots\urcorner, \ulcorner\_\_\urcorner, \ulcorner--\urcorner) = \ulcorner\ldots--\ldots\urcorner$ to indicate the numerical function-- where the odd semi-parentheses indicate the numerical coding. (The standard notation is a pair of raised corners, but these do not exist on my typewriter.)

What Gödel did is very similar. Instead of handling English text, he treated a formal mathematical language; and, instead of using binary strings, he coded on ordinary numbers. In doing this, he noted an interesting phenomenon: A formula ϕv with the free variable v would have a numerical code. This number n would have a name $\overline{n}$ in the language and this name could be substituted back into ϕv. Thus, ϕ could refer to itself. If one is dealing with a formal theory powerful enough to perform routine computations-- such as the word processor's *subst*-function-- this self-reference can be formalised:

<u>1.1. DIAGONALISATION LEMMA</u>. For any formula ψ with only the variable v free, there is a sentence ϕ such that $\vdash \phi \leftrightarrow \psi(\ulcorner\phi\urcorner)$.

This is easily proven-- modulo a tiny explanation. I am assuming here a theory about numbers (and possibly other entities). For such a theory to be good, it must have names $\overline{n}$ for the numbers n. By the theory's computational capability, in particular its word processing prowess, it has a means of coding these things and a substitution function $sub(\cdot,\cdot)$ that satisfies the following: $sub(\ulcorner\phi v_0\urcorner, \overline{n}) = \ulcorner\phi\overline{n}\urcorner$, i.e. *sub* gives us the code of the result of replacing v_0 by the n-th numeral in ϕv_0 (the string v_0 being assumed fixed in a more comprehensive function *subst*).

Proof of 1.1: Let ψv be given and let θv_0 be $\psi(sub(v_0,v_0))$, $m = \ulcorner\theta v_0\urcorner$, and $\phi = \theta\overline{m}$. Then notice

$$\phi = \theta\overline{m} = \psi(sub(\overline{m},\overline{m})),$$

whence
$$\begin{aligned} \vdash \phi &\leftrightarrow \psi(sub(\overline{m},\overline{m})) \\ &\leftrightarrow \psi(sub(\ulcorner\theta v_0\urcorner,\overline{m})) \\ &\leftrightarrow \psi(\ulcorner\theta\overline{m}\urcorner) \\ &\leftrightarrow \psi(\ulcorner\phi\urcorner). \end{aligned}$$

QED

Ever since Epimenides, self-reference has been suspect. But in a formal context we have nothing to fear; the vexing "I am lying" simply cannot be expressed:

<u>1.2. GÖDEL-TARSKI THEOREM</u>. There is no formal truth definition, i.e. there is no formula $Tr(\cdot)$ such that, for all sentences ϕ of the formal language in question,

$$\vdash \phi \leftrightarrow Tr(\ulcorner\phi\urcorner).$$

Proof: Suppose $Tr(\cdot)$ existed. Apply the Diagonalisation Lemma to obtain a sentence ϕ such that

$$\vdash \phi \leftrightarrow \sim Tr(\ulcorner\phi\urcorner).$$

If the basic property of the truth definition also held, we would have

$$\vdash \phi \leftrightarrow Tr(\ulcorner\phi\urcorner),$$

whence $\quad \vdash \phi \leftrightarrow \sim\phi,$

which quickly yields a contradiction within our theory (which I implicitly assume to be consistent).

QED

The Gödel-Tarski Theorem is cute, but it is not much fun to apply self-reference to a formula only to conclude that the formula in question doesn't exist. It is far more profitable to see what self-reference can accomplish in existing contexts. Gödel proved his First Incompleteness Theorem by finding such a context:

1.3. FIRST INCOMPLETENESS THEOREM. There is a "definition" of provability, i.e. a formula $Pr(\cdot)$ such that, for all sentences ψ,

i) (completeness) $\vdash \psi \Rightarrow \vdash Pr(\ulcorner\psi\urcorner)$

ii) (soundness) $\vdash Pr(\ulcorner\psi\urcorner) \Rightarrow \vdash \psi$.

Moreover, if we let $\vdash \phi \leftrightarrow \sim Pr(\ulcorner\phi\urcorner)$, then

i) $\nvdash \phi$

ii) $\nvdash \sim\phi$.

Proof: I will skip for now the difficult part-- the construction of the formula $Pr(\cdot)$. But I will note that the existence of such should not be too surprising: Formal derivations follow strict rules and the correctness of such a derivation is a computational matter; the verification that y codes a proof of a sentence coded by x is a computational procedure. Hence, this can be expressed adequately within the language by some formula $Prov(y,x)$. $Pr(x)$ is defined by $\exists y\, Prov(y,x)$.

Given $Pr(\cdot)$ satisfying the completeness and soundness conditions, the rest is easy. First choose ϕ by the Diagonalisation Lemma so that $\vdash \phi \leftrightarrow \sim Pr(\ulcorner\phi\urcorner)$.

i) Suppose $\vdash \phi$. By completeness, $\vdash Pr(\ulcorner\phi\urcorner)$. But, by definition of ϕ, $\vdash \sim\phi$, a contradiction.

ii) Suppose $\vdash \sim\phi$, i.e. (by choice of ϕ) $\vdash Pr(\ulcorner\phi\urcorner)$. By soundness we get $\vdash \phi$ and a contradiction. QED

To reiterate, Gödel's First Incompleteness Theorem destroyed Hilbert's Programme. If we assume the above carried out for the "meaningful" finite system S, i.e. $\vdash$ denotes provability in S and $Pr(\cdot)$ defines the codes of theorems of S, then Gödel's sentence ϕ is a meaningful sentence unprovable in S. Since ϕ asserts its unprovability and is unprovable, we conclude that it is true. Hence it is a theorem of our transfinite T and T is not conservative over S, much less provably so in S.

The first conclusion of the First Incompleteness Theorem, that ϕ is unprovable, depends only on the consistency of the theory in question. Thus, if *Con* denotes this consistency, a formalisation would yield

$$\vdash Con \rightarrow \sim Pr(\ulcorner \phi \urcorner),$$

i.e. $\vdash Con \rightarrow \phi,$

whence $\nvdash Con.$

<u>1.4. SECOND INCOMPLETENESS THEOREM</u>. Letting $Con = \sim Pr(\ulcorner \Lambda \urcorner)$, where Λ is any convenient absurdity (e.g. $\overline{0} = \overline{1}$), we have: $\nvdash Con$.

The Second Incompleteness Theorem is essentially a formalisation of the First Incompleteness Theorem; but, despite what one often reads, it is not proven by formalising the *proof* of the First Incompleteness Theorem. To avoid such excessive labour, Bernays introduced *Derivability Conditions*-- formalisations of a few basic properties of the "proof predicate" $Pr(\cdot)$. With these, the Second Incompleteness Theorem reduces rather quickly to the First.

Bernays was justly proud of his Derivability Conditions, nowadays termed the *Hilbert-Bernays* Derivability Conditions. They certainly did their job of making the proof of the Second Incompleteness Theorem printable. Nonetheless, they are unrelentingly ugly and inelegant:

HB1. $\vdash \phi \rightarrow \psi \;\; \Rightarrow \;\; \vdash Pr(\ulcorner \phi \urcorner) \rightarrow Pr(\ulcorner \psi \urcorner)$

HB2. $\vdash Pr(\ulcorner \sim\phi v \urcorner) \rightarrow Pr(sub(\ulcorner \sim\phi v \urcorner, x))$

HB3. $\vdash fx = \overline{0} \;\rightarrow\; Pr(sub(\ulcorner fv = \overline{0} \urcorner, x))$, for each primitive recursive term f.

Condition *HB1* is a generalisation of the adequacy condition,

$$\vdash \phi \;\; \Rightarrow \;\; \vdash Pr(\ulcorner \phi \urcorner),$$

used in the proof of the First Incompleteness Theorem. The other two conditions are moderately bizarre. *HB2* offers a uniform expression of the assertion that, if $\sim\phi v$ is derivable with v a free variable, then each instance $\sim\phi\overline{n}$ is also derivable. This is clear enough-- even for unnegated formulae. *HB3* is an odd way of asserting that the theory computes primitive recursive functions-- the functions used in syntactic encoding. A slightly more reasonable version would allow computations to be simulated for arbitrary outputs:

$$\vdash fx = y \rightarrow Pr(sub_2(\ulcorner fx = y \urcorner, x, y)),$$

(where sub_2 denotes the obvious two-fold substitution function).

One thing is clear: The Hilbert-Bernays Derivability Conditions were produced for the purpose of analysing a proof rather than a concept. A major step in the direction of this latter type of analysis was made in 1954 by Martin Hugo Löb when he streamlined the Hilbert-Bernays Derivability Conditions for the purpose of applying them to a new proof. These new *Löb Derivability Conditions*, or, more simply, *the* Derivability Conditions, are unquestionably more elegant:

D1. $\vdash \phi \Rightarrow \vdash Pr(\ulcorner \phi \urcorner)$

D2. $\vdash Pr(\ulcorner \phi \urcorner) \wedge Pr(\ulcorner \phi \rightarrow \psi \urcorner) \rightarrow Pr(\ulcorner \psi \urcorner)$

D3. $\vdash Pr(\ulcorner \phi \urcorner) \rightarrow Pr(\ulcorner Pr(\ulcorner \phi \urcorner) \urcorner)$.

Not being one for subtleties, I will explain the greater elegance of *D1-D3* over *HB1-HB3*: There are no extraneous features in *D1-D3*-- no substitutions, no general functions. Conditions *D1-D3* offer a partial analysis of $Pr(\cdot)$ mentioning only $Pr(\cdot)$. As an analysis of the provability of sentences, it mentions only sentences; no free variables occur. Where *HB1-HB3* are formulae of the predicate calculus, *D1-D3* are sentences of *modal* propositional logic. But I am getting ahead of myself... the immediate point is to derive the Second Incompleteness Theorem from the First.

Proof of 1.4: Choose ϕ so that $\vdash \phi \leftrightarrow \sim Pr(\ulcorner \phi \urcorner)$ and let Con be $\sim Pr(\ulcorner \Lambda \urcorner)$. We shall show $\vdash Con \rightarrow \phi$, whence the underivability of ϕ (by 1.3) will yield the underivability of Con.

Observe

$$\vdash \phi \leftrightarrow \sim Pr(\ulcorner \phi \urcorner) \Rightarrow \vdash \sim\phi \leftrightarrow Pr(\ulcorner \phi \urcorner)$$

$$\Rightarrow \vdash Pr(\ulcorner \sim\phi \urcorner) \leftrightarrow Pr(\ulcorner Pr(\ulcorner \phi \urcorner) \urcorner), \quad (1)$$

by a few applications of *D1*, *D2*; similarly,

$$\vdash \phi \wedge \sim\phi \rightarrow \Lambda \Rightarrow \vdash Pr(\ulcorner \phi \urcorner) \wedge Pr(\ulcorner \sim\phi \urcorner) \rightarrow Pr(\ulcorner \Lambda \urcorner). \quad (2)$$

But, by *D3*,

$$\vdash Pr(\ulcorner \phi \urcorner) \rightarrow Pr(\ulcorner Pr(\ulcorner \phi \urcorner) \urcorner). \quad (3)$$

Now (1) and (2) combine to yield

$$\vdash Pr(\ulcorner\phi\urcorner) \to Pr(\ulcorner\phi\urcorner) \wedge Pr(\ulcorner\sim\phi\urcorner),$$

whence (2) yields $\vdash Pr(\ulcorner\phi\urcorner) \to Pr(\ulcorner\Lambda\urcorner)$,

i.e. $\vdash \sim Pr(\ulcorner\Lambda\urcorner) \to \sim Pr(\ulcorner\phi\urcorner)$,

i.e. $\vdash Con \to \phi$. QED

Although it was the *First* and not the Second Incompleteness Theorem that destroyed Hilbert's Programme and although the Second is more-or-less a mere formalisation of the First, the Second Incompleteness Theorem is not unimportant. The First Incompleteness Theorem tells us that any true, sufficiently strong theory is necessarily incomplete-- there are recognisably true sentences undecided by the theory; the Second Incompleteness Theorem provides a meaningful instance of such: The very consistency of the theory cannot be proven in the theory. There is a third result completing the picture of theoretical impotence-- Löb's Theorem. Consistency is an expression of faith in the system which the Second Incompleteness Theorem asserts the system cannot prove; Löb's Theorem generalises this by characterising provable instances of a more general expression of faith:

1.5. LÖB'S THEOREM. Let ψ be any sentence. Then:

$$\vdash Pr(\ulcorner\psi\urcorner) \to \psi \quad \text{iff} \quad \vdash \psi.$$

Proof: The right-to-left implication is trivial.

To prove the left-to-right implication, choose ϕ so that $\vdash \phi \leftrightarrow. Pr(\ulcorner\phi\urcorner) \to \psi$ and observe

$$\begin{aligned}
\vdash \phi \leftrightarrow. Pr(\ulcorner\phi\urcorner) \to \psi \quad &\Rightarrow \quad \vdash \phi \to. Pr(\ulcorner\phi\urcorner) \to \psi \\
&\Rightarrow \quad \vdash Pr(\ulcorner\phi \to. Pr(\ulcorner\phi\urcorner) \to \psi\urcorner), \text{ by } D1 \\
&\Rightarrow \quad \vdash Pr(\ulcorner\phi\urcorner) \to Pr(\ulcorner Pr(\ulcorner\phi\urcorner) \to \psi\urcorner), \text{ by } D2 \\
&\Rightarrow \quad \vdash Pr(\ulcorner\phi\urcorner) \to. Pr(\ulcorner Pr(\ulcorner\phi\urcorner)\urcorner) \to Pr(\ulcorner\psi\urcorner), \text{ by } D2 \\
&\Rightarrow \quad \vdash Pr(\ulcorner\phi\urcorner) \to Pr(\ulcorner\psi\urcorner), \text{ by } D3 \\
&\Rightarrow \quad \vdash Pr(\ulcorner\phi\urcorner) \to \psi, \qquad (*)
\end{aligned}$$

by assumption $\vdash Pr(\ulcorner\psi\urcorner) \to \psi$. Now use the definition of ϕ:

$$\begin{aligned}
\vdash Pr(\ulcorner\phi\urcorner) \to \psi \quad &\Rightarrow \quad \vdash \phi \\
&\Rightarrow \quad \vdash Pr(\ulcorner\phi\urcorner), \text{ by } D1 \\
&\Rightarrow \quad \vdash \psi, \text{ by } (*).
\end{aligned}$$

QED

The above was Löb's proof of his theorem. There are two other proofs. The cuter of the two appeals to the Second Incompleteness Theorem: If $\nvdash \psi$, then $\sim\psi$ can be added as a new axiom. Since the Second Incompleteness Theorem depended only on the consistency of the theory, it applies to this consistent extension by $\sim\psi$,

$$\nvdash \sim\psi \rightarrow Con_{\sim\psi}. \qquad (*)$$

But $Con_{\sim\psi} = \sim Pr(\ulcorner \sim\psi \rightarrow \Lambda \urcorner)$ is equivalent to $\sim Pr(\ulcorner \psi \urcorner)$ and one can contrapose (*) to conclude $\nvdash Pr(\ulcorner \psi \urcorner) \rightarrow \psi$.

This slick reduction of Löb's Theorem to the Second Incompleteness Theorem (for finite extensions) is very pleasing and it has been popularly exposited of late. It was multiply discovered; the earliest reported discovery is Saul Kripke's in 1967.

The remaining proof is Kreisel's variant of Löb's. In it, one uses the fixed point (as I shall be calling self-referential sentences) ϕ to $Pr(\ulcorner \phi \rightarrow \psi \urcorner)$, i.e. one assumes $\vdash \phi \leftrightarrow Pr(\ulcorner \phi \rightarrow \psi \urcorner)$. I urge the reader to attempt this proof on his own now as it is an important one. But, if the reader finds such propositional derivations too confusing, he shouldn't worry: We will encounter it in Chapter 1 and again, in a slightly different context, in Chapter 4.

One of the nice things about Kreisel's fixed point is that it rather effortlessly yields the formalised version of Löb's Theorem:

1.6. FORMALISED LÖB'S THEOREM. Let ψ be any sentence. Then:

$$\vdash Pr(\ulcorner Pr(\ulcorner \psi \urcorner) \rightarrow \psi \urcorner) \rightarrow Pr(\ulcorner \psi \urcorner).$$

I defer the proof to Chapter 1.

There has been no result on $Pr(\cdot)$ since Löb published his proof in 1955. There is a reason for this: The Derivability Conditions and Löb's Theorem (Formalised or not) tell the complete story of $Pr(\cdot)$.

2. SELF-REFERENCE

When I say that the Derivability Conditions and Löb's Theorem tell the whole story of $Pr(\cdot)$, I do not, of course, mean that there is nothing else interesting to say about $Pr(\cdot)$; I mean only to imply that these properties account for all other properties of $Pr(\cdot)$. One such property is an analysis of self-referential sentences.

2.1. EXAMPLE. (Gödel's sentence). Let $\vdash \phi_1 \leftrightarrow \sim Pr(\ulcorner \phi_1 \urcorner)$. By the proof of the Second Incompleteness Theorem,

$$\vdash Con \rightarrow \phi_1.$$

But ϕ_1 asserts the unprovability of something (namely itself); so it immediately implies consistency. Thus,

$$\vdash \phi_1 \leftrightarrow Con.$$

2.2. EXAMPLE. (Henkin's sentence). In 1952, Leon Henkin published a small query: By Gödel's work, any sentence asserting its own unprovability is unprovable; what about sentences asserting their own provability? I.e., consider any sentence ϕ_2 satisfying $\vdash \phi_2 \leftrightarrow Pr(\ulcorner \phi_2 \urcorner)$. Löb's Theorem answers this question immediately:

$$\vdash \phi_2 \leftrightarrow Pr(\ulcorner \phi_2 \urcorner) \quad \Rightarrow \quad \vdash Pr(\ulcorner \phi_2 \urcorner) \rightarrow \phi_2 \quad \Rightarrow \quad \vdash \phi_2.$$

Thus, e.g., $\vdash \phi_2 \leftrightarrow \overline{0} = \overline{0}$

2.3. EXAMPLE. (Kreisel's sentence). Let $\vdash \phi_3 \leftrightarrow Pr(\ulcorner \phi_3 \rightarrow \psi \urcorner)$ for some sentence ψ. By the Formalised Löb Theorem,

$$\vdash Pr(\ulcorner Pr(\ulcorner \psi \urcorner) \rightarrow \psi \urcorner) \leftrightarrow Pr(\ulcorner \psi \urcorner),$$

and $Pr(\ulcorner \psi \urcorner)$ is an example of such a ϕ_3. In 1973, Angus Macintyre and Harry Simmons showed that, in fact,

$$\vdash \phi_3 \leftrightarrow Pr(\ulcorner \psi \urcorner).$$

2.4. EXAMPLE. (Löb's sentence). Let $\vdash \phi_4 \leftrightarrow . Pr(\ulcorner \phi_4 \urcorner) \rightarrow \psi$ for some sentence ψ. Macintyre and Simmons have shown: $\vdash \phi_4 \leftrightarrow . Pr(\ulcorner \psi \urcorner) \rightarrow \psi$.

These examples can be multiplied endlessly. The fact is that they are not atypical, but rather are instances of a general phenomenon: All legitimate modally expressible fixed points are unique and explicitly definable.

3. THINGS TO COME

I have twice used the word "modal" in referring to $Pr(\cdot)$ and it is now time to explain this usage. Basically, it is simple: Consider the sentences of the language of S (or T, or whatever) as propositions and let $p, q, r, \ldots$ be propositional variables ranging over them; consider provability as necessity-- i.e. $Pr(\cdot)$ is $\Box$. The result is a formal interpretation of modal logic. Modulo some assumptions on S (or T ...), the modal schemata provable in S (or T) are axiomatised by the modal translates of the Derivability Conditions and the Formalised Löb's Theorem. Moreover, this modal logic-- which I follow Albert Visser in denoting PRL (for *Pr*ovability *L*ogic)-- suffices for the modal analysis of self-referential sentences arising from this modal context. Part I of this monograph, consisting of three chapters, is devoted to the study of PRL: In the first of these chapters, the modal language is introduced, the "provability interpretation" explained, the axioms set out and various syntactic matters pursued, including the above cited modal analysis of self-reference. The second chapter is devoted to semantic (set theoretically semantic) matters-- the definition of Kripke models, completeness with respect to them, and the less syntactically involved discussion of self-reference. The final chapter concerns the intended semantics (i.e. the provability interpretation). The main result is Solovay's Completeness Theorem, by which PRL *is* the logic of provability. Various refinements and applications are also discussed.

If it were permissible to refer to such a recent development as such, I would say that Part I is the "classical core" of the modal study of self-reference. It is, at least, the modal analysis of the "classical" results in self-reference. There are other aspects of self-reference amenable to modal analysis. Parts II and III are devoted to two of these.

Already when he proved PRL adequate for the provability interpretation, Solovay observed that PRL is also the logic of certain other predicates. He also noted that, by adding some new axioms, it becomes the logic of certain predicates with more properties. Not every interesting predicate satisfies all the laws of PRL, however, and the modal logic of such may not be as rich as PRL. In such a case, it can happen that the predicate in question is amenable to study if one studies it in conjunction with *Pr(·)*. If one adds to this the reflexion that the relationship between the two predicates is fully as interesting as the individual predicates, it should not surprise the reader to find that Part II is devoted to multi-modal generalisations of the modal analysis of Part I. Chapter 4, the first chapter of Part II, studies a few specific multi-modal (generally: bi-modal) logics. These include a weak system SR in which to analyse self-reference in $\Box$ and the new operator ∇, PRL_1 in which the new operator Δ is the provability predicate of an RE theory extending our base theory, the logic PRL_{ZF} extending PRL_1 by axioms asserting the extension is powerful relative to the base, and another system MOS to be explained later. While SR is merely a weak base theory with many interpretations desired, PRL_1, PRL_{ZF}, and MOS have intended interpretations and are given model theories and completeness proofs with respect to these model theories and intended interpretations.

Syntactically, the presence of the box, $\Box$, in the modal analyses of the new operators Δ and ∇ in Chapter 4 is initially dictated by the need for a modal law of the form,

$$\Box(A \to B) \wedge \nabla A \to \nabla B,$$

in the absence of the validity of

$$\nabla(A \to B) \wedge \nabla A \to \nabla B.$$

Model-theoretically, this presence is acutely felt: The completeness of PRL_1 and MOS for their intended interpretations is established by reduction to the uni-modal case, i.e. to Solovay's Completeness Theorem for PRL. The main theorem of Chapter 5 purports to explain this. The fixed point algebras of Chapter 5 were introduced for the purpose of providing a general setting for the study of *extensional* self-reference. The chief theorem about them asserts that finite such algebras are subalgebras of those generated by the box of PRL; thus, in a weak sense, any modal operator having an adequate finite algebraic modelling reduces to the box.

Given the weakly-established limitation on a modal analysis of extensional self-reference, the question arises: What can be said about non-extensional self-reference? Part III is devoted to the beginnings of the answer to this question. Chapter 6 presents Guaspari's and Solovay's modal analysis of Rosser sentences.

I will explain Rosser sentences in more detail in Chapter 6. Nonetheless, a few words are in order here: Although one is primarily interested in true theories, the requirement of soundness in the second half of the First Incompleteness Theorem (1.3.ii) is a technical weakness. In 1936, J. Barkley Rosser circumvented this requirement by introducing a new twist: Instead of asserting its own unprovability as Gödel's sentence did, Rosser's sentence asserted that, if it were provable, so was its negation-- in fact, its negation had an even earlier proof. This use of the ordering of the natural numbers-- hence of numerical codes of syntactic objects such as proofs-- was the key to many applications of self-reference to the study of formal systems. In studying the literature in the 1970s I realised that all of these basic applications of self-reference used fixed points of one simple form. This form and some of its applications are discussed in Chapter 7 along with generalisations of more recent instances of self-reference.

Thus, we see in outline the contents of the present monograph and an explanation of the trisection into Parts I, II, and III. There are other explanations of the trichotomy and other reasonable subdivisions of the material. Parts I and II cover the smooth theory-- those instances of self-reference that are extremely well-behaved: The fixed points are unique up to provable equivalence; they are explicitly definable; and they have common explicit definitions. Beyond the basic results (e.g., Gödel's and Löb's Theorems in the case of *Pr(·)*) these self-referential sentences have no interesting known applications. The Rosser sentences serve as examples of the failure of the modal analysis to extend to the general case: The fixed points need be neither unique nor explicitly definable. A second theme in self-reference-- metamathematical application-- is finally taken up in the last Chapter. Although prior to 1970 most work in self-reference followed this theme, it is only briefly touched on here-- not because of any lack of interest, but because of the immense uniformity of this aspect of the subject. There is a third theme to self-reference-- the use of self-reference to define functions and predicates by recursion. To date there is no modal theory of this type of self-reference and it is not studied here. It does occur though-- as a tool in proving the completeness of PRL and such with respect to the intended interpretations.

This explains all the rest of the book other than what is left of this chapter. The rest of this chapter merely standardises the formal theory PRA the provability predicate of which is to be referred to as *Pr(·)* and presents a few details of the formal development within PRA of the syntactic coding underlying Gödel's Diagonalisation Lemma and the ensuing theory. The reader who is familiar with or simply not interested in these details can easily skip ahead to Chapter 1; only once in a while should he need to refer back to this material.

4. THE THEORY PRA

The predicate $Pr(\cdot)$ is supposed to represent provability in a formal system S. Mathematically, the choice of the theory S is not particularly delicate-- any reasonably strong theory will do. Socially, the matter appears to be more delicate: Proof theorists, who believe in analysing individual proofs, believe a theory of strings directly discussing syntax to be the proper choice; many modern logicians, confusing set theory with foundations, believe one should choose a theory of hereditarily finite sets; and traditionally one has referred to Peano Arithmetic. My own, rather Pythagorean, belief is that number lies at the heart of mathematics and the incompleteness of arithmetic is the most interesting result. However, there are technical disadvantages to working within Peano Arithmetic. First, one has to perform a preliminary coding just to verify that one can do the sort of coding necessary. Second, unless one extends the language, one does not have function symbols for the definable functions used in the encoding and the resultant circumlocutions tend to be unreadable. This practically dictates choosing an extension of Peano Arithmetic with a goodly stock of functions. However, one does not need quite so powerful a theory and I choose for my standard (a variant of) the theory PRA of *P*rimitive *R*ecursive *A*rithmetic. The primitive recursive functions are ideal for encoding recursively defined syntactic objects (such as derivations) and the amount of induction in PRA is only a tiny bit more than that needed for proofs by induction on the inductive generation of syntactic objects.

Modulo specification of certain initial functions and certain innocuous closure properties, the class of primitive recursive functions is just that of those functions of natural numbers generated by recursion:

4.1. DEFINITION. A function $f:\omega^n \rightarrow \omega$ of natural numbers is *primitive recursive* if it can be defined after finitely many steps by means of the following rules:

F1. $Z(x) = 0$ *Zero*

F2. $S(x) = x + 1$ *Successor*

$F3^n_i$.	$P^n_i(x_1,\ldots,x_n) = x_i$	$(1 \le i \le n)$	*Projection*
$F4^n_m$.	$f(x_1,\ldots,x_n) = g(h_1(x_1,\ldots,x_n),\ldots,h_m(x_1,\ldots,x_n))$		*Composition*
$F5^n$.	$f(0,x_1,\ldots,x_n) = g(x_1,\ldots,x_n)$ $f(x+1,x_1,\ldots,x_n) = h(f(x,\vec{x}),x,\vec{x})$,		*Primitive Recursion*

where $\vec{x}$ abbreviates the sequence $x_1, \ldots, x_n$.

The functions Z, S and P^n_i are the initial functions from which all others are to be generated. Projection and Composition yield closure under explicit definition: We can define new functions by permuting variables, adding dummy variables, diagonalising (i.e. going from $g(x,y)$ to $g(x,x) = g(P^1_1(x),P^1_1(x)))$, etc. The schema $F5^n$ of Primitive Recursion is only one of many possible types of recursion, some of which reduce to $F5^n$ and some of which yield new functions. Generally, simpler recursions reduce to $F5^n$ via explicit definability; e.g. the recursion,

$$f(0,\vec{x}) = g(\vec{x})$$

$$f(x+1,\vec{x}) = h(f(x,\vec{x})),$$

yields a primitive recursive function since one can replace h by

$$h_1(y,x,\vec{x}) = h(P^{n+2}_1(y,x,\vec{x})).$$

Thus, e.g., addition is primitive recursive since

$$x + 0 = A(0,x) = x$$

$$x + (y + 1) = A(y+1,x) = S(A(y,x)) = (x + y) + 1.$$

Similarly, multiplication is primitive recursive.

Our goal in the present section is not, however, to generate a lot of primitive recursive functions, but rather to define a formal system of Primitive Recursive Arithmetic in which to generate these functions and prove their properties.

Every formal theory has a formal language, rules of term and formula construction, and basic axioms and rules of inference. First, we specify the language of PRA:

VARIABLES: $v_0, v_1, \ldots$

CONSTANT: $\overline{0}$

FUNCTION SYMBOLS: $\overline{f}$ for each (primitive recursive definition of a) primitive recursive function f

RELATION SYMBOL: $=$

PROPOSITIONAL CONNECTIVES: $\sim, \wedge, \vee, \rightarrow$

QUANTIFIERS: $\forall, \exists$

One point of the above needs clarification: If one regards functions *extensionally*-- say, as sets of ordered pairs-- then a given function can be generated by the schemata of primitive recursion in many ways and one cannot effectively decide if two generating principles yield the same function. Thus, we associate function symbols not to functions but to their rules of generation, thereby allowing an obvious choice of axioms. If, e.g., f is defined from g, h by $F5^n$ and g, h have been assigned function symbols $\overline{g}$, $\overline{h}$, respectively, we assign f the function symbol $\overline{f}$ and will add axioms (cf. 4.9, below),

$$\overline{f}(\overline{0}, v_1, \ldots, v_n) = \overline{g}(v_1, \ldots, v_n)$$
$$\overline{f}(\overline{S}v_0, v_1, \ldots, v_n) = \overline{h}(\overline{f}(v_0, \vec{v}), v_0, \vec{v}),$$

where $\vec{v}$ abbreviates $v_1, \ldots, v_n$.

Before discussing axioms, however, we should discuss the rules of term and formula formation. As these are just the standard rules, there would seem to be no need to discuss them here. There isn't; in the next section, however, when we encode syntax we will have to be quite explicit on these points. The reader who is not keen on such details is invited to skip ahead to Definition 4.9, where the non-logical axioms for PRA are given.

4.2. DEFINITIONS. i. The set of *terms* of the language of PRA is inductively defined by:

a. $\overline{0}$ is a term; each v_i is a term

b. if $\overline{f}$ is an n-ary function symbol and $t_1, \ldots, t_n$ are terms, $\overline{f}t_1 \ldots t_n$ is a term.

ii. The set of *formulae* of the language of PRA is inductively defined by:

a. if t_1, t_2 are terms, $=t_1t_2$ is a formula

b. if ϕ, ψ are formulae, so are $\sim\phi$, $\wedge\phi\psi$, $\vee\phi\psi$, and $\rightarrow\phi\psi$

c. if ϕ is a formula and v a variable, then $\exists v\phi$ and $\forall v\phi$ are also formulae.

The use of Polish notation is dictated by the desire for unique readability.

In actual practice, we use the usual infix notation, parentheses, and parentheses-avoiding conventions.

Another point worth clarifying: The functions S, Z are unary; P_i^n is n-ary; if f is introduced via $F4_m^n$ it is n-ary; and if f is introduced by $F5^n$ it is $(n+1)$-ary. The arity of $\overline{f}$ is that of f.

The axioms of PRA tesserachotomise: There are propositional axioms, quantifier axioms, equality axioms, and non-logical axioms. There are also three rules of inference-- modus ponens and two generalisation rules.

4.3. DEFINITION. The propositional axioms of PRA are the following schemata:

i. $\phi \rightarrow (\psi \rightarrow \phi)$

ii. $(\phi \rightarrow (\psi \rightarrow \chi)) \rightarrow ((\phi \rightarrow \psi) \rightarrow (\phi \rightarrow \chi))$

iii. $\phi \wedge \psi \rightarrow \phi$

iv. $\phi \wedge \psi \rightarrow \psi$

v. $\phi \rightarrow (\psi \rightarrow \phi \wedge \psi)$

vi. $\phi \rightarrow \phi \vee \psi$

vii. $\psi \rightarrow \phi \vee \psi$

viii. $(\phi \rightarrow \chi) \rightarrow ((\psi \rightarrow \chi) \rightarrow (\phi \vee \psi \rightarrow \chi))$

ix. $(\phi \rightarrow \psi) \rightarrow ((\phi \rightarrow \sim\psi) \rightarrow \sim\phi)$

x. $\sim\sim\phi \rightarrow \phi$.

It can be shown that axiom schemata 4.3.i-x, together with modus ponens, generate all tautologies. To generate all logical truths, we must add some logical axioms and rules. Before giving these we must carefully define the notions of free and bound occurrences of a variable.

4.4. DEFINITIONS. We inductively define the notions i. *v occurs in t*, ii. *v occurs in ϕ*, iii. *v has a free occurrence in ϕ*, and iv. *v has a bound occurrence in ϕ*:

i. a. v occurs in v

b. if v occurs in (at least) one of $t_1, \ldots, t_n$, and $\overline{f}$ is an n-ary function symbol, then v occurs in $\overline{f}t_1 \ldots t_n$;

ii. a. v occurs in $t_1 = t_2$ just in case v occurs in one of t_1, t_2

b. if v occurs in one of ϕ, ψ, then v occurs in $\phi \wedge \psi$, $\phi \vee \psi$, and $\phi \rightarrow \psi$

c. if v occurs in ϕ, then v occurs in $\sim\phi$, $\exists v^*\phi$, and $\forall v^*\phi$, for any variable v^*;

iii. a. v has a free occurrence in $t_1 = t_2$ just in case v occurs in one of t_1, t_2

b. if v has a free occurrence in one of ϕ, ψ, then v has a free occurrence in $\phi \wedge \psi$, $\phi \vee \psi$, and $\phi \rightarrow \psi$

c. if v has a free occurrence in ϕ, then v has a free occurrence in $\sim\phi$, $\exists v^*\phi$, and $\forall v^*\phi$, for any variable v^* other than v;

iv. a. v has a bound occurrence in $\exists v\phi$ and $\forall v\phi$

b. if v has a bound occurrence in one of ϕ, ψ, then v has a bound occurrence in $\phi \wedge \psi$, $\phi \vee \psi$, and $\phi \rightarrow \psi$

c. if v has a bound occurrence in ϕ, then v has a bound occurrence in $\sim\phi$, $\exists v^*\phi$, and $\forall v^*\phi$, for any variable v^*.

Somewhat more important than these notions is the derivative one of substitutability:

4.5. DEFINITION. We inductively define the relation, *t is substitutable for v in ϕ*, as follows:

i. if ϕ is atomic (i.e. $t_1 = t_2$), then t is substitutable for v in ϕ

ii. if t is substitutable for v in ϕ, ψ, then t is substitutable for v in $\sim\phi$, $\phi \wedge \psi$, $\phi \vee \psi$, and $\phi \rightarrow \psi$

iii. if ϕ is $\exists v^*\psi$ or $\forall v^*\psi$, then t is substitutable for v in ϕ iff either

a. v does not have a free occurrence in ϕ

or b. v^* does not occur in t and t is substitutable for v in ψ.

With these rather hideous technicalities, we can give the usual axioms for the quantifiers:

4.6. DEFINITION. The quantifier axioms for **PRA** are the following schemata:

i. $\forall v\, \phi v \rightarrow \phi t$

ii. $\phi t \rightarrow \exists v\, \phi v$,

where, in both cases, t is substitutable for v in ϕv.

The rules of inference are now also statable:

4.7. DEFINITION. The rules of inference for PRA are modus ponens and two schemata of generalisation:

i. From ϕ, $\phi \to \psi$ derive ψ

ii. From $\phi v \to \psi$ derive $\exists v\, \phi v \to \psi$, provided v has no free occurrence in ψ

iii. From $\psi \to \phi v$ derive $\psi \to \forall v\, \phi v$, provided v has no free occurrence in ψ.

The axioms and rules thus far given suffice for purely logical purposes: A sentence ϕ follows semantically from a set Γ of sentences iff ϕ is derivable from Γ by means of these axioms and rules of inference. We have but to specify the non-logical axioms Γ of concern to us. These are the semi-logical equality axioms and the properly non-logical axioms of primitive recursion and induction.

4.8. DEFINITION. The equality axioms of PRA are the following:

i. $v_0 = v_0$

ii. $v_0 = v_1 \to v_1 = v_0$

iii. $v_0 = v_1 \wedge v_1 = v_2 \to v_0 = v_2$

iv. $v_i = v^* \to \bar{f}(v_1, \ldots, v_i, \ldots, v_n) = \bar{f}(v_1, \ldots, v^*, \ldots, v_n)$,

where $1 \le i \le n$ and $\bar{f}$ is an n-ary function symbol.

Finally, we can give the non-logical axioms of PRA:

4.9. DEFINITION. The non-logical axioms of PRA are the following:

a. (Initial Functions)

i. $\bar{Z}v_0 = \bar{0}$

ii. $\sim(\bar{0} = \bar{S}v_0)$; $\bar{S}v_0 = \bar{S}v_1 \to v_0 = v_1$

iii. $\bar{P}^n_i(v_1, \ldots, v_n) = v_i$, for $1 \le i \le n$.

b. (Derived Functions)

i. $\bar{f}(v_1, \ldots, v_n) = \bar{g}(\bar{h}_1(\vec{v}), \ldots, \bar{h}_m(\vec{v}))$, where f is defined from $g, h_1, \ldots, h_m$ by composition $F4^n_m$

ii. $\bar{f}(\bar{0}, v_1, \ldots, v_n) = \bar{g}(v_1, \ldots, v_n)$

$\bar{f}(\bar{S}v_0, v_1, \ldots, v_n) = \bar{h}(\bar{f}(v_0, \vec{v}), v_0, \vec{v})$, where f is defined from g, h by primitive recursion $F5^n$.

c. (Induction)

$\phi\bar{0} \wedge \forall v\big(\phi v \to \phi(\bar{S}v)\big) \to \forall v\, \phi v$, where ϕv has the form $\exists v_n\big(\bar{f}(v, v_0, \vec{v}) = \bar{0}\big)$.

These axioms are fairly self-explanatory. Only the odd restriction on induction requires explanation. The obvious choice is full induction, i.e. the full schema of induction applying to *all* formulae. The resulting theory is a version of *P*eano *A*rithmetic, PA, and is the theory with which Gödel's Incompleteness Theorems have come to be associated. PA is, however, very powerful-- much more powerful than is necessary for the encoding of syntax. The system PRA, with its primitive recursive functions to capture the simple recursive definitions of syntactic notions (as e.g. Definitions 4.2, 4.4, and 4.5, above) and its induction on primitive recursive relations, has just the right level of strength for the sort of theory of self-reference we wish to study in this monograph. I have cheated and allowed, for convenience sake, a little more induction than is necessary; quantifier-free induction, i.e. induction restricted to quantifier-free formulae, suffices for the "classical theory of incompleteness". (In the literature, the name "PRA" generally attaches itself to this weaker theory with only quantifier-free induction. The system described herein has, however, no greater proof theoretic strength and proves exactly the same universal assertions as this weaker theory and, thus, I proclaim myself justified in the nominal liberty I have taken.)

We now have the system PRA-- we've specified its language, axioms, and rules of inference. All that remains to complete the ritual is to give a formal definition of a (formal) derivation. Following this, we can actually begin to work with(in) the system.

4.10. DEFINITION. A *formal derivation* in the system PRA is a sequence $\phi_0, \phi_1, \ldots, \phi_k$ of formulae in the language of PRA satisfying: Each ϕ_i

i. is an axiom of PRA

or ii. follows from two formulae ϕ_j, ϕ_l, where $j, l < i$ by the rule 4.7.i of modus ponens,

or iii. follows from one formula ϕ_j, where $j < i$, by one of the rules 4.7.ii and 4.7.iii of generalisation.

The primitive recursive encoding of the syntax of PRA will be outlined in the next section. For now, I would simply like to show that the axioms and rules given are strong enough to compute the primitive recursive functions. To this end, we first need a definition:

4.11. DEFINITION. Let n be a natural number. The *numeral* $\overline{n}$ is the term $\overline{S}\ldots\overline{S}\overline{0}$ consisting of n $\overline{S}$'s followed by a $\overline{0}$, i.e. $\overline{0}$ is the constant $\overline{0}$ and $\overline{n+1}$ is $\overline{S}\overline{n}$.

4.12. THEOREM. Let f be an n-ary primitive recursive function and let $\overline{f}$ be the function symbol representing it in PRA. For all $k_1, \ldots, k_n, k \in \omega$,

$$f(k_1,\ldots,k_n) = k \implies \mathrm{PRA} \vdash \overline{f}(\overline{k}_1,\ldots,\overline{k}_n) = \overline{k}.$$

To save space in presenting the axioms, I gave free variable formulations with specific free variables. We will need substitution instances of these:

4.13. LEMMA. Let ϕv be given and let t be substitutable for v in ϕ. If $\mathrm{PRA} \vdash \phi v$, then $\mathrm{PRA} \vdash \phi t$.

Proof: Let ψ be the sentence $\overline{0} = \overline{0} \wedge \overline{0} = \overline{0} \rightarrow \overline{0} = \overline{0}$ and extend a derivation of ϕv as follows:

0.	ϕv	derivation assumed
1.	$\phi v \rightarrow (\psi \rightarrow \phi v)$	axiom 4.3.i
2.	$\psi \rightarrow \phi v$	0, 1, modus ponens (4.7.i)
3.	$\psi \rightarrow \forall v \phi v$	2, generalisation (4.7.iii)
4.	ψ	axiom 4.3.iii
5.	$\forall v \phi v$	3, 4, modus ponens
6.	$\forall v \phi v \rightarrow \phi t$	axiom 4.6.i
7.	$\phi t.$	5, 6, modus ponens

Thus, we obtain a derivation of ϕt by appending a few lines to any derivation of ϕt.

QED

Proof of 4.12: The proof is by induction on the generation of f by the generating rules *F1*-*F5*.

F1: From the axiom,

$$\overline{Z}v_0 = \overline{0},$$

one concludes via the Lemma, $\mathrm{PRA} \vdash \overline{Z}\overline{k}_1 = \overline{0}$ for all $k_1 \in \omega$.

F2: The identity axiom, $v_0 = v_0$, entails via the Lemma, $\mathrm{PRA} \vdash \overline{S}\overline{k}_1 = \overline{k_1+1}$, for all

$k_1 \in \omega$-- since $\overline{k_1+1}$ is $S\overline{k}_1$.

$F3^n_i$: Again, the axiom, $\overline{P}^n_i(v_1,\ldots,v_n) = v_i$, entails $\mathsf{PRA} \vdash \overline{P}^n_i(\overline{k}_1,\ldots,\overline{k}_n) = \overline{k}_i$, for all $k_1,\ldots,k_n \in \omega$.

$F4^n_m$: Suppose f is defined from g, h_i by composition, $h_i(k_1,\ldots,k_n) = l_i$ for $1 \le i \le m$ and $g(l_1,\ldots,l_m) = k$, so that $f(k_1,\ldots,k_n) = k$. By the induction hypothesis,

$$\mathsf{PRA} \vdash \overline{h}_i(\overline{k}_1,\ldots,\overline{k}_n) = \overline{l}_i, \tag{1}$$

for $1 \le i \le m$, and

$$\mathsf{PRA} \vdash \overline{g}(\overline{l}_1,\ldots,\overline{l}_m) = \overline{k}. \tag{2}$$

Now we also have the axiom,

$$\overline{f}(v_1,\ldots,v_n) = \overline{g}(\overline{h}_1(v_1,\ldots,v_n),\ldots,\overline{h}_m(v_1,\ldots,v_n)),$$

whence n applications of the Lemma yield

$$\mathsf{PRA} \vdash \overline{f}(\overline{k}_1,\ldots,\overline{k}_n) = \overline{g}(\overline{h}_1(\overline{k}_1,\ldots,\overline{k}_n),\ldots,\overline{h}_m(\overline{k}_1,\ldots,\overline{k}_n)). \tag{3}$$

Again, a little propositional logic and the application of the Lemma to equality axioms yield

$$\mathsf{PRA} \vdash \bigwedge \overline{h}_i(\overline{k}_1,\ldots,\overline{k}_n) = \overline{l}_i \rightarrow \\ \rightarrow \overline{g}(\overline{h}_1(\overline{k}_1,\ldots,\overline{k}_n),\ldots,\overline{h}_m(\overline{k}_1,\ldots,\overline{k}_n)) = \overline{g}(\overline{l}_1,\ldots,\overline{l}_m). \tag{4}$$

Now, (1), (4) and modus ponens yield

$$\mathsf{PRA} \vdash \overline{g}(\overline{h}_1(\overline{k}_1,\ldots,\overline{k}_n),\ldots,\overline{h}_m(\overline{k}_1,\ldots,\overline{k}_n)) = \overline{g}(\overline{l}_1,\ldots,\overline{l}_m),$$

which with (3) and the appropriate substitution instance of equality axiom 4.8.iii,

$$v_0 = v_1 \wedge v_1 = v_2 \rightarrow v_0 = v_2,$$

yields

$$\mathsf{PRA} \vdash \overline{f}(\overline{k}_1,\ldots,\overline{k}_n) = \overline{g}(\overline{l}_1,\ldots,\overline{l}_m).$$

This, (2), and another substitution instance of 4.8.iii yields

$$\mathsf{PRA} \vdash \overline{f}(\overline{k}_1,\ldots,\overline{k}_n) = \overline{k}.$$

$F5^n$: Let f be defined from g, h by primitive recursion,

$$f(0,\vec{x}) = g(\vec{x})$$

$$f(x+1,\vec{x}) = h(f(x,\vec{x}),x,\vec{x}).$$

We prove by a subsidiary induction on m that

$$f(m,k_1,\ldots,k_n) = k \Rightarrow \mathsf{PRA} \vdash \overline{f}(\overline{m},\overline{k}_1,\ldots,\overline{k}_n) = \overline{k}.$$

Basis. $m = 0$. As usual, we have substitution instances of free-variable axioms:

$$\mathsf{PRA} \vdash \overline{f}(\overline{0},\overline{k}_1,\ldots,\overline{k}_n) = \overline{g}(\overline{k}_1,\ldots,\overline{k}_n)$$

$$\mathsf{PRA} \vdash \overline{f}(\overline{0},\vec{\overline{k}}) = \overline{g}(\vec{\overline{k}}) \wedge \overline{g}(\vec{\overline{k}}) = \overline{k} \rightarrow \overline{f}(\overline{0},\vec{\overline{k}}) = \overline{k}.$$

We also have the induction hypothesis:

$$\mathsf{PRA} \vdash \overline{g}(\overline{k}_1,\ldots,\overline{k}_n) = \overline{k}.$$

Simple propositional logic yields from these the desired conclusion:

$$\mathsf{PRA} \vdash \overline{f}(\overline{0},\overline{k}_1,\ldots,\overline{k}_n) = \overline{k}.$$

Induction step. Again, we have substitution instances of free-variable axioms and the induction hypotheses:

$$\mathsf{PRA} \vdash \overline{f}(\overline{Sm},\overline{k}_1,\ldots,\overline{k}_n) = \overline{h}(\overline{f}(\overline{m},\overline{k}_1,\ldots,\overline{k}_n),\overline{m},\overline{k}_1,\ldots,\overline{k}_n)$$

$$\mathsf{PRA} \vdash \overline{f}(\overline{m},\overline{k}_1,\ldots,\overline{k}_n) = \overline{l}, \text{ say,}$$

$$\mathsf{PRA} \vdash \overline{h}(\overline{l},\overline{m},\overline{k}_1,\ldots,\overline{k}_n) = \overline{k}.$$

The equality axioms and propositional logic quickly yield

$$\mathsf{PRA} \vdash \overline{f}(\overline{Sm},\overline{k}_1,\ldots,\overline{k}_n) = \overline{k}.$$ QED

I suppose the above proof is overly detailed. It certainly is from the point of view of understanding the truth of the Theorem. For our later purpose of formalising the proof within PRA, it is quite sketchy. Formalise this? Whatever for? Theorem 4.12 is the technical result underlying the first Derivability Condition *D1*; condition *D3* reduces to its formalisation.

5. ENCODING SYNTAX IN PRA

Back in section 3, I announced that the remainder of the Chapter would include "a few details of the formal development within PRA of the syntactic encoding underlying" the theory behind the Incompleteness Theorems. This was followed in section 4 by a somewhat disgustingly detailed description of syntax. In the present section, finally, I shall omit vast quantities of details. However, I shall also include many details. It might help if I first explain what I shall and shall not do.

By way of explanation, let me cite a simple example. Addition is defined by primitive recursion:

$$\left.\begin{aligned} x + 0 &= x \\ x + Sy &= S(x + y). \end{aligned}\right\} \qquad (*)$$

Formalising this definition within PRA would be done as follows: Let $\overline{g}$, $\overline{h}$ be terms built up from those for the projection and successor functions so that

$$\mathsf{PRA} \vdash \overline{g}(x) = x$$

$$\mathsf{PRA} \vdash \overline{h}(z,y,x) = \overline{S}z,$$

and introduce $\overline{f}$ with the axioms

$$\overline{f}(\overline{0},x) = \overline{g}(x)$$

$$\overline{f}(\overline{S}y,x) = \overline{h}(\overline{f}(y,x),y,x).$$

One would then observe that the function symbol $\overline{f}_1$ defined by

$$\overline{f}_1(x,y) = \overline{f}(y,x) \quad \text{(i.e. } \overline{f}(\overline{P}^2_2(x,y),\overline{P}^2_1(x,y)))$$

satisfies the recursion (*) and, hence, defines addition in the sense that $\overline{f}_1$ satisfies the recursion defining addition in the standard model $\mathcal{N} = (\omega;+,\cdot,0,')$. By Theorem 4.12, we know also that this defines addition in the sense that

$$m + n = k \Rightarrow \mathsf{PRA} \vdash \overline{f}_1(\overline{m},\overline{n}) = \overline{k}$$

(with the consistency of PRA yielding the converse as well). What we might not *know*, but is true and will be assumed and left unproved, is that $\overline{f}_1$ satisfies in PRA the familiar laws of addition; e.g., writing $x + y$ for $\overline{f}_1(x,y)$, we do not bother proving basic facts (although we might use them) like commutativity,

$$x + y = y + x,$$

associativity,

$$x + (y + z) = (x + y) + z$$

or, when we have defined multiplication, distributivity,

$$x\cdot(y + z) = x\cdot y + x\cdot z.$$

More generally, the details I shall present will be those exhibiting definitions and establishing definability; the details I omit are the enumeration of basic properties. Although such an enumeration of propositions and proofs is routine, it is also time and space consuming and I leave such details to the reader either to work out for himself or to locate in the literature. (Elliot Mendelson's textbook is a good source for some of these neglected proofs.)

Now, about notation: It is important to distinguish between distinct objects when the distinction matters; it is equally important not to let one's notation proliferate when distinctions do not matter. In the last section, we had to distinguish between a primitive recursive function f and the function symbol $\overline{f}$

denoting it. In much of the present section, we don't need to make this distinction. Thus, in the sequel, we shall let "f" stand for both f and $\overline{f}$. Heuristically, the reader can view much of the ensuing discussion as taking place within PRA rather than being about PRA.

Enough discussion! Let's get to work.

We begin with some examples of primitive recursive functions.

5.1. EXAMPLES. The following functions are primitive recursive; moreover, they provably have their defining properties in PRA:

i. $K(\vec{x}) = k$, any number k — *Constant*

ii. $A(x,y) = x + y$ — *Addition*

iii. $M(x,y) = x \cdot y$ — *Multiplication*

iv. $E(x,y) = x^y$ — *Exponentiation*

v. $pd(x) = \begin{cases} x - 1, & x > 0 \\ 0, & x = 0 \end{cases}$ — *Predecessor*

The predecessor function is defined by recursion:

$$pd(0) = 0$$

$$pd(Sx) = x.$$

vi. $x \dot{-} y = \begin{cases} x - y, & x \geq y \\ 0, & x < y. \end{cases}$ — *Cut-Off Subtraction*

Cut-off subtraction is defined by iterating the predecessor function:

$$x \dot{-} 0 = x$$

$$x \dot{-} Sy = pd(x \dot{-} y).$$

vii. $sg(x) = \begin{cases} 0, & x = 0 \\ 1, & x > 0 \end{cases}$ — *Signum*

viii. $\overline{sg}(x) = \begin{cases} 1, & x = 0 \\ 0, & x > 0. \end{cases}$ — *Signum Complement*

The signum function and it complement can be defined by recursion,

$$sg(0) = 0 \qquad \overline{sg}(0) = 1$$

$$sg(Sx) = 1; \qquad \overline{sg}(Sx) = 0,$$

or by cut-off subtraction,

$$\overline{sg}(x) = 1 \dot{-} x, \qquad\qquad sg(x) = 1 \dot{-} \overline{sg}(x).$$

(The conflict between the somewhat firmly entrenched $\overline{sg}$ notation and that of the previous section is another reason for my dropping the overlining convention of that section. Another common bit of notation we will encounter, $\tilde{f}$, for the course-of-values function associated with f would, with the overlining convention, lead, not to ambiguity, but to ugliness: $\overline{\tilde{f}}$.)

The functions sg and $\overline{sg}$ are logical rather than numerical. To understand their use, we introduce the notion of a primitive recursive relation.

5.2. DEFINITION. A relation $R \subseteq \omega^n$ is *primitive recursive* (sometimes: *PR*) if its *representing function*,

$$\chi_R(\vec{x}) = \begin{cases} 0, & R(\vec{x}) \\ 1, & \sim R(\vec{x}), \end{cases}$$

is a primitive recursive function.

By Examples 5.1.vii and 5.1.viii, the relations of being 0 and of being positive are primitive recursive. Our next goal is to give a listing of more examples of primitive recursive relations. We set about this task in a rather pedestrian manner.

5.3. LEMMA. (Definition by Cases). Let g_1, g_2, and h be primitive recursive functions and define f by

$$f(\vec{x}) = \begin{cases} g_1(\vec{x}), & h(\vec{x}) = 0 \\ g_2(\vec{x}), & h(\vec{x}) \neq 0. \end{cases}$$

Then: f is primitive recursive.

Proof: $f(\vec{x}) = g_1(\vec{x}) \cdot \overline{sg}(h(\vec{x})) + g_2(\vec{x}) \cdot sg(h(\vec{x}))$. QED

5.4. COROLLARY. The relation of equality is primitive recursive.

Proof: $h(x,y) = |x - y| = (x \dot{-} y) + (y \dot{-} x)$. QED

The logical rôle played by sg and $\overline{sg}$ is further illustrated by the following.

5.5. REMARK. The class of primitive recursive relations is closed under complement, intersection, and union.

Proof: If χ_R, χ_S are given,

$$\chi_{\sim R}(\vec{x}) = \overline{sg}(\chi_R(\vec{x}))$$

$$\chi_{R \wedge S}(\vec{x}) = sg(\chi_R(\vec{x}) + \chi_S(\vec{x}))$$

$$\chi_{R \vee S}(\vec{x}) = \chi_R(\vec{x}) \cdot \chi_S(\vec{x}).$$

QED

More interesting is the observation that, as disjunction corresponds to multiplication, bounded existential quantification corresponds to bounded iterated multiplication: Let $R(y,\vec{x})$ be given. Then

$$\chi_{\exists y \le x R(y,\vec{x})}(x,\vec{x}) = \prod_{y \le x} \chi_R(y,\vec{x}),$$

where $\exists y \le x\, \phi$ means $\exists y (y<x \wedge \phi)$. Further,

$$\chi_{\forall y \le x R(y,\vec{x})}(x,\vec{x}) = \chi_{\sim \exists y \le x \sim R}(x,\vec{x}),$$

where $\forall y \le x\, \phi$ means $\forall y (y \le x \rightarrow \phi)$.

5.6. LEMMA. i. Let $g(y,\vec{x})$ be primitive recursive. Then f_1, f_2 are primitive recursive, where

$$f_1(x,\vec{x}) = \sum_{y \le x} g(y,\vec{x})$$

$$f_2(x,\vec{x}) = \prod_{y \le x} g(y,\vec{x}).$$

ii. If $R(y,\vec{x})$ is a primitive recursive relation, then the relations S, T are also primitive recursive, where

$$S(x,\vec{x}) \leftrightarrow \exists y \le x\, R(y,\vec{x})$$

$$T(x,\vec{x}) \leftrightarrow \forall y \le x\, R(y,\vec{x}).$$

I leave the proof to the reader.

The closure of the class of primitive recursive relations under bounded quantification is a powerful tool in exposing the primitive recursiveness of various relations. From arithmetic, for example, we have the following:

5.7. EXAMPLES. The following relations are primitive recursive:

i. $x|y$ *(x divides y)*

ii. $x < y$; $x \le y$

iii. $Res(x,y) = z$ *(the residue of x after division by y is z)*

iv. $z = [x/y]$ *(z is the greatest integer in x/y)*

v. $x \in exp(y)$ *(x occurs as an exponent in the dyadic expansion of y, i.e. the x-th digit of y written in binary is a 1)*

Proofs: i. $x|y \leftrightarrow \exists z \le y\ (x \cdot z = y)$. (Here and below I use the hidden lemma: If R, f_1, ..., f_n are primitive recursive, so is $R(f_1(\vec{x}),\ldots,f_n(\vec{x}))$. This follows by the closure of the class of primitive recursive functions under composition.)

ii. $x \le y \leftrightarrow \exists z \le y(z = x)$

$x < y \leftrightarrow \exists z \le y(z = x + 1)$

iii. $Res(x,y) = z \leftrightarrow \exists w \le x(x = y \cdot w + z \wedge z < y)$

iv. $z = [x/y] \leftrightarrow y \cdot z \le x < y(z + 1)$

v. $x \in exp(y) \leftrightarrow y = \ldots + 2^x + \ldots$

$\leftrightarrow Res(y, 2^{x+1}) = 2^x + \ldots$

$\leftrightarrow [Res(y, 2^{x+1})/2^x] = 1.$ QED

This last relation, $x \in exp(y)$, is a very important one. The map *set*, defined by

$$set(0) = \{\ \}$$

$$set(n+1) = \{set(m):\ m \in exp(n+1)\}$$

identifies the set of natural numbers with the hereditarily finite sets, i.e. the sets of finite rank in ordinary set theory. This means that we have at our disposal the coding capability of finite set theory; in particular, finite sequences can be treated as finite functions the domains of which are finite ordinals ...

We will not take the set theoretic route. Although set theoretic encoding is, perhaps, more familiar than arithmetic encoding, it is yet encoding and we would still have to verify its primitive recursive nature. Arithmetic encoding affords us a short cut. It is also, in my opinion, more interesting.

The relation $x \in exp(y)$ gives a one-one correspondence between natural numbers and finite sets thereof: The finite set D_x with so-called *canonical index* x is

$$D_0 = \{\ \},\quad \text{if } x = 0$$

$$D_x = \{x_0,\ldots,x_{n-1}\},\quad \text{if } x = 2^{x_0} + \ldots + 2^{x_{n-1}} \text{ where } x_0 < \ldots < x_{n-1}.$$

Finite sets can be put into one-one correspondence with finite sequences of natural numbers. In fact, a finite set is, when listed in order, a finite strictly increasing sequence of natural numbers. If we restrict our attention to *positive* integers, a correspondence between sequences of positive integers and strictly increasing such sequences is readily given: The sequence,

$$(a_0,\ldots,a_{n-1}),$$

can be associated with the strictly increasing sequence,

$$(a_0,a_0 + a_1,\ldots,a_0 + a_1 + \ldots + a_{n-1});$$

and, conversely, a strictly increasing sequence,

$$(s_0,\ldots,s_{n-1}), \quad s_0 < s_1 < \ldots < s_{n-1},$$

can readily be decoded into an ordinary sequence,

$$(s_0,s_1 - s_0,\ldots,s_{n-1} - s_{n-2}).$$

With the possibility that 0 could occur among the a_i's and that the resulting increasing sequence might not be strict, for natural numbers we modify the procedure: To encode a sequence as a strictly increasing sequence, we add a lot of 1's: $(a_0,\ldots,a_{n-1}) \rightarrow (a_0,a_0 + a_1 + 1,a_0 + a_1 + a_2 + 2,\ldots,a_0+\ldots+a_{n-1}+(n-1))$; to decode a strictly increasing sequence, we subtract such:

$$(s_0,\ldots,s_{n-1}) \rightarrow (s_0,s_1 - s_0 - 1,\ldots,s_{n-1} - s_{n-2} - 1).$$

Since the increasing sequences are just sets, this gives an encoding S_x of finite sequences of natural numbers:

$$S_0 = (\ )$$

$$S_x = (x_0,\ldots,x_{n-1}),$$

where $D_x = \{x_0,x_0 + x_1 + 1, \ldots, x_0 + \ldots + x_{n-1} + n - 1\}$ for $x > 0$; conversely, if $x = 2^{x_0} + \ldots + 2^{x_{n-1}}$ with $x_0 < x_1 < \ldots < x_{n-1}$, then

$$S_x = (x_0,x_1 - x_0 - 1,\ldots,x_{n-1} - x_{n-2} - 1).$$

The correspondence $x \rightarrow S_x$ is a one-one correspondence between the set of natural numbers and the set of finite sequences thereof. This fact cannot be proven in PRA; indeed, it cannot even be expressed therein. What can be done is, upon identifying each number x with its corresponding sequence S_x, the important operations on finite sequences can be primitive recursively simulated within PRA and all important properties of such sequences and operations can be proven within this theory. What are these operations? There are three: finding the length, projecting onto a coördinate, and concatenating sequences.

Finding the length of the sequence S_x identified with x is relatively easy: The length of S_0 (= $(\ )$) is 0 and of $S_x = (x_0,\ldots,x_{n-1})$ is n; but

$$x = 2^{x_0} + 2^{x_0+x_1+1} + \ldots + 2^{x_0+\ldots+x_{n-1}+n-1},$$

whence n is the number of 1's in the binary expansion of x. Thus,

$$lh(x) = \sum_{y \leq x} \overline{sg}(\chi_{exp}(y,x)),$$

where χ_{exp} is the representing function of the relation $y \in exp(x)$. Hence we see that $lh(x)$ is a primitive recursive function.

To verify the primitive recursiveness of the projection and concatenation functions, we need a definition and a lemma.

5.8. DEFINITION. (Bounded μ-Operator). Let $g(y,\vec{x})$ be a function. We define

$$f(x,\vec{x}) = \mu y < x\,(g(y,\vec{x}) = 0)$$

by $f(x,\vec{x})$ = the least $y < x$ such that $g(y,\vec{x}) = 0$, if such a y exists, and $f(x,\vec{x}) = x$, otherwise.

5.9. LEMMA. If $g(y,\vec{x})$ is primitive recursive, then so is

$$f(x,\vec{x}) = \mu y < x\,(g(y,\vec{x}) = 0).$$

Proof: First define the auxiliary function,

$$f_1(x,\vec{x}) = \begin{cases} 0, & \exists y \leq x\,(g(y,\vec{x}) = 0) \\ 1, & \sim\exists y \leq x\,(g(y,\vec{x}) = 0), \end{cases}$$

by cases. Then observe that

$$f(x,\vec{x}) = \sum_{y < x} f_1(y,\vec{x}).$$

QED

With Lemma 5.9, we readily obtain the primitive recursiveness of the projection functions:

$$(x)_0 = \mu y \leq x\,(y \in exp(x))$$

$$(x)_{i+1} = \mu y \leq x\,(y \in exp(x) \wedge y > (x)_i) \dot{-} ((x)_i + 1).$$

Finally, concatenation is also fairly simple. Let

$$S_x = (x_0,\ldots,x_{n-1}), \quad S_y = (y_0,\ldots,y_{m-1})$$

be non-trivial. The code of

$$S_x * S_y = (x_0,\ldots,x_{n-1},y_0,\ldots,y_{m-1})$$

is

$$2^{x_0} + 2^{x_0+x_1+1} + \ldots + 2^{x_0+x_1+\ldots+x_{n-1}+n-1} +$$

$$+ 2^{x_0+\ldots+x_{n-1}+y_0+n} + \ldots + 2^{x_0+\ldots+x_{n-1}+y_0+\ldots+y_{m-1}+n+m-1},$$

which is

$$x + 2^{x_0 + \ldots + x_{n-1} + n} \cdot y.$$

Thus, if we define

$$E(x) = x_0 + \ldots + x_{n-1} + n = \sum_{z < lh(x)} ((x)_z + 1),$$

we have

$$x*y = \begin{cases} 0, & x = 0 \wedge y = 0 \\ y, & x = 0 \wedge y \neq 0 \\ x, & x \neq 0 \wedge y = 0 \\ x + 2^{E(x)} \cdot y, & x \neq 0 \wedge y \neq 0. \end{cases}$$

We have now reached the point that most proof theorists consider the natural place to begin-- a theory of finite sequences. For, most syntactic objects are finite sequences and the syntactic coding is readily established once we can project and concatenate and such. We have only one last lemma to prove before we encode syntax. This lemma establishes the primitive recursiveness of functions defined by a slightly more complicated recursion-- the type used in the inductive definitions of syntactic concepts in section 4.

5.10. LEMMA. (Course-of-Values Recursion). Let g, h be primitive recursive and define f by

$$f(0,\vec{x}) = g(\vec{x})$$

$$f(x+1,\vec{x}) = h(\tilde{f}(x,\vec{x}),x,\vec{x}),$$

where $\tilde{f}(x,\vec{x}) = (f(0,\vec{x}),\ldots,f(x,\vec{x}))$ is the *course-of-values* function associated with f. Then: f is primitive recursive.

Proof: First observe that $\tilde{f}$ is primitive recursive:

$$\tilde{f}(0,\vec{x}) = (g(\vec{x})) \quad (= 2^{g(\vec{x})})$$

$$\tilde{f}(x+1,\vec{x}) = \tilde{f}(x,\vec{x})*(h(\tilde{f}(x,\vec{x}),x,\vec{x})).$$

Then define f explicitly: $f(x,\vec{x}) = (\tilde{f}(x,\vec{x}))_x$. QED

At last! We can begin the encoding of syntax within PRA. We start by assigning numerical codes to basic syntactic symbols. These codes are just sequence numbers, denoted $\ulcorner s \urcorner$ for a syntactic symbol s.

Some of these are quite easy to give:

CONSTANT. $\ulcorner \bar{0} \urcorner = (0)$

RELATION SYMBOL. $\ulcorner = \urcorner = (1)$

PROPOSITIONAL CONNECTIVES. $\ulcorner \sim \urcorner = (2)$; $\ulcorner \wedge \urcorner = (3)$; $\ulcorner \vee \urcorner = (4)$; $\ulcorner \rightarrow \urcorner = (5)$

QUANTIFIERS. $\ulcorner \forall \urcorner = (6)$; $\ulcorner \exists \urcorner = (7)$

VARIABLES. $\ulcorner v_i \urcorner = (8,i)$

FUNCTION SYMBOLS. These are a bit more complex. The constant, relation symbol, connectives and quantifiers were each given a number as a code. There are only finitely many of these and there is no difficulty with them. A number is the code of, say, $\rightarrow$, iff it is (5) (i.e. $2^5 = 32$). The code of a variable v_i was more complicated-- it contained first a component, 8, to identify it as a variable and then a component, i, to tell which variable it was. Codes for function symbols must give more information-- first, a component, 9, telling us it is intended to code a function symbol, and then a list of additional pertinent information such as the arity of the function and the schema used to define it. We define these codes inductively:

	Function	*Index*
$F1.$	$Z(x) = 0$	$(9,1,1)$
$F2.$	$S(x) = x + 1$	$(9,2,1)$
$F3_i^n.$	$P_i^n(\vec{x}) = x_i$	$(9,3,n,i)$
$F4_m^n.$	$f(\vec{x}) = g(h_1(\vec{x}),\ldots,h_m(\vec{x}))$	$(9,4,n,m,(g^*,h_1{}^*,\ldots,h_m{}^*))$
$F5^n.$	$f(0,\vec{x}) = g(\vec{x})$ $f(x+1,\vec{x}) = h(f(x,\vec{x}),x,\vec{x})$,	$(9,5,n+1,g^*,h^*)$

where g^*, $h_1{}^*$, $\ldots$, $h_m{}^*$ in $F4_m^n$ and g^*, h^* in $F5^n$ are the codes assigned to the functions g, h_1, $\ldots$, h_m and g, h, respectively.

Before we go on to encode more complex syntactic objects, we must check the primitive recursiveness of the encoding thus far given. For example, the set of codes of variables is primitive recursive:

$$x \in Var \text{ iff } lh(x) = 2 \wedge (x)_0 = 8;$$

alternatively:

$$x \in Var \text{ iff } \exists y \le x\,(x = (8)*(y)),$$

where we use the easy inequality, $(z)_w \le z$, for all w, z.

The primitive recursiveness of the set of function symbols is less easily established. The function symbols are generated by a course-of-values recursion and their recognisability depends on such. Writing down the representing function of the relation in question is facilitated by first listing a few related primitive recursive properties. First, we can define the codes of initial functions outright:

$$Init(x) \leftrightarrow x = (9,1,1) \vee x = (9,2,1) \vee \exists n{\le}x\, \exists i{\le}n\, \big(1 \le i \wedge x = (9,3,n,i)\big).$$

Next, we can assert the code of a function f generated by $F4$ to have the right form and to relate properly to the codes g^*, h_1^*, ..., h_m^* occurring in the last component:

$$\begin{aligned} F4(x) \leftrightarrow\ & lh(x) = 5 \wedge (x)_0 = 9 \wedge (x)_1 = 4 \wedge (x)_2 > 0 \wedge (x)_3 > 0 \wedge \\ & \wedge\ lh((x)_4) = (x)_3 + 1 \wedge (((x)_4)_0)_2 = (x)_3 \wedge \\ & \wedge\ \forall i < (x)_3 \big((((x)_4)_{i+1})_2 = (x)_2 \big). \end{aligned}$$

$F4(x)$ asserts x to have the form $(9,4,n,m,(a,b_1,\ldots,b_m))$, where, should $a,b_1,\ldots,b_m$ also be function codes, a codes an m-ary function and $b_1,\ldots,b_m$ code n-ary functions. All that is missing is the inductive assumption that $a,b_1,\ldots,b_m$ are codes... Similarly, we can assert the code of a function f generated by $F5$ to have the right form:

$$\begin{aligned} F5(x) \leftrightarrow\ & lh(x) = 5 \wedge (x)_0 = 9 \wedge (x)_1 = 5 \wedge (x)_2 > 1 \wedge \\ & \wedge ((x)_3)_2 = (x)_2 \dot{-} 1 \wedge ((x_4)_2 = (x)_2 + 1. \end{aligned}$$

$F5(x)$ asserts x to have the form $(9,5,n{+}1,a\ b)$, where, should a,b also be codes, the arities match: The function with code a is n-ary; that with code b is $n{+}2$-ary.

With these abbreviations, we quickly define the representing function of the class of codes of function symbols by course-of-values recursion:

$$f(x) = \begin{cases} 0, & Init(x) \\ 0, & F4(x) \wedge \forall i \le (x)_3 \big(f(((x)_4)_i) = 0 \big) \\ 0, & F5(x) \wedge f((x)_3) = 0 \wedge f((x)_4) = 0 \\ 1, & otherwise. \end{cases}$$

Thus,

$$x \in FncSym: \quad f(x) = 0$$

is a primitive recursive relation.

The next step is to mimic the inductive definitions (4.2) of terms and formulae to generate codes for such. There are two approaches we could take. For example, a term

$$\bar{f}v_0\bar{g}v_0v_1$$

can be viewed as a sequence and given the code

$$(\ulcorner\bar{f}\urcorner, \ulcorner v_0\urcorner, \ulcorner\bar{g}\urcorner, \ulcorner v_0\urcorner, \ulcorner v_1\urcorner),$$

or, perhaps less obviously, it can be viewed as a tree,

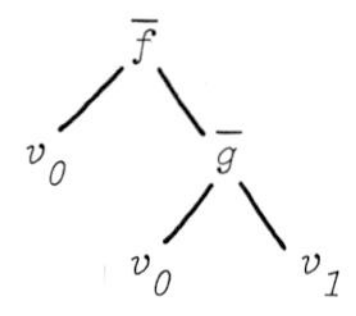

and given the code

$$(\ulcorner\bar{f}\urcorner, \ulcorner v_0\urcorner, (\ulcorner\bar{g}\urcorner, \ulcorner v_0\urcorner, \ulcorner v_1\urcorner)).$$

Gödel originally chose the first alternative; in metamathematics, one generally chooses the latter nowadays.

Thus, we define the codes of terms as follows:

i. $\ulcorner\bar{0}\urcorner$, $\ulcorner v_i\urcorner$ are codes of $\bar{0}$, v_i, respectively

ii. if $\bar{f}$ is an n-ary function symbol and $t_1,\ldots,t_n$ are terms with codes $\ulcorner t_1\urcorner,\ldots,\ulcorner t_n\urcorner$, respectively, then

$$\ulcorner\bar{f}t_1\ldots t_n\urcorner = (\ulcorner\bar{f}\urcorner, \ulcorner t_1\urcorner,\ldots,\ulcorner t_n\urcorner).$$

The representing function of the set of codes is defined by course-of-values recursion:

$$f(x) = \begin{cases} 0, & x = \ulcorner\bar{0}\urcorner \\ 0, & x \in Var \\ 0, & (x)_0 \in FncSym \wedge lh(x) = ((x)_0)_1 + 1 \wedge \forall i{<}lh(x)\big(f((x)_{i+1} = 0\big) \\ 1, & otherwise. \end{cases}$$

(The garbage in the third clause asserts x to be of the form $(\ulcorner\bar{f}\urcorner, \ulcorner t_1\urcorner,\ldots,\ulcorner t_n\urcorner)$ and $\bar{f}$ to be n-ary.) Thus,

$$x \in Term:\quad f(x) = 0$$

is primitive recursive.

Formulae are given codes inductively as follows:

i. if t_1,t_2 are terms, $\ulcorner{=}t_1t_2\urcorner = (\ulcorner{=}\urcorner, \ulcorner t_1\urcorner, \ulcorner t_2\urcorner)$

ii. if ϕ,ψ are formulae,

$$\ulcorner \sim\phi \urcorner = (\ulcorner \sim \urcorner, \ulcorner \phi \urcorner) \qquad \ulcorner \phi \wedge \psi \urcorner = (\ulcorner \wedge \urcorner, \ulcorner \phi \urcorner, \ulcorner \psi \urcorner)$$

$$\ulcorner \phi \vee \psi \urcorner = (\ulcorner \vee \urcorner, \ulcorner \phi \urcorner, \ulcorner \psi \urcorner) \qquad \ulcorner \phi \to \psi \urcorner = (\ulcorner \to \urcorner, \ulcorner \phi \urcorner, \ulcorner \psi \urcorner)$$

iii. if ϕ is a formula and v a variable,

$$\ulcorner \forall v\phi \urcorner = (\ulcorner \forall \urcorner, \ulcorner v \urcorner, \ulcorner \phi \urcorner) \qquad \ulcorner \exists v\phi \urcorner = (\ulcorner \exists \urcorner, \ulcorner v \urcorner, \ulcorner \phi \urcorner).$$

Again, the representing function of the set of codes of formulae is readily given by a course-of-values recursion:

$$f(x) = \begin{cases} 0, & \exists y \le x \, \exists z \le x \left(y \in Term \wedge z \in Term \wedge x = (\ulcorner = \urcorner, y, z) \right) \\ 0, & \exists y < x \left(f(y) = 0 \wedge x = (\ulcorner \sim \urcorner, y) \right) \\ 0, & \exists y < x \exists z < x \left(f(y) = 0 \wedge f(z) = 0 \wedge \right. \\ & \quad \left. \wedge \left(x = (\ulcorner \wedge \urcorner, y, z) \vee x = (\ulcorner \vee \urcorner, y, z) \vee x = (\ulcorner \to \urcorner, y, z) \right) \right) \\ 0, & \exists y < x \exists z < x \left(f(z) = 0 \wedge y \in Var \wedge \right. \\ & \quad \left. \wedge \left(x = (\ulcorner \forall \urcorner, y, z) \vee x = (\ulcorner \exists \urcorner, y, z) \right) \right) \\ 1, & otherwise. \end{cases}$$

Hence,

$$x \quad Fmla: \qquad f(x) = 0$$

is primitive recursive.

It may, at this point, be prudent to collect a list of such primitive recursive relations:

5.11. LIST. The following relations are primitive recursive:

	RELATION	*EXPLANATION*
i.	$x \in Var$	*"x is the code of a variable"*
ii.	$x \in FncSym$	*"x is the code of a function symbol"*
iii.	$x \in Term$	*"x is the code of a term"*
iv.	$x \in Fmla$	*"x is the code of a formula"*
v.	$x \in Var_T(y)$	*"x is the code of a variable occurring in the term with code y"*
vi.	$x \in Var_F(y)$	*"x is the code of a variable occurring in the formula with code y"*
vii.	$x \in FV(y)$	*"x is the code of a free variable of the formula with code y"*
viii.	$x \in BV(y)$	*"x is the code of a bound variable of the formula with code y"*
ix.	$x \in Sent$	*"x is the code of a sentence (i.e. a*

		formula with no free variables)"
x.	x *sub for* y *in* z	"x *is the code of a term substitutable for the variable with code* y *in the formula with code* z".

We have already verified the primitive recursiveness of the relations i-iv; that of the relations v-viii and x is verified similarly and I omit the details; the primitive recursiveness of ix is readily established:

$$x \in Sent \;\leftrightarrow\; x \in Fmla \wedge \forall y < x \left(y \notin FV(x) \right).$$

The last relation, that of substitutability of a term for a variable in a formula, was introduced for the purpose of properly stating the logical axioms. To recognise these primitive recursively, we need also a substitution function which yields the code $\ulcorner \phi t \urcorner$ from $\ulcorner \phi v \urcorner$, $\ulcorner v \urcorner$, and $\ulcorner t \urcorner$. There are actually two such functions, one for substitution inside terms and one for substitution inside formulae. Both are defined by course-of-values recursion.

We first define

$$subst^T(\ulcorner t_1(v) \urcorner; \ulcorner v \urcorner, \ulcorner t_2 \urcorner) \;=\; t_1(t_2),$$

the result of substituting t_2 for v in $t_1(v)$:

$$subst^T(x;y,z) \;=\; \begin{cases} x, & x \in Var \wedge y \in Var \wedge z \in Term \wedge x \neq y \\ z, & x \in Var \wedge y \in Var \wedge z \in Term \wedge x = y \\ (x)_0 * \underset{i < lh(x)}{*} (subst^T((x)_i;y,z)), & \text{if: } \\ & x \in Term \wedge y \in Var \wedge z \in Term \wedge x \notin Var \\ 0, & otherwise, \end{cases}$$

where $\underset{i<y}{*} f(i)$ is defined by

$$\underset{i<0}{*} f(i) \;=\; (\;)$$

$$\underset{i<y+1}{*} f(i) \;=\; \left(\underset{i<y}{*} f(i) \right) * f(y).$$

(The reader should verify that the third clause in the definition of $subst^T$ yields:

$$subst^T(\ulcorner f t_1 \ldots t_n \urcorner; \ulcorner v \urcorner, \ulcorner t \urcorner) \;=$$

$$=\; (\ulcorner f \urcorner, subst^T(\ulcorner t_1 \urcorner; \ulcorner v \urcorner, \ulcorner t \urcorner), \ldots, subst^T(\ulcorner t_n \urcorner; \ulcorner v \urcorner, \ulcorner t \urcorner)).)$$

Using substitution within terms we define substitution within formulae,

$$subst(\ulcorner \phi v \urcorner; \ulcorner v \urcorner, \ulcorner t \urcorner) \;=\; \ulcorner \phi t \urcorner,$$

as follows:

$$subst(x;y,z) = \begin{cases} ((x)_0, subst^T((x)_1;y,z), subst^T((x)_2;y,z)), & \text{if} \\ \quad x \in Fmla \wedge \exists w_1 w_2 \leq x\,(x = (\ulcorner = \urcorner, w_1, w_2)) \wedge \\ \quad \wedge\, y \in Var \wedge z \in Term \\ (\ulcorner \sim \urcorner, subst((x)_1;y,z)), & \text{if} \\ \quad x \in Fmla \wedge y \in Var \wedge z \in Term \wedge (x)_0 = \ulcorner \sim \urcorner \\ \vdots \\ etc. \end{cases}$$

I leave to the reader the odious task of completing the definition. (In doing so, the reader should bear in mind that

$$subst(\ulcorner \forall v_i \phi \urcorner; \ulcorner v_j \urcorner, \ulcorner t \urcorner) = (\ulcorner \forall \urcorner, \ulcorner v_i \urcorner, subst(\ulcorner \phi \urcorner; \ulcorner v_j \urcorner, \ulcorner t \urcorner))$$

if $i \neq j$, but

$$subst(\ulcorner \forall v_i \phi \urcorner; \ulcorner v_i \urcorner, \ulcorner t \urcorner) = \ulcorner \forall v_i \phi \urcorner.)$$

Armed with the substitution function, we can primitive recursively define the set of logical axioms. First, there are the propositional axioms, which are most easily defined if we introduce some abbreviations:

$$\dot{\sim} x = (\ulcorner \sim \urcorner, x) \qquad x \mathbin{\dot\wedge} y = (\ulcorner \wedge \urcorner, x, y)$$

$$x \mathbin{\dot\vee} y = (\ulcorner \vee \urcorner, x, y) \qquad x \mathbin{\dot\rightarrow} y = (\ulcorner \rightarrow \urcorner, x, y).$$

With this, "x codes a propositional axiom" (Cf. 4.3.) is defined by

$$\begin{aligned} PropAx(x) \leftrightarrow\ & \exists yzw \leq x\,(x = y \mathbin{\dot\rightarrow} (z \mathbin{\dot\rightarrow} y) \vee \\ & \vee\, x = (y \mathbin{\dot\rightarrow} (z \mathbin{\dot\rightarrow} w)) \mathbin{\dot\rightarrow} ((y \mathbin{\dot\rightarrow} w) \mathbin{\dot\rightarrow} (z \mathbin{\dot\rightarrow} w)) \vee \\ & \vee \ldots \vee\, x = (\dot\sim\dot\sim y) \mathbin{\dot\rightarrow} y). \end{aligned}$$

(The "..." indicates 7 missing disjuncts.)

The quantifier axioms (4.6) are defined by:

$$\begin{aligned} QuantAx(x) \leftrightarrow\ & \exists yzw \leq x\,(y \in Var \wedge z \in Fmla \wedge w \in Term \wedge w \text{ sub for } y \text{ in } z \\ & \wedge\, (x = (\ulcorner \forall \urcorner, y, z) \mathbin{\dot\rightarrow} subst(z;y,w) \vee \\ & \vee\, x = subst(z;y,w) \mathbin{\dot\rightarrow} (\ulcorner \exists \urcorner, y, z))). \end{aligned}$$

The equality axioms (4.8) consist of three individual axioms with specific codes, say e_1, e_2, e_3, and the schema:

$$v_i = v_0 \to \overline{f}v_1 \ldots v_i \ldots v_n = \overline{f}v_1 \ldots v_0 \ldots v_n,$$

where $\overline{f}$ is n-ary. First, define

$$SimpTerm(x) \leftrightarrow \exists fn \leq x \bigl(f \in FnSym \wedge (f)_1 = n \wedge lh(x) = n+1 \wedge (x)_0 = f \wedge$$
$$\wedge \forall i < n \bigl((x)_{i+1} = (8,i+1) \bigr) \bigr).$$

$SimpTerm(x)$ asserts $x = \ulcorner \overline{f}v_1 \ldots v_n \urcorner$ for an n-ary function symbol $\overline{f}$. Define also the abbreviation

$$x \doteq y: \quad (\ulcorner = \urcorner, x, y).$$

Then:

$$EqAx(x) \leftrightarrow x = e_1 \vee x = e_2 \vee x = e_3 \vee \exists iyz \leq x \bigl(x = y \dot{\to} z \wedge$$
$$\wedge\ y = (8,i) \doteq (8,0) \wedge \exists w \leq z \bigl(Simpterm(w) \wedge$$
$$\wedge\ z = w \doteq subst^T(w;(8,i),(8,0)) \bigr) \bigr).$$

I leave as an exercise to the reader the exhibition as primitive recursive of the set of nonlogical axioms (4.4), say $NonLogAx(x)$. Given this, one can finally define

$$Ax(x) \leftrightarrow PropAx(x) \vee QuantAx(x) \vee EqAx(x) \vee NonLogAx(x).$$

To define provability, one must also express the rules of inference (4.7). First, "x follows from y, z by modus ponens" is expressed by:

$$MP(x;y,z) \leftrightarrow z = y \dot{\to} x.$$

There are two generalisation rules:

$$Gen_1(x;y) \leftrightarrow \exists vwz \leq x \bigl(v \in Var \wedge w \in Fmla \wedge z \in Fmla \wedge v \notin FV(z) \wedge$$
$$\wedge\ y = w \dot{\to} z \wedge x = (\ulcorner \exists \urcorner, v, w) \dot{\to} z \bigr)$$

$$Gen_2(x;y) \leftrightarrow \exists vwz \leq x \bigl(v \in Var \wedge w \in Fmla \wedge z \in Fmla \wedge v \notin FV(z) \wedge$$
$$\wedge\ y = w \dot{\to} z \wedge x = w \dot{\to} (\ulcorner \forall \urcorner, v, z) \bigr).$$

(Cf. 5.11 for FV.) We can combine these:

$$Gen(x;y) \leftrightarrow Gen_1(x;y) \vee Gen_2(x;y).$$

A formal derivation is, of course, a sequence of formulae $\phi_0, \ldots, \phi_n$, each member of which is either an axiom or follows from earlier entries in the sequence by a rule of inference. We assign to the sequence $\phi_0, \ldots, \phi_n$ the code $(\ulcorner \phi_0 \urcorner, \ulcorner \phi_1 \urcorner, \ldots, \ulcorner \phi_n \urcorner)$. We immediately see the primitive recursiveness of the relation "y codes a derivation of the formula with code x":

$$Prov(y,x) \leftrightarrow (y)_{lh(y) \dot{-} 1} = x \wedge \forall i < lh(y) \bigl(Ax((y)_i) \vee$$
$$\vee\ \exists jk < i\, MP((y)_i;(y)_j,(y)_k) \vee \exists j < i\, Gen((y)_i;(y)_j) \bigr).$$

With this we can also define, but *not* primitive recursively, the relation "x codes a provable formula":

$$Pr(x) \leftrightarrow \exists y Prov(y,x).$$

Before proceeding, let us pause to collect the functions and relations we have shown primitive recursive:

5.12. SECOND LIST. The following are primitive recursive:

	FUNCTION or RELATION	*EXPLANATION*
i.	$subst^T(x;y,z)$	*"the result of substituting the term coded by z for the variable coded by y in the term coded by x"*
ii.	$subst(x;y,z)$	*"the result of substituting the term coded by z for the variable coded by y in the formula coded by x"*
iii.	$PropAx(x)$	*"x codes a propositional axiom"*
iv.	$QuantAx(x)$	*"x codes a quantifier axiom"*
v.	$EqAx(x)$	*"x codes an equality axiom"*
vi.	$NonLogAx(x)$	*"x codes a nonlogical axiom"*
vii.	$Ax(x)$	*"x codes an axiom"*
viii.	$MP(x:y,z)$	*"the formula with code x follows by modus ponens from the formulae with codes y, z"*
ix.	$Gen(x;y)$	*"the formula with code x follows by a generalisation rule from the formula with code y"*
x.	$Prov(y,x)$	*"y codes a proof of the formula with code x"*

And, of course, in addition to these there is the non-primitive recursive

$$Pr(x) = \exists y Prov(y,x).$$

Now, primitive recursive relations behave, by Theorem 4.12, exactly as they should: If $R \subseteq \omega^n$ is primitive recursive, one has, for all $k_1,\ldots,k_n \in \omega$

$$Rk_1\ldots k_n \text{ holds} \Rightarrow \chi_R(k_1,\ldots,k_n) = 0$$
$$\Rightarrow \mathrm{PRA} \vdash \overline{\chi}_R(\overline{k}_1,\ldots,\overline{k}_n) = \overline{0}$$
$$\Rightarrow \mathrm{PRA} \vdash \overline{R}(\overline{k}_1,\ldots,\overline{k}_n),$$

assuming, of course, the representation $\overline{\chi}_R(v_1,\ldots,v_n) = \overline{0}$ for $\overline{R}v_1\ldots v_n$. Similarly,

$$Rk_1\ldots k_n \text{ fails} \Rightarrow \chi_R(k_1,\ldots,k_n) = 1$$

$$\Rightarrow \text{PRA} \vdash \bar{\chi}_R(\bar{k}_1,\ldots,\bar{k}_n) = \bar{1}$$
$$\Rightarrow \text{PRA} \vdash \sim\bar{R}(\bar{k}_1,\ldots,\bar{k}_n).$$

When we existentially quantify such a relation (as in the case of $Pr(x)$), we get only

$$\begin{aligned} \exists k\,(R(k,k_1,\ldots,k_n) \text{ holds}) &\Rightarrow \exists k\,(\chi_R(k,k_1,\ldots,k_n) = 0) \\ &\Rightarrow \exists k\,(\text{PRA} \vdash \bar{\chi}_R(\bar{k},\bar{k}_1,\ldots,\bar{k}_n) = \bar{0}) \\ &\Rightarrow \text{PRA} \vdash \exists v\,(\bar{\chi}_R(v,\bar{k}_1,\ldots,\bar{k}_n) = \bar{0}) \\ &\Rightarrow \text{PRA} \vdash \exists v\bar{R}(v,\bar{k}_1,\ldots,\bar{k}_n). \end{aligned}$$

We get the converse by the fact that PRA only proves true assertions (a fact that fails for many other theories). But we get nothing about the relation $\sim\exists vR$. However, what we have is enough to establish the following:

5.13. LEMMA. (*D1*). For any formula ϕ,

$$\text{PRA} \vdash \phi \Rightarrow \text{PRA} \vdash Pr(\ulcorner\phi\urcorner).$$

(It is bad form to change notation like this, but I have given up on using the capitals of the Orator typing ball on this rented typewriter. Henceforth, I shall stick to the lower case, which prints legibly without the need for touching up. If you keep this secret with me, the publisher will never find out and get mad at me.)

Proof: If $\text{PRA} \vdash \phi$, there is a proof $\phi_1,\ldots,\phi_n = \phi$ of ϕ. Thus

$$\overline{Prov}((\ulcorner\phi_1\urcorner,\ldots,\ulcorner\phi_n\urcorner),\ulcorner\phi\urcorner) \text{ is true,}$$

whence

$$\text{PRA} \vdash \overline{Prov}((\ulcorner\phi_1\urcorner,\ldots,\ulcorner\phi_n\urcorner),\ulcorner\phi\urcorner),$$

whence

$$\text{PRA} \vdash \exists v\overline{Prov}(v,\ulcorner\phi\urcorner),$$

i.e.

$$\text{PRA} \vdash \overline{Pr}(\ulcorner\phi\urcorner).$$

QED

(In this proof I have carefully distinguished between the relations *Prov*, *Pr* and their representing formulae $\overline{Prov}$, $\overline{Pr}$. As I announced earlier, it is important to distinguish between each when discussing their relations to each other. From here on, it will not be necessary to distinguish between *Prov* and $\overline{Prov}$ or between *Pr* and $\overline{Pr}$. I thus use the simpler notation for both.)

The Second Derivability Condition is also easily established:

5.14. LEMMA. (*D2*). For any formulae ϕ, ψ

$$\text{PRA} \vdash Pr(\ulcorner\phi\urcorner) \wedge Pr(\ulcorner\phi \to \psi\urcorner) \to Pr(\ulcorner\psi\urcorner).$$

(In fact: $\text{PRA} \vdash \forall v_0 v_1 \big(Fmla(v_0) \wedge Fmla(v_1) \to (Pr(v_0) \wedge Pr(v_0 \dot{\to} v_1) \to Pr(v_1))\big)$)

For,

$$Prov(v_0, \ulcorner\phi\urcorner) \wedge Prov(v_1, \ulcorner\phi \to \psi\urcorner) \to Prov(v_0 * v_1 * (\ulcorner\psi\urcorner), \ulcorner\psi\urcorner)$$

is readily verified within PRA.

The Third Derivability Condition is not as easy. We must first introduce one more primitive recursive function. Recall from 4.11 that the numeral $\overline{n}$ is the term defined by recursion:

$\overline{0}$ is the constant $\overline{0}$

$\overline{n+1}$ is $S\overline{n}$.

The function mapping $\overline{n}$ to $\ulcorner\overline{n}\urcorner$ is thus defined by recursion:

$$num(0) = \ulcorner\overline{0}\urcorner = (0) = 2^0 = 1$$

$$num(x+1) = (\ulcorner\overline{S}\urcorner, num(x))$$

$$(= 2^{20992} + 2^{20993+num(x)}).$$

(I never said this was an *efficient* encoding!)

To give a nice statement of the Lemma from which the Third Derivability Condition derives, we introduce the notation $\ulcorner\phi\dot{v}_0 \ldots \dot{v}_{n-1}\urcorner$. This is defined recursively on n-- outside PRA-- as follows:

$$\ulcorner\phi\dot{v}_0\urcorner = subst(\ulcorner\phi v_0\urcorner; (8,0), num(v_0))$$

$$\ulcorner\phi\dot{v}_0 \ldots \dot{v}_k\urcorner = subst(\ulcorner\phi\dot{v}_0 \ldots \dot{v}_{k-1} v_k\urcorner; (8,k), num(v_k)),$$

i.e. $\ulcorner\phi\dot{v}_0 \ldots \dot{v}_{n-1}\urcorner$ denotes the code, not of the formula $\phi v_0 \ldots v_{n-1}$, but of the formula resulting from substituting numerals $\overline{v}_0, \ldots, \overline{v}_{n-1}$ for the variables: Replacing v_i by k_i would yield $\ulcorner\phi\overline{k}_1 \ldots \overline{k}_{n-1}\urcorner$.

In this notation, the crucial lemma is the following:

5.15. LEMMA. Let $\overline{f}$ be an n-ary primitive recursive function symbol. There is a function g, depending on f, and such that

$$\text{PRA} \vdash \overline{f}v_0 \ldots v_{n-1} = v_n \to Prov(\overline{g}v_0 \ldots v_{n-1}, \ulcorner\overline{f}\dot{v}_0 \ldots \dot{v}_{n-1} = \dot{v}_n\urcorner).$$

5.16. COROLLARY. ($D3$). Let ϕ be any formula. Then:

$$\text{PRA} \vdash Pr(\ulcorner\phi\urcorner) \to Pr(\ulcorner Pr(\ulcorner\phi\urcorner)\urcorner).$$

The derivation of 5.15 is a formalisation of that of Theorem 4.12. I do not give it here because I have been rather loose in the exposition in the present section of the encoding; however, I hope I have given enough detail in the proof of Theorem 4.12 and in the encoding of syntax to allow the reader to construct the function g and to see that the induction involved in the proof of Theorem 4.12 is of the type allowable within PRA.

The derivation of Corollary 5.16 from Lemma 5.15 is also similar to that of Lemma 5.13 from Theorem 4.12 and I omit it.

With the Derivability Conditions explained, the only thing left to account for in our discussion in section 1 of the Incompleteness Theorems is the Diagonalisation Lemma 1.1. To obtain this, we merely use $subst(x;y,z)$ and $num(x)$ again to define

$$sub(y,x) \quad = \quad subst(y;(8,0),x).$$

Thus, if ϕv_0 has v_0 free,

$$sub(\ulcorner\phi v_0\urcorner,v_1) \quad = \quad \ulcorner\phi\dot{v}_1\urcorner$$

and

$$sub(\ulcorner\phi v_0\urcorner,n) \quad = \quad \ulcorner\phi\overline{n}\urcorner.$$

Letting $\overline{sub}$ represent sub in PRA, we derive the Diagonalisation Lemma as usual: Let ψv have v free and let $\theta v_0 \; = \; \psi(\overline{sub}(v_0,v_0))$, $m \; = \; \ulcorner\theta v_0\urcorner$, and $\phi = \theta\overline{m}$. Since $m = \ulcorner\theta v_0\urcorner$, $sub(\ulcorner\theta v_0\urcorner,\ulcorner\theta v_0\urcorner) = \ulcorner\theta\overline{m}\urcorner = \ulcorner\phi\urcorner$ and

$$\mathrm{PRA}\vdash \overline{sub}(\ulcorner\theta v_0\urcorner,\ulcorner\theta v_0\urcorner) \quad = \quad \ulcorner\phi\urcorner.$$

Thus

$$\mathrm{PRA}\vdash \psi(\overline{sub}(\ulcorner\theta v_0\urcorner,\ulcorner\theta v_0\urcorner)) \quad \leftrightarrow \quad \psi(\ulcorner\phi\urcorner),$$

i.e.

$$\begin{aligned} \mathrm{PRA}\vdash \theta(\ulcorner\theta v_0\urcorner) &\leftrightarrow \psi(\ulcorner\phi\urcorner) \\ \vdash \theta\overline{m} &\leftrightarrow \psi(\ulcorner\phi\urcorner), \text{ since } \ulcorner\theta v_0\urcorner \textit{ is } m \\ \vdash \phi &\leftrightarrow \psi(\ulcorner\phi\urcorner), \text{ since } \phi \textit{ is } \theta\overline{m}. \end{aligned}$$

A seemingly small, but important, point: The fixed point ϕ to ψv is of the form ψt for some primitive recursive term t with no free variables. In particular, if ψ has other free variables, ϕ shares exactly these; i.e. we have proven the following stronger result:

<u>5.17. DIAGONALISATION THEOREM</u>. Let $\psi v v_0 \ldots v_{n-1}$ have exactly $v,v_0,\ldots,v_{n-1}$ free. There is a formula $\phi v_0 \ldots v_{n-1}$ with exactly $v_0,\ldots,v_{n-1}$ free and such that

$$\mathrm{PRA}\vdash \phi v_0 \ldots v_{n-1} \quad \leftrightarrow \quad \psi(\ulcorner\phi v_0 \ldots v_{n-1}\urcorner,v_0,\ldots,v_{n-1}).$$

This strengthened result extends beyond the type of self-reference we shall examine in this monograph, but not beyond that we will need to use as a tool in our narrower examination.

6. ADDITIONAL ARITHMETIC PREREQUISITES

Except for the fine points of some proofs, the discussion of the preceding sections is a complete account of the Incompleteness Theorems and their metatheory. In the sequel, however, we will want to go beyond this. Doing so will require a deeper understanding of Arithmetic. I propose, in the present section, to outline the foundations of this understanding. The reader who has had his fill of such detail can skip this section for now-- the material will not be needed until Chapter 3. (Indeed, the reader might wish to consider much of the present section as an appendix to Chapter 3.)

There is a number of things to discuss. The most immediate application of Arithmetic in the sequel will occur in Chapter 3, where we will want to define a function by self-reference. Thus, we can begin with a discussion of functionality.

The proof of Theorem 4.12, hence of the important Third Derivability Condition, rested on the fact that the more-or-less normal computation of a primitive recursive function falls short of being a formal proof only insofar as it is based on computational, rather than logical, rules of inference. To obtain a formal proof, one has but to pad the computation with a number of implications and applications of modus ponens. A greatly simplified proof of Theorem 4.12, and thus of the Third Derivability Condition, could be had by more directly accepting computations as derivations. Be that as it may, the fact is that, just as we exhibited the primitive recursive encodability of derivations, we can exhibit that of computations.

A computation of a value of a function, say,

$$f(k_1,\ldots,k_n) \;=\; k,$$

must contain certain information: an index of f, the sequence of arguments $(k_1,\ldots,k_n)$, the value k, and (occasionally) information verifying this value-- i.e. side computations. To compute an initial function, one needs only the basic information:

$((9,1,1),(x),0)$ computes $Z(x) = 0$

$((9,2,1),(x),x+1)$ computes $S(x) = x + 1$

$((9,3,n.i),(x_1,\ldots,x_n),x_i)$ computes $P_i^n(x_1,\ldots,x_n) = x_i$.

(Recall $(9,1,1) = \ulcorner\overline{Z}\urcorner$, etc.) For functions defined by $F4$ and $F5$, an additional component contains the side computations: If

$$f(\vec{x}) = g(h_1(\vec{x}),\ldots,h_m(\vec{x})),$$

a computation of

$$f(x_1,\ldots,x_n) = y$$

is a quadruple

$$((9,4,n,m,(g^*,h_1{}^*,\ldots,h_m{}^*)),(x_1,\ldots,x_n),y,(c_0,c_1,\ldots,c_m)),$$

where $c_0,c_1,\ldots,c_m$ are computations of appropriate values of $g,h_1,\ldots,h_m$. Similarly, if f is defined from g,h by primitive recursion $F5^n$, a computation of

$$f(x,x_1,\ldots,x_n) = y$$

is a quadruple

$$((9,5,n+1,g^*,h^*),(x,x_1,\ldots,x_n),y,(c_0,c_1)),$$

where either

i. $x = 0$ and $c_0 = c_1$ is a computation showing $g(\vec{x}) = y$

or ii. $x > 0$ and c_0 is a computation of $f(x \dot{-} 1,\vec{x})$ and c_1 is one of $h(f(x \dot{-} 1,\vec{x}),x\dot{-}1,\vec{x})$.

The usual course-of-values recursion yields a primitive recursive relation $Comp(x,y,z,w)$ such that, for any primitive recursive function f, any $x_1,\ldots,x_n,y \in \omega$ and any $z \in \omega$,

$Comp(\ulcorner\overline{f}\urcorner,(\overline{x}_1,\ldots,\overline{x}_n),\overline{y},\overline{z})$ is true iff z codes a computation verifying

$$f(x_1,\ldots,x_n) = y.$$

Moreover, a careful check shows that, if $f(x_1,\ldots,x_n) = y$, a computation z verifying this can be found primitive recursively from $\ulcorner f\urcorner,x_1,\ldots,x_n$. This can be proven in PRA:

6.1. THEOREM. There is a primitive recursive relation $Comp$ such that, for any primitive recursive n-ary function symbol $\overline{f}$,

$$\text{PRA} \vdash \overline{f}v_0\ldots v_{n-1} = v_n \leftrightarrow \exists v Comp(\ulcorner\overline{f}\urcorner,(v_0,\ldots,v_{n-1}),v_n,v).$$

I propose not to prove this. Instead, I restate it:

6.2. COROLLARY. For each n, there is a primitive recursive relation C_n such that, for all n-ary primitive recursive function symbols $\overline{f}$,

$$\mathrm{PRA} \vdash \overline{f}v_0 \ldots v_{n-1} = v_n \leftrightarrow \exists v C_n(v; \ulcorner\overline{f}\urcorner, v_0, \ldots, v_{n-1}, v_n).$$

To this it is easy to add existential quantifiers:

6.3. COROLLARY. For each n, there is a primitive recursive relation D_n such that, for all $(n+1)$-ary primitive recursive function symbols $\overline{f}$,

$$\mathrm{PRA} \vdash \exists v_n (\overline{f}v_0 \ldots v_n = \overline{0}) \leftrightarrow \exists v D_n(v; \ulcorner \exists v_n \overline{f}v_0 \ldots v_n = \overline{0}\urcorner, v_0, \ldots, v_{n-1})$$

Proof: Observe,

$$\begin{aligned} \exists v_n(\overline{f}v_0 \ldots v_n = \overline{0}) &\leftrightarrow \exists v_n v_{n+1} C_{n+1}(v_{n+1}; \ulcorner\overline{f}\urcorner, v_0, \ldots, v_n, \overline{0}) \\ &\leftrightarrow \exists v \exists v_n v_{n+1} \le v C_{n+1}(v_{n+1}; \ulcorner\overline{f}\urcorner, v_0, \ldots, v_n, \overline{0}) \\ &\leftrightarrow \exists v D_n(v; \ulcorner \exists v_n \overline{f}v_0 \ldots v_n = \overline{0}\urcorner, v_0, \ldots, v_n), \end{aligned}$$

since $\ulcorner\overline{f}\urcorner$ can be recovered primitive recursively from $\ulcorner \exists v_n \overline{f}v_0 \ldots v_n = \overline{0}\urcorner$ $(= (\ulcorner\exists\urcorner, \ulcorner v_n\urcorner, (\ulcorner=\urcorner, (\ulcorner\overline{f}\urcorner, \ulcorner v_0\urcorner, \ldots, \ulcorner v_n\urcorner), (\ulcorner\overline{0}\urcorner))))$ and the class of primitive recursive relations is closed under bounded existential quantification. QED

What is the point of this? As we saw in section 1, there is no truth definition within PRA for the entire language. Corollary 6.3 gives us a partial truth definition for an expressive fragment of the language. Moreover, the definition belongs to the fragment.

6.4. DEFINITION. A formula $\phi v_0 \ldots v_{n-1}$ is an *RE-formula* if it has the form,

$$\exists v_n(\overline{f}v_0 \ldots v_{n-1}v_n = \overline{0}),$$

for some primitive recursive function symbol $\overline{f}$.

Thus we have, for each n, an RE-formula with $(n+1)$ free variables defining truth (or, more accurately: satisfaction) for RE-formulae in n free variables. Using some effective closure properties, we can generalise this.

6.5. DEFINITION. The class of Σ_1-*formulae* is defined inductively as follows:

i. if $\overline{f}$ is an n-ary primitive recursive function symbol and $v_{i_0}, \ldots, v_{i_{n-1}}$ are variables,

$$\overline{f}v_{i_0} \ldots v_{i_{n-1}} = \overline{0}$$

is Σ_1,

ii. if ϕ, ψ are Σ_1, v is a variable, and t is a term not containing v, then $\phi \wedge \psi$, $\phi \vee \psi$, $\exists v \le t\phi$, $\forall v \le t\phi$, and $\exists v\phi$ are Σ_1.

6.6. THEOREM. Let $n > 0$ be given. There is a Σ_1-formula $Sat_n(v, v_0, \ldots, v_{n-1})$ such that, for all Σ_1-formulae $\phi v_0 \ldots v_{n-1}$ with exactly $v_0, \ldots, v_{n-1}$ free,

$$\mathsf{PRA} \vdash \phi v_0 \ldots v_{n-1} \leftrightarrow Sat_n(\ulcorner \phi \urcorner, v_0, \ldots, v_{n-1}).$$

Proof sketch: We see that the primitive recursive relation 6.5.i is *RE* by the simple device of introducing a dummy variable. We next observe the *RE*-relations to be closed (up to provable equivalence) under the closure conditions 6.5.ii: Write $\phi = \exists v\phi' v$, $\psi = \exists v\psi' v$, ϕ', ψ' both primitive recursive relations.

$$\begin{aligned} \wedge: \quad \phi \wedge \psi \quad &= \quad \exists v\phi' v \wedge \exists v\psi' v \\ &\leftrightarrow \quad \exists vv^*(\phi' v \wedge \psi' v^*) \\ &\leftrightarrow \quad \exists v' \exists v \le v' \exists v^* \le v'(\phi' v \wedge \psi' v^*), \end{aligned}$$

provably in PRA. But $\exists v \le v' \exists v^* \le v'(\phi' v \wedge \psi' v^*)$ is itself primitive recursive, so expressible in the form $\overline{f} v_0 \ldots v_{n-1} v = \overline{0}$.

$$\begin{aligned} \vee: \quad \phi \vee \psi \quad &= \quad \exists v\phi' v \vee \exists v\psi' v \\ &\leftrightarrow \quad \exists v(\phi' v \vee \psi' v) \end{aligned}$$

and the disjunction of primitive recursive relations is primitive recursive.

$\exists v^*$, $\exists v^* \le t$: Take the latter case:

$$\begin{aligned} \exists v^* \le t \exists v\phi' \quad &\leftrightarrow \quad \exists v^* \exists v(v^* \le t \wedge \phi') \\ &\leftrightarrow \quad \exists v' \exists v^* \le v' \exists v \le v'(v^* \le t \wedge \phi'), \end{aligned}$$

and we recognise this to be *RE*.

$\forall v^* \le t$: We have but to establish the equivalence,

$$\forall v^* \le t \exists v\phi' v^* v \leftrightarrow \exists v \forall v^* \le t\phi'(v^*, (v)_{v^*}),$$

as the right-hand-side is readily recognised to be *RE*. The equivalence is proven by a simple induction on t; it is, however, an important induction: It is the first induction in which we use the full power of *RE*-induction postulated in Axiom 4.9.c.

Changing notation slightly, what we wish to prove is

$$\forall v_0 \le v_1 \exists v_2 \phi(v_0, v_2) \rightarrow \exists v \forall v_0 \le v_1 \phi(v_0, (v)_{v_0}),$$

where v_1 does not occur in ϕ and ϕ is a primitive recursive formula. We wish to prove this by induction on v_1. This formula is, however, not *RE*. To get around this, we assume

$$\forall v_0 \le v_1 \exists v_2 \phi v_0 v_2 \qquad (*)$$

and prove

$$\exists v \forall v_0 \le v_3 (v_3 \le v_1 \rightarrow \phi(v_0, (v)_{v_0})),$$

which is readily seen to be equivalent to an *RE*-formula, by induction on v_3.

For $v_3 = 0$, since $\exists v_2 \phi(\overline{0}, v_2)$, choose such a v_2 and set $v = (v_2)$.

For $v_3 + 1 \le v_1$, assume

$$\forall v_0 \le v_3 (v_3 \le v_1 \rightarrow \phi(v_0, (v)_{v_0})).$$

But also, by (*), for some v_2 $\phi(v_3+1, v_2)$ holds. Let $v^* = v*(v_2)$ and observe

$$\forall v_0 \le v_3 + \overline{1} (v_3 + 1 \le v_1 \rightarrow \phi(v_0, (v^*)_{v_0})).$$

This completes the induction and therewith the verification that the class of relations is closed under bounded universal quantification.

Thus, we see that, up to an equivalence provable in PRA, the classes of *RE* and Σ_1-formulae coincide. I claim further that this coincidence is primitive recursively effective, i.e. there is a primitive recursive function f such that, for any Σ_1-formula ϕ, $f(\ulcorner\phi\urcorner)$ is the code of an *RE* formula ψ such that

$$\text{PRA} \vdash \phi \leftrightarrow \psi.$$

I leave to the reader the construction of the function f on the basis of the obviously effective closure proof I have just given.

Using f, the construction of Sat_n is very easy:

$$Sat_n(v, v_0, \ldots, v_{n-1}) \quad = \quad \exists v_n D_n(v_n; \overline{f}(v), v_0, \ldots, v_{n-1}).$$

(To be pedantically correct, one should write

$$Sat_n(v, v_0, \ldots, v_{n-1}) \quad = \quad \exists v_n \exists v_{n+1} (\overline{f}(v) = v_{n+1} \wedge D_n(v_n; v_{n+1}, v_0, \ldots, v_{n-1})),$$

which is literally Σ_1.) Then observe, if $\phi \in \Sigma_1$ has exactly $v_0, \ldots, v_{n-1}$ free and $\ulcorner\psi\urcorner = f(\ulcorner\phi\urcorner)$,

$$\begin{aligned} \text{PRA} \vdash Sat_n(\ulcorner\phi\urcorner, v_0, \ldots, v_{n-1}) &\leftrightarrow \exists v_n D_n(v_n; \ulcorner\psi\urcorner, v_0, \ldots, v_{n-1}) \\ &\leftrightarrow \psi v_0 \ldots v_{n-1}, \quad \text{by 6.3} \\ &\leftrightarrow \phi v_0 \ldots v_{n-1}. \end{aligned}$$

QED

(The reader uncomfortable with the function f can take heart: What is really of importance here is the equivalence, effective or otherwise, of Σ_1- with *RE* formulae. Any application we make of this Theorem can be replaced by one of Corollary 6.3 and a specific instance of the equivalence of Σ_1- with *RE* formulae.)

A few pages back I announced that we would be discussing functionality first and I appear to have strayed a bit off course. I haven't really: The functions we want are those with Σ_1-graphs and our already announced ability to define functions by self-reference will be an application of Theorem 6.6 and the Diagonalisation Theorem.

6.7. DEFINITION. A function $F:\omega^n \to \omega$ is *recursive* (or: *general recursive*) if its graph is Σ_1, i.e. if there is a Σ_1-formula $\phi v_0 \ldots v_{n-1} v_n$ such that, for all $k_0, \ldots, k_{n-1}, k$

$$F(k_0, \ldots, k_{n-1}) = k \quad \text{iff} \quad \phi \overline{k}_0 \ldots \overline{k}_{n-1} \overline{k} \text{ is true.}$$

We will often be interested in *partial* functions, i.e. functions defined on subsets of ω^n.

6.8. DEFINITION. A function $F:A \to \omega$, for some $A \subseteq \omega^n$, is *partial recursive* if its graph is Σ_1, i.e. if there is a Σ_1-formula $\phi v_0 \ldots v_{n-1} v_n$ such that, for all $k_0, \ldots, k_{n-1}, k$

$$F(k_0, \ldots, k_{n-1}) = k \quad \text{iff} \quad \phi \overline{k}_0 \ldots \overline{k}_{n-1} \overline{k} \text{ is true.}$$

Of course, not every Σ_1-formula defines a partial recursive function. However, just as every relation contains a function, every Σ_1-formula defines a relation containing a Σ_1-function.

6.9. SELECTION THEOREM. Let $\phi v_0 \ldots v_{n-1} v_n$ be Σ_1. There is a Σ_1-formula $Sel(\phi)$ with exactly the same free variables and such that

i. $Sel(\phi) v_0 \ldots v_n \to \phi v_0 \ldots v_n$

ii. $Sel(\phi) v_0 \ldots v_{n-1} v_n \wedge Sel(\phi) v_0 \ldots v_{n-1} v^* \to v_n = v^*$

iii. $\exists v_n \phi v_0 \ldots v_{n-1} v_n \to \exists v_n Sel(\phi) v_0 \ldots v_{n-1} v_n$.

Moreover, this is provable in PRA.

Proof sketch: Without loss of generality, we can assume ϕ is *RE*, say

$$\phi v_0 \ldots v_n : \quad \exists v (f v v_0 \ldots v_n = \overline{0})$$

The obvious candidate for $Sel(\phi)$ is one choosing the least v_n:

$$\exists v(fvv_0 \ldots v_n = \overline{0}) \wedge \forall v_{n+1} \leq v_n \sim \exists v(fvv_0 \ldots v_{n-1} v_{n+1} = \overline{0}).$$

Unfortunately, this formula is not Σ_1. To get a Σ_1-formula, we must minimise the pair (v, v_n). Thus, we first define ϕ^*:

$$\phi^* vv_0 \ldots v_{n-1}: \quad f((v)_0, v_0, \ldots, v_{n-1}, (v)_1) = \overline{0}.$$

We then minimise v in ϕ^*:

$$\phi^{**} vv_0 \ldots v_{n-1}: \quad \phi^* vv_0 \ldots v_{n-1} \wedge \forall v^* \leq v \sim \phi^* v^* v_0 \ldots v_{n-1}.$$

$Sel(\phi)$ is defined from ϕ^{**} by reading v_n off v:

$$Sel(\phi) v_0 \ldots v_{n-1} v_n: \quad \exists v\left(\phi^{**} vv_0 \ldots v_{n-1} \wedge (v)_1 = v_n\right).$$

$Sel(\phi)$ is certainly Σ_1. It is not hard to see that it satisfies i-iii: The Least Number Principle assures the existence of a unique least $v = ((v)_0, v_n)$ satisfying ϕ^*, provided anything at all satisfies ϕ-- for any given $v_0, \ldots, v_{n-1}$.

What is not immediately clear is that the proof can be carried out with the minimal induction available in PRA. But this is not too hard to verify: The Least Number Principle applied to ψv,

$$\exists v \psi v \rightarrow \exists v\left(\psi v \wedge \forall v^* < v \sim \psi v^*\right),$$

is just the contrapositive to the Strong Form of Induction applied to $\sim\psi$,

$$\forall v\left(\forall v^* < v \sim \psi v^* \rightarrow \sim \psi v\right) \rightarrow \forall v \sim \psi v,$$

which follows from ordinary induction applied to $\forall v^* < v \sim \psi v^*$,

$$\forall v^* < \overline{0} \sim \psi v^* \wedge \forall v\left(\forall v^* < v \sim \psi v^* \rightarrow \forall v^* < v+1 \sim \psi v^*\right) \rightarrow \forall v \forall v^* < v \sim \psi v^*.$$

Hence, for ψ a *PR* relation (or even a negation of a Σ_1-formula), $\forall v^* < v \sim \psi v^*$ is *PR* (respectively, Σ_1) and the *RE*-induction of PRA yields the Least Number Principle for ψ. In the application at hand, we want to apply the Least Number Principle to ϕ^*, which is *PR*. (A little later we will say something more about the kind of induction available in PRA.) QED

Notice that, if ϕ already defines a partial function, then ϕ is equivalent to $Sel(\phi)$; if ϕ is provably functional, i.e. if

$$\text{PRA} \vdash \phi v_0 \ldots v_{n-1} v_n \wedge \phi v_0 \ldots v_{n-1} v^* \rightarrow v_n = v^*$$

then ϕ is provably equivalent to $Sel(\phi)$. Moreover, the implication just made is provable in PRA. Hence, the graphs $Sel(\phi)$ (provably) contain all and only the graphs of partial recursive functions. Hence, we can use the codes $\ulcorner\phi\urcorner$ of Σ_1-formulae to code partial recursive functions. Via Theorem 6.6, we can, in fact, extend this to allowing arbitrary natural numbers to code partial recursive functions.

6.10. DEFINITION. Let $n > 0$ and let x be arbitrary. The n-ary partial recursive function F_x^n with code x is the partial recursive function with graph defined by

$$Sel\left(Sat_{n+1}(\overline{x}, v_0, \ldots, v_{n-1}, v_n)\right).$$

The importance of codes has already been demonstrated in the linguistic case. The importance for functions is similar. For now, we have two immediate applications.

6.11. UNIVERSAL PARTIAL FUNCTIONS. For each $n \geq 1$, there is an $(n+1)$-ary partial recursive function U_n such that, for all $x, x_0, \ldots, x_{n-1}$,

$$U_n(x, x_0, \ldots, x_n-1) \simeq F_x(x_0, \ldots, x_{n-1}),$$

where "$\simeq$" asserts that either both sides are defined and equal or neither side is defined.

Proof: Let U_n be the function with graph

$$Sel\left(Sat_{n+1}(v, v_0, \ldots, v_{n-1})\right).$$

QED

6.12. RECURSION THEOREM. Let $G:\omega \to \omega$ be recursive. For each $n > 0$, there is an $x \in \omega$ such that

$$F_{G(x)}^n = F_x^n.$$

Proof: By the Diagonalisation Theorem 5.17, there is a formula $\phi v_0 \ldots v_n$ such that

$$\phi v_0 \ldots v_n \leftrightarrow Sel\left(Sat_{n+1}(G(\ulcorner\phi\urcorner), v_0, \ldots, v_{n-1}, v_n)\right),$$

i.e.

$$\phi v_0 \ldots v_n \leftrightarrow Sel\left(\exists v\left(Sat_{n+1}(v, v_0, \ldots, v_n) \wedge \psi(\ulcorner\phi\urcorner, v)\right)\right),$$

where ψ defines the graph of G. Let $x = \ulcorner\phi\urcorner$ and observe

$$F_x^n x_0 \ldots x_{n-1} = y \leftrightarrow F_{G(x)}^n x_0 \ldots x_{n-1} = y.$$

QED

Again, these Theorems are schematically proven in PRA.

As a sample application of the existence of a universal partial function, let me cite the following:

6.13. COROLLARY. There is a partial recursive function F of (say) one variable having no recursive extension.

Proof: Let $F(x) \simeq U(x,x) + 1$, with $\simeq$ as in the statement of Theorem 6.11.

Evidently, the graph of F is Σ_1 and F is partial recursive. Suppose G were a recursive extension of F. Let G have code x_0, whence

$$G(x) = U(x_0,x)$$

for all $x \in \omega$. But then $U(x_0,x_0)$ is defined, whence $F(x_0)$ is defined and

$$G(x_0) = F(x_0) = U(x_0,x_0) + 1 = G(x_0) + 1,$$

a contradiction. QED

Corollary 6.13 and Theorem 6.11 jointly explain our interest in *partial* recursive functions: We want, for various reasons, universal functions. By Theorem 6.11, we can have them if we accept partial functions; by Corollary 6.13, we cannot have them for total functions. (We might view Corollary 6.13 as an analogue to the Gödel-Tarski Theorem 1.2.)

The applications of the Recursion Theorem are many. The application we will make most frequently is in the definition of functions by self-reference. We can illustrate this here by showing closure under primitive recursion. (With all of our coding apparatus, we could directly get a Σ_1-definition of a function defined by primitive recursion from two recursive functions, but such an approach is less elegant than appeal to the Recursion Theorem. Moreover, in Chapter 3, the recursion we will use will not be as straightforward and we will need to apply the Recursion Theorem.)

6.14. APPLICATION. Let $G:\omega \to \omega$, $H:\omega^3 \to \omega$ be recursive functions and define $F:\omega^2 \to \omega$ by

$$F(0,y) = G(y)$$

$$F(x+1,y) = H(F(x,y),x,y).$$

Then: F is recursive.

Proof: Let $\phi_0 v_0 v_1$ be arbitrary, $\psi v_0 v_1$ define the graph of G, and $\chi v_0 v_1 v_2$ define the graph of H. Consider the formula

$$\Phi(\phi_0): \ (v_0 = \overline{0} \wedge \psi v_0 v_1) \vee (v_0 > \overline{0} \wedge \exists v(\phi_0(v_0 \dot{-} 1, v) \wedge \chi v v_0 v_1)).$$

Evidently, $\ulcorner \Phi(\phi_0) \urcorner$ is a primitive recursive function of $\ulcorner \phi_0 \urcorner$, say

$$\ulcorner \Phi(\phi_0) \urcorner = K(\ulcorner \phi_0 \urcorner).$$

Choose x_0 by the Recursion Theorem so that $F^2_{K(x_0)} = F^2_{x_0}$ and observe, for $F = F^2_{x_0}$,

$$F(x,y) = z \leftrightarrow (x = 0 \wedge G(y) = z) \vee (x > 0 \wedge H(F(x\dot{-}1,y),x,y) = z),$$

i.e. $F(0,y) = G(y)$

$$F(x+1,y) = H(F(x,y),x,y),$$

provided, of course, F is defined-- which can be proven by Σ_1-induction. QED

This last proof brings us back to induction. This has been an issue already: In the case of Lemma 5.15, we skirted the issue by not giving the proof; but the statement of the result,

$$fv_0 \ldots v_{n-1} = v_n \rightarrow Pr(gv_0 \ldots v_{n-1}, \ulcorner f\dot{v}_0 \ldots \dot{v}_{n-1} = \dot{v}_n \urcorner),$$

clearly calls for an induction-- a rather messy one taking into account the intricacies of g. With the RE-induction available in PRA, we could have proven instead the less informative, but equally useful,

$$\exists v\big(fv_0 \ldots v_{n-1} = v_n \rightarrow Pr(v, \ulcorner f\dot{v}_0 \ldots \dot{v}_{n-1} = \dot{v}_n \urcorner)\big).$$

The details of such an induction are a bit simpler. This is more-or-less what we assumed in stating Theorem 6.1 without proof. Nonetheless, in Theorem 6.1, this was only a convenience: We could have replaced the existential quantifier therein by a primitive recursive witnessing function and relied on PR-induction. It is, as announced during the proof, Theorem 6.6 that first made essential use of full RE-induction. The equivalence proven by this induction,

$$\forall v_0 < v_1 \exists v\, \phi(v_0,v) \leftrightarrow \exists v \forall v_0 < v_1\, \phi(v_0,(v)_{v_0}),$$

implies RE-induction in the face of mere PR-induction. Insofar as we will be using the equivalence, established in the course of proving Theorem 6.6, of Σ_1- with RE-formulae, we will be depending on the full power of the induction available to us. Indeed, our dependence on RE-induction will be more explicit than this in the sequel. In fact, at one point we will go beyond this. Thus, we pause to say a few words about how much induction is available in PRA.

6.15. DEFINITIONS. i. The classes of Σ_n-, Π_n-formulae, for $n > 0$, are defined (outside PRA) inductively as follows:

a. Σ_1 has already been defined; Π_1 consists of the negations of Σ_1-formulae

b. if ϕv is Π_n, then $\exists v \phi v$ is Σ_{n+1}; if ϕ is Σ_{n+1}, $\sim\phi$ is Π_{n+1}.

ii. Given a class Γ of formulae, we define Γ-*Induction* (Γ-*Ind*) to be the schema,

$$\phi\overline{0} \wedge \forall v(\phi v \to \phi(Sv)) \to \forall v\phi v,$$

for $\phi \in \Gamma$; we define the Γ-*Least Number Principle* (Γ-*LNP*) to be the schema

$$\exists v\phi v \to \exists v(\phi v \wedge \forall v^* < v \sim\phi v^*).$$

We have already seen, in the proof of Theorem 6.9, that Σ_1-*Ind* implies Π_1-*LNP*. The same sort of argument establishes that, modulo Σ_n-*Ind* and Π_n-*Ind*, respectively, Σ_n and Π_n are closed under bounded quantification and that Σ_n-*Ind* implies Π_n-*LNP* and Π_n-*Ind* implies Σ_n-*LNP*. If we define PRA$^-$ to be the subtheory of PRA obtained by further restricting induction to *PR*-formulae, we can state the following:

6.16. FACTS. Over PRA$^-$ the following schemata are equivalent:

i. Σ_n-*Ind*

ii. Π_n-*Ind*

iii. Σ_n-*LNP*

iv. Π_n-*LNP*.

The implications i $\Rightarrow$ iv and ii $\Rightarrow$ iii have already been discussed and their converses offer nothing new. The equivalence i $\Leftrightarrow$ ii is more interesting. We prove i $\Rightarrow$ ii in the case $n = 1$. The proof for $n > 1$ is identical, modulo the closure of Σ_n under bounded quantification.

Proof of i $\Rightarrow$ ii for $n = 1$: Assume Σ_1-*Ind* and suppose, for $\psi v \in \Sigma_1$, the instance

$$\sim\psi\overline{0} \wedge \forall v(\sim\psi v \to \sim\psi Sv) \to \forall v\sim\psi v$$

of Π_1-*Ind* fails. This means

$$\sim\psi\overline{0}, \quad \forall v(\sim\psi v \to \sim\psi Sv), \quad \exists v\psi v. \qquad (*)$$

Choose v_0 so that ψv_0. We get a contradiction by using Σ_1-*Ind* on the variable v in $\psi(v_0 \dot{-} v)$ to prove $\psi\overline{0}$: $\psi(v_0 \dot{-} \overline{0})$ follows immediately on assumption ψv_0. The induction step, $\psi(v_0 \dot{-} Sv)$ follows from the central conclusion of (*), since $S(v_0 \dot{-} Sv) = v_0 \dot{-} v$, unless we have already reached $v = v_0$, in which case $v_0 \dot{-} Sv = v_0 \dot{-} v$. Thus, one concludes $\forall v\psi(v_0 \dot{-} v)$, and, in particular, $\psi\overline{0}$, the desired contradiction. QED

What Facts 6.16 tell us about PRA is that, since Σ_1-formulae are equivalent to *RE*-formulae and PRA has *RE*-induction, PRA has Σ_1- and Π_1-induction, as well as the Σ_1- and Π_1- Least Number Principles. In fact, as Harvey Friedman has shown, PRA proves induction on all boolean combinations of Σ_1-formulae.

<u>6.17. THEOREM</u>. PRA$\vdash$ *Bool*(Σ_1)-*Ind*, where *Bool*(Σ_1) is the class of boolean combinations of Σ_1-formulae.

The proof of this is nontrivial and would carry us too far afield, whence I omit it.

The question might arise in the reader's mind of why we don't just ignore all the difficulties inherent in keeping track of how much induction we have and how much we need by assuming full induction and working with PA instead of PRA. Indeed, the reader can do this easily; only one result needs to be reformulated, but the reformulation is the natural version anyway. My main concern in choosing PRA over PA is simply to use the appropriate amount of induction.

There is, in Chapter 4, one result where, by cheating slightly, we obtain for PRA a slightly cleaner statement of the result. The reason we can do this is that PRA is, unlike PA, almost a finitely axiomatised theory; it is (equivalent to) a definitional extension of a finitely axiomatised theory. In Chapter 4, we use this fact to pretend that PRA *is* a finitely axiomatised theory. Although I do not wish to go into too much detail as we will not really need to make much use of this, I will say that the difference between PRA and PA can be explained by induction and the partial truth definitions available from Theorem 6.6.

Let us first consider PRA. We don't need *all* the primitive recursive functions to do what we have so far done; finitely many suffice for the encoding machinery we have constructed. With these finitely many, we can still obtain a Σ_1-truth definition for the restricted language, define the notion of a recursive function, prove the Selection and Recursion Theorems, and, *à la* Application 6.14, prove the closure of the class of recursive functions under primitive recursion. Axiomatically, this means we only need finitely many primitive recursive function symbols and finitely many recursion equations in our linguistically restricted reformulation of PRA. Finally, the schema of Σ_1-induction can be replaced by the single instance

$$\forall \ulcorner\phi v\urcorner \in \Sigma_1\big(Tr_{\Sigma_1}(\ulcorner\phi\overline{0}\urcorner) \wedge \forall v_0(Tr_{\Sigma_1}(\ulcorner\phi\dot{v}_0\urcorner) \rightarrow Tr_{\Sigma_1}(\ulcorner\phi S\dot{v}_0\urcorner)) \rightarrow \forall v_0 Tr_{\Sigma_1}(\ulcorner\phi\dot{v}_0\urcorner)\big).$$

Hence, PRA is equivalent to a theory, say PRA$_{fin}$, in a finite sublanguage of that of PRA and possessing only finitely many non-logical axioms. Moreover, this equivalence is very strong: PRA is a definitional extension of PRA_{fin}.

The same cannot be done for PA. Certainly, the language can be restricted. Indeed, for both PRA and PA, at the cost of some messiness, one need not have any more function symbols than $S,+,\cdot$. Moreover, the induction schema for PA can be replaced, via partial truth definitions, by a single instance of Σ_1-*Ind*, a single instance of Σ_2-*Ind*,...-- an infinity of such instances. This infinity cannot be dispensed with. In fact, for $n \geq 1$, PRA + Σ_{n+1}-*Ind* proves the consistency of PRA + Σ_n-*Ind*. In fact, a lot more holds.

In order for us to state formally and to appreciate any result of the sort just cited, we must digress to announce that PRA is not the only theory with a decently encodable syntax: The encoding in PRA of the syntax of PRA is merely exemplary; the syntax of many theories can also be encoded in PRA (and, assuming the theories strong, also in themselves). For example, PA shares the same language as PRA. The coding of syntactic objects summarised in List 5.11 is still correct for PA. Of those syntactic codings summarised in 5.12, the primitive recursive predicate *NonLog(x)* must be changed to admit all instances of induction and not merely the *RE*-instances accepted in PRA. The ensuing predicates in 5.12 are built up from *NonLog(x)* and, once we've made this one change, we've made them all. In the end, we get a predicate $Pr_{PA}(x)$ representing provability in PA. If we reformulate PRA or PA in a smaller language or if we choose to deal with another theory, like ZF set theory, in a completely different language, we must go back and redo some of the other syntactic encoding, but this is usually routine.

A theory T for which the predicate $Pr_T(x)$ defining its set of theorems is an *RE*-predicate is called an *RE-theory*. If an *RE*-theory T is "sufficiently strong", we can derive for it all of the incompleteness results we obtained for PRA. The hitch is, of course, what do we mean by "sufficiently strong"? The simplest definition is that T "contains" PRA. Again, I put "contains" in quotes. The simplest notion of containment, T $\supseteq$ PRA, is that the language of T contains that of PRA and the set of theorems of T includes all theorems of PRA. ZF does not satisfy this condition and we clearly like to think ZF contains arithmetic. There are two simple ways in which it does: Firstly, there is an *interpretation* of PRA in ZF; secondly, this interpretation yields a *definitional extension* ZF_{PRA} of ZF such that $ZF_{PRA} \supseteq$ PRA. I shall state the following for simple extensions and leave to the reader the discussion of theories like ZF.

6.18. THEOREM. Let $T \supseteq$ PRA be a consistent *RE* theory.

i. (Löb's Derivability Conditions). For any sentences ϕ, ψ,

a. $T \vdash \phi \Rightarrow \mathrm{PRA} \vdash Pr_T(\ulcorner\phi\urcorner)$

b. $\mathrm{PRA} \vdash Pr_T(\ulcorner\phi\urcorner) \wedge Pr_T(\ulcorner\phi \rightarrow \psi\urcorner) \rightarrow Pr_T(\ulcorner\psi\urcorner)$

c. $\mathrm{PRA} \vdash Pr_T(\ulcorner\phi\urcorner) \rightarrow Pr_{PRA}(\ulcorner Pr_T(\ulcorner\phi\urcorner)\urcorner)$

d. $\mathrm{PRA} \vdash Pr_T(\ulcorner\phi\urcorner) \rightarrow Pr_T(\ulcorner Pr_T(\ulcorner\phi\urcorner)\urcorner)$.

ii. (Gödel's Incompleteness Theorem). Let $\mathrm{PRA} \vdash \phi \leftrightarrow \sim Pr_T(\ulcorner\phi\urcorner)$.

a. $T \nvdash \phi$

b. if T is Σ_1-sound, i.e. if T proves only true Σ_1-sentences, $T \nvdash \sim\phi$

c. $T \nvdash Con_T$, i.e. $T \nvdash \sim Pr_T(\ulcorner \overline{0} = \overline{1}\urcorner)$

iii. (Löb's Theorem). For any sentence ϕ,

a. $T \vdash Pr_T(\ulcorner\phi\urcorner) \rightarrow \phi \Rightarrow T \vdash \phi$

b. $\mathrm{PRA} \vdash Pr_T(\ulcorner Pr_T(\ulcorner\phi\urcorner) \rightarrow \phi\urcorner) \rightarrow Pr_T(\ulcorner\phi\urcorner)$.

Getting back to our discussion of PA vs. PRA, recall that we mentioned the following:

6.19. THEOREM. For $n \geq 1$,

$$\mathrm{PRA} + \Sigma_{n+1}\text{-}Ind \vdash Con(\mathrm{PRA} + \Sigma_n\text{-}Ind).$$

Hence, $\mathrm{PRA} + \Sigma_n\text{-}Ind \nvdash \Sigma_{n+1}\text{-}Ind$.

The underivability assertion follows from the positive assertion via 6.18.ii.c, i.e. Gödel's Second Incompleteness Theorem for $\mathrm{PRA} + \Sigma_n\text{-}Ind$. The positive assertion is too deep to be proven here. For our purposes, the crucial thing is that, over PRA, PA is axiomatised by an infinite sequence of strictly stronger and stronger sentences and thus, even in the finite language of PRA_{fin}, cannot be finitely axiomatised over PRA.

The relation between PRA and PA is deeper than this. Although I really cannot prove Theorem 6.19 here, I should like to say that its proof yields more: With Σ_{n+1}-*Ind*, one proves that any Σ_{n+1}-consequence of Σ_n-*Ind* is *true*. In particular, PA is strong enough to prove that all consequences of PRA are true, i.e. PA proves the soundness of PRA.

6.20. DEFINITIONS. There are two schematic representations of the soundness of an *RE* theory $T \supseteq$ PRA: *Local* and *Uniform Reflexion*. Local reflexion is the schema,

$$Rfn(T):\quad Pr_T(\ulcorner\phi\urcorner) \to \phi, \text{ all sentences } \phi.$$

Uniform reflexion is the schema,

$$RFN(T):\quad \forall v\big(Pr_T(\ulcorner\phi\dot{v}\urcorner) \to \phi v\big),$$

where ϕv is a formula with only v free.

The two reflexion schemata are quite a bit different in strength. Indeed, one can show

$$T + RFN(T) \vdash RFN(T + Rfn(T)),$$

in particular,

$$T + RFN(T) \vdash Con(T + Rfn(T)),$$

whence the uniform schema is strictly stronger than the local one. That both schemata attempt and appear to express soundness and yet they are not equivalent is another incompleteness result. Be that as it may, in proof theory, the stronger schema of uniform reflexion is the more central one. In the sequel, where propositional modal logic is the issue, our ability to simulate local reflexion by the schema

$$\Box A \to A,$$

and our inability to offer any better expression of uniform reflexion will make local reflexion the relevant schema. This will occur in Chapter 4. For now, we merely note that, with the concept of reflexion, we can state exactly the relation between PRA and PA:

6.21. THEOREM. (Kreisel and Levy). PA and PRA + *RFN*(PRA) are equivalent, i.e. for any arithmetic sentence ϕ,

$$\text{PA} \vdash \phi \quad \text{iff} \quad \text{PRA} + RFN(\text{PRA}) \vdash \phi.$$

In particular, it follows that

$$\text{PA} \vdash Rfn(\text{PRA}),$$

and PA is much stronger than PRA. Other pairs of comparable difference in strength are, say, ZF and PA,

$$\mathrm{ZF} \vdash Rfn(\mathrm{PA}),$$

and $\mathrm{ZFI} = \mathrm{ZF} + \exists\kappa(\kappa$ is an inaccessible cardinal$)$ and ZF,

$$\mathrm{ZFI} \vdash Rfn(\mathrm{ZF}).$$

The discussion (due to Timothy Carlson) of the bi-modal logic of provabilities in such pairs is a central topic in Chapter 4, below. All we need to know in this discussion is that certain theories T_1 are much stronger than certain theories T_0; indeed, we discuss the pair given by $T_1 = \mathrm{ZF}$, $T_0 = \mathrm{PRA}$. For this pair, the provability of $Rfn(\mathrm{PRA})$ in ZF is unremarkable: Merely formalise in ZF the usual proof of soundness of PRA, using the fact that the truth definition for the arithmetic language is expressible in the set theoretic one. It is with pairs like PA and PRA, where, by the Gödel-Tarski Theorem (1.2, above), we cannot express truth for the common language, that the result becomes remarkable.

Let me quickly summarise a few points (and, perhaps, announce one or two additional facts):

i. PRA can almost be finitely axiomatised. This fact will be of minor importance in Chapter 4, below, where we will remark that, if T extends PRA, it can be assumed that T provably does so:

$$\mathrm{PRA} \vdash Pr_{PRA_{fin}}(\ulcorner\phi\urcorner) \rightarrow Pr_T(\ulcorner\phi\urcorner)$$

for any sentence ϕ.

ii. PA proves uniform reflexion for PRA. In fact, PA does so for *any* finite subtheory T_0. It is not the case that every extension of PA is provably an extension. Thus, in Chapter 4, when we wish to discuss results about extensions of PA, we must speak of theories *provably* extending PA.

iii. This reflexive property of PA is shared by ZF, which also proves uniform reflexion for its finite subsystems. GB, the finitely axiomatised class theory extending ZF, does not (by Gödel's Second Incompleteness Theorem) have this property. When one gets away from the usual provability predicates, there arise distinct issues in self-reference for these two types of theories-- reflexive vs. finitely axiomatised theories. We will touch on this (just barely) in Chapter 7, below.

Of what has just been discussed about reflexion principles, the only thing used in the sequel is the fact that some theories are strong enough to prove local reflexion for others. This will only be needed in Chapter 4 when we discuss Carlson's characterisation of the bi-modal logic of the two provability predicates of the two theories. Throughout most of the sequel, such arithmetic facts will merely be cited as they are needed. The reader interested primarily in the modal logics can simply accept these facts when they are offered and ignore much of the present Chapter. The reader primarily interested in arithmetic will find, in addition to these citations of results of the present Chapter, numerous exercises complementing the text where modal techniques are inadequate. For some of these, he will need just a couple of additional facts. First, an application of Theorem 5.15 yields the following often used generalisation of *D3*:

6.22. DEMONSTRABLE Σ_1-COMPLETENESS. Let $\phi v_0 \ldots v_{n-1}$ be Σ_1 with free variables as shown. Then: $\mathsf{PRA} \vdash \phi v_0 \ldots v_{n-1} \rightarrow Pr(\ulcorner \phi \dot{v}_0 \ldots \dot{v}_{n-1} \urcorner)$.

Proof: By the proof of Theorem 6.6, there is an *RE*-formula $\exists v(fvv_0 \ldots v_{n-1} = \overline{0})$ such that

$$\mathsf{PRA} \vdash \phi v_0 \ldots v_{n-1} \leftrightarrow \exists v(fvv_0 \ldots v_{n-1} = \overline{0}) \qquad (1)$$

and

$$\mathsf{PRA} \vdash Pr(\ulcorner \phi \dot{v}_0 \ldots \dot{v}_{n-1} \leftrightarrow \exists v(fv\dot{v}_0 \ldots \dot{v}_{n-1} = \overline{0}) \urcorner), \qquad (2)$$

by *D1*. Observe:

$$\begin{aligned} \mathsf{PRA} \vdash \phi v_0 \ldots v_{n-1} &\rightarrow \exists v(fvv_0 \ldots v_{n-1} = \overline{0}), \text{ by (1)} \\ &\rightarrow \exists v Pr(\ulcorner fv\dot{v}_0 \ldots \dot{v}_{n-1} = \overline{0} \urcorner), \text{ by 5.15} \\ &\rightarrow Pr(\ulcorner \exists v(fv\dot{v}_0 \ldots \dot{v}_{n-1} = \overline{0}) \urcorner), \text{ by } D1, D2 \\ &\rightarrow Pr(\ulcorner \phi \dot{v}_0 \ldots \dot{v}_{n-1} \urcorner), \text{ by (2).} \end{aligned}$$

QED

Theorem 6.22 gives us an essential difference between Π_1- and Σ_1-sentences. If ϕ is Π_1, one can have $\phi + Con(\mathsf{PRA} + \phi)$ and $\phi + Con(\mathsf{PRA} + \sim\phi)$ both consistent; $\phi + Con(\mathsf{PRA} + \sim\phi)$ is inconsistent if ϕ is Σ_1. This observation, suitably formalised, is the following:

6.23. COROLLARY. Let T be a consistent *RE* extension of PRA. Over PRA, the following are equivalent:

i. Con_T

ii. $Rfn_{\Pi_1}(T)$

iii. $RFN_{\Pi_1}(T)$

where the subscript "Π_1" indicates the restriction of the given schema to $\phi \in \Pi_1$.

Proof: Obviously iii $\Rightarrow$ ii $\Rightarrow$ i. To prove i $\Rightarrow$ iii, let ϕv Π_1 and observe:

$$\mathrm{PRA} \vdash \sim\phi \rightarrow Pr_T(\ulcorner\sim\phi\dot{v}\urcorner), \text{ by 6.22} \quad (1)$$

$$\vdash \sim Pr_T(\ulcorner\sim\phi\dot{v}\urcorner) \rightarrow \phi v. \quad (2)$$

But,
$$\mathrm{PRA} + Con_T \vdash Pr_T(\ulcorner\phi\dot{v}\urcorner) \rightarrow \sim Pr_T(\ulcorner\sim\phi\dot{v}\urcorner)$$

$$\vdash Pr_T(\ulcorner\phi\dot{v}\urcorner) \rightarrow \phi v, \text{ by (2).}$$

QED

A tiny remark: (1) follows from 6.22 easily for T *provably* extending PRA. Since T provably extends PRA_{fin}, it actually follows for all T and we don't need to assume the provability of the extension. This sort of subtlety pops up again in Chapter 4.

A corollary to 6.23 of interest is the following:

6.24. COROLLARY. Let T be a consistent *RE* extension of PRA. Let π be a Π_1-sentence. Then: $T \vdash \pi \Rightarrow \mathrm{PRA} + Con_T \vdash \pi$.

Proof: Observe,

$$T \vdash \pi \Rightarrow \mathrm{PRA} \vdash Pr_T(\ulcorner\pi\urcorner)$$

$$\Rightarrow \mathrm{PRA} + Rfn_{\Pi_1}(T) \vdash \pi$$

$$\Rightarrow \mathrm{PRA} + Con_T \vdash \pi, \text{ by 6.23.}$$

QED

With this result we have come full circle. It was an observation such as this (although unformalised) that led Hilbert to stress the importance of consistency: If, say, $\mathrm{PRA} \vdash Con_T$, then PRA proves all Π_1-consequences of T and the extra power of T is unnecessary. As Gödel noted in the remarks cited at the beginning of this Chapter, however, such a conservation result only holds for a restricted class of formulae: Let $T = \mathrm{PRA} + \sim Con_{PRA}$ and observe:

$$T \vdash \sim Con_{PRA},$$

but $\mathrm{PRA} + Con_{PRA} \vdash Con_T$, by Gödel's Second Incompleteness Theorem

yet $\mathrm{PRA} + Con_{PRA} \nvdash \sim Con_{PRA}$,

by the soundness of PRA.

Part I
The Logic of Provability

Chapter 1
Provability as Modality

Although the idea of provability as modal necessity is hardly novel, the serious study of the modal logic of provability did not get underway until the 1970s. There were early flirtations with the idea, but they never amounted to anything: About the time his Incompleteness Theorems gave him instant fame (among mathematicians and philosophers), Gödel wrote up a short note on an embedding of intuitionistic logic into modal logic. The idea behind the embedding was simple: Intuitionistic truth is defined in terms of proof and provability is necessity. Composition yields the embedding. Gödel never bothered (so far as anyone knew before his death-- his Nachlass may, or may not, reveal otherwise) to connect this modal view of provability with his self-referential one.

Richard Montague, in a paper published in 1963, made the connexion between $Pr(\cdot)$ as a box and self-reference. However, he was not really interested in the same $Pr(\cdot)$ as we are. Specifically, he partially modally analysed the First Incompleteness Theorem as follows: If $Pr(\cdot)$ is a predicate in a language admitting self-reference then one cannot have

$$\vdash \phi \Rightarrow \vdash Pr(\ulcorner \phi \urcorner)$$

$$\vdash Pr(\ulcorner \phi \urcorner) \rightarrow \phi$$

holding simultaneously. If Pr represents provability within the given formal system, since

$$\vdash \phi \Rightarrow \vdash Pr(\ulcorner \phi \urcorner)$$

readily holds, the (weak) Gödelian conclusion is that

$$\nvdash Pr(\ulcorner \phi \urcorner) \rightarrow \phi.$$

However, as I said, Montague wanted to apply this argument to different predicates $Pr_T(\cdot)$. He took for his basic system a theory like Peano Arithmetic, PA, or

Zermelo-Fraenkel Set Theory, ZF, which is strong enough to prove the Reflexion Schema,

$$Pr_T(\ulcorner\phi\urcorner) \to \phi,$$

for each finite subtheory T and drew from his modal representation of Gödel's Theorem the conclusion

$$\vdash\phi \;\not\Rightarrow\; \vdash Pr_T(\ulcorner\phi\urcorner),$$

i.e. $\vdash\phi \;\not\Rightarrow\; T\vdash\phi,$

i.e. PA and ZF are not finitely axiomatisable.

I don't know when or by whom the idea of studying $Pr(\cdot)$ by modal means first arose. In 1973 Angus Macintyre and Harry Simmons were the first to publish anything of substance on the matter and Simmons wrote a follow-up. By 1975, work on the modal logic of provability was well underway in Italy, the Netherlands, and the United States. And this work has spread: I know of contributions from Eastern Europe (Bulgaria, Czechoslovakia, and the Soviet Union), West Germany, and Israël. Such popularity attests to the coherence of the subject.

But what is the subject? In my (I fear soon to be customarily) elliptic manner, I have been discussing the modal interpretation of provability without defining it. So let me define it: What we have in mind are functions * from the modal language (rigorously defined in section 1, below) into the arithmetic language (rigorously defined in Chapter 0, section 4, above) which preserve boolean operations and satisfy also:

$$(\Box A)^* = Pr(\ulcorner A^*\urcorner).$$

The modal theory PRL, axiomatised by the modal inverse images of the Derivability Conditions and the Formalised Löb's Theorem (cf. section 2, below, for a formal definition), proves all modal laws of $Pr(\cdot)$, i.e. PRL proves all modal sentences A for which PRL$\vdash A^*$ for all arithmetic interpretations *. PRL is, thus, the "Logic of Provability".

The present chapter devotes itself to syntactic matters. In section 1 immediately following, a system BML of *Basic Modal Logic* is presented and a few results about it are derived. There are two reasons for beginning with BML: In the next chapter on model theory, it will be convenient to first derive a strong completeness theorem for BML and then look at the derivative model theory for PRL. The second reason is of more immediate significance: BML serves as a convenient neutral theory over which to prove the equivalence of a number of alternative axiomatisations of PRL. These alternatives are presented in section 2 and the equivalence proofs are then begun. The most important parts of these equivalence proofs are also the most important syntactic goals of the chapter. These are the proofs within PRL of the uniqueness and explicit definability of fixed points, which proofs occupy section 3. The chapter finishes somewhat anti-climactically in section 4 with a discussion on avoiding the rule of necessitation; this material is merely technical preparation for Chapter 2.

1. A SYSTEM OF BASIC MODAL LOGIC

Through all of Part I, our basic modal language will be given by:

PROPOSITIONAL VARIABLES: $p, q, r, \ldots$

TRUTH VALUES: t, f

PROPOSITIONAL CONNECTIVES: $\sim, \wedge, \vee, \rightarrow$

MODAL OPERATOR: $\Box$.

Sentences (which we denote by capital Roman letters $A, B, C, \ldots$) are constructed in the usual manner and parentheses are avoided whenever possible by the usual conventions on precedence along with the modal convention that $\Box$ is given minimal scope. Thus, e.g., $\Box A \wedge B$ reads $(\Box A) \wedge B$.

1.1. DEFINITION. BML is the modal theory with axioms and rules of inference as follows:

AXIOMS. *A1.* All (boolean) tautologies

A2. $\Box A \wedge \Box(A \rightarrow B) \rightarrow \Box B$

A3. $\Box A \rightarrow \Box\Box A$

RULES. *R1.* $A, A \rightarrow B / B$ *(Modus Ponens)*

R2. $A / \Box A$. *(Necessitation)*

The modal system BML is our system of *B*asic *M*odal *L*ogic over which we wish to study the system PRL of *Pr*ovability *L*ogic. The system BML has been studied by modal logicians and is commonly known as K4, the "4" indicating that the logic extends an even more basic system K by the addition of axiom *A3*. Model theoretically, K would be a more reasonable choice of minimal system than BML; proof theoretically, BML is the obvious choice: Axioms *A1* and rule *R1* are mere logic; the nonlogical axioms and rule are modal simulations of the Derivability Conditions-- *A2* is *D2*, *A3* is *D3*, and *R2* is *D1*. In short, BML axiomatises those properties of $Pr(\cdot)$ that do not depend on the Diagonalisation Lemma.

In the next section we will consider several ways of modally simulating the Diagonalisation Lemma and begin the proof of their equivalence over BML. For now, we have a few syntactic preliminaries to dispose of. We must, as in beginning logic, exhibit a few useful modal tautologies (i.e. derive a few sentences in BML), prove a Deduction Theorem, and prove a Substitution Lemma. Unfortunately, unlike the situation in beginning logic, we cannot do all of this:

1.2. REMARK ON THE DEDUCTION THEOREM. The Deduction Theorem, i.e. closure under the rule,

From $A \vdash B$ conclude $\vdash A \rightarrow B$,

fails. In fact, it fails badly: By *R2* one has $A \vdash \Box A$; but one certainly doesn't have $\vdash A \rightarrow \Box A$. The model theory of the next chapter will readily yield the underivability of the instance $p \rightarrow \Box p$; bearing in mind the provability interpretation, an arithmetical counterexample is also easy: $\mathrm{PRA} \nvdash Con \rightarrow Pr(\ulcorner Con \urcorner)$. For, the Formalised Second Incompleteness Theorem (cf. Exercise 2.3.iii, below) yields $\mathrm{PRA} \vdash Con \rightarrow \sim Pr(\ulcorner Con \urcorner)$. However, *R2* is the only obstacle to the Deduction Theorem: Any derivation *not* invoking the rule *R2* can be viewed as a derivation within the propositional calculus from strange new propositional variables and special axioms, whence we may freely apply the Deduction Theorem for the propositional calculus. We will do so without mention in the sequel. (For more on *R2*, cf. section 4, below.)

The other preparatory tasks are performable. First, there is the list of modal tautologies:

1.3..LEMMA. i. $\mathrm{BML} \vdash \Box(A \wedge B) \leftrightarrow \Box A \wedge \Box B$

ii. $\mathrm{BML} \vdash \Box A \vee \Box B \rightarrow \Box(A \vee B)$

iii. $\mathrm{BML} \vdash \Box(A \rightarrow B) \rightarrow. \Box A \rightarrow \Box B$

iv. $\mathsf{BML} \vdash \Box(A \leftrightarrow B) \rightarrow. \Box A \leftrightarrow \Box B$

v. $\mathsf{BML} \vdash \Box f \rightarrow \Box A$

vi. $\mathsf{BML} \vdash \sim\Box f \leftrightarrow. \Box A \rightarrow \sim\Box\sim A$.

I leave these to the reader as exercises.

Implications ii-v cannot in general be reversed. Easy counterexamples are possible with the model theory of Chapter 2. More sophisticated counterexamples are possible via arithmetic interpretations and the Incompleteness Theorems. For example, Gödel's sentence ϕ such that

$$\mathsf{PRA} \vdash \phi \leftrightarrow \sim Pr(\ulcorner\phi\urcorner)$$

provides a counterexample to the converse to 1.3.ii. For, suppose

$$\mathsf{PRA} \vdash Pr(\ulcorner\phi \vee \sim\phi\urcorner) \rightarrow Pr(\ulcorner\phi\urcorner) \vee Pr(\ulcorner\sim\phi\urcorner).$$

Since the antecedent is derivable (by *D1*), it follows that

$$\mathsf{PRA} \vdash Pr(\ulcorner\phi\urcorner) \vee Pr(\ulcorner\sim\phi\urcorner).$$

Now the soundness of PRA yields

$$\mathsf{PRA} \vdash \phi \quad \text{or} \quad \mathsf{PRA} \vdash \sim\phi,$$

contradicting the First Incompleteness Theorem, by which the chosen ϕ is undecided.

Our third immediate goal is the Substitution Lemma. This requires a few more modal tautologies, which in turn require some preparation. First, a lemma:

1.4. LEMMA. Let ML be a modal system extending BML and closed under *R2* (e.g. ML = BML or ML = PRL). Then:

$$\mathsf{ML} \vdash \Box A \rightarrow B \quad \Rightarrow \quad \mathsf{ML} \vdash \Box A \rightarrow \Box B.$$

Proof: $\mathsf{ML} \vdash \Box A \rightarrow B \Rightarrow \mathsf{ML} \vdash \Box(\Box A \rightarrow B)$, by *R2*

$\Rightarrow \mathsf{ML} \vdash \Box\Box A \rightarrow \Box B$, by 1.3.iii

$\Rightarrow \mathsf{ML} \vdash \Box A \rightarrow \Box B$, by *A3*. QED

Substitution outside a modal context requires mere equivalence; inside a modal context it will require necessary equivalence. Hence, general substitution will require equivalence and necessary equivalence. This leads to the following:

1.5. DEFINITION. The *strong box* $\boxed{s}$ is defined by

$$\boxed{s}A = A \wedge \Box A.$$

The strong box is a modal operator in its own right:

1.6. LEMMA. $\mathrm{BML}(\Box) \vdash \mathrm{BML}(\boxed{s})$, i.e.

i. $\mathrm{BML} \vdash \boxed{s} A \wedge \boxed{s}(A \to B) \to \boxed{s} B$

ii. $\mathrm{BML} \vdash \boxed{s} A \to \boxed{s}\boxed{s} A$

iii. $\mathrm{BML} \vdash A \Rightarrow \mathrm{BML} \vdash \boxed{s} A$.

I leave the proof to the reader. I note also that 1.6.iii holds for any $\mathrm{ML} \supseteq \mathrm{BML}$ closed under $R2$:

$$\mathrm{ML} \vdash A \Rightarrow \mathrm{ML} \vdash \boxed{s} A.$$

A quick corollary is the strongly boxed analogue to Lemmas 1.3 and 1.4.

1.7. COROLLARY. i. $\mathrm{BML} \vdash \boxed{s}(A \wedge B) \leftrightarrow \boxed{s} A \wedge \boxed{s} B$

ii. $\mathrm{BML} \vdash \boxed{s} A \vee \boxed{s} B \to \boxed{s}(A \vee B)$

iii. $\mathrm{BML} \vdash \boxed{s}(A \to B) \to . \boxed{s} A \to \boxed{s} B$

iv. $\mathrm{BML} \vdash \boxed{s}(A \leftrightarrow B) \to . \boxed{s} A \leftrightarrow \boxed{s} B$

v. $\mathrm{BML} \vdash \boxed{s} A \to B \Rightarrow \mathrm{BML} \vdash \boxed{s} A \to \boxed{s} B$.

Again, 1.7.v holds for $\mathrm{ML} \supseteq \mathrm{BML}$ closed under $R2$. (Query: Why aren't the analogues to 1.3.v and 1.3.vi listed?)

In addition to the properties inherited from $\Box$, $\boxed{s}$ has some of its own.

1.8. LEMMA. i. $\mathrm{BML} \vdash \boxed{s} A \to A$

ii. $\mathrm{BML} \vdash \boxed{s} A \leftrightarrow \boxed{s}\boxed{s} A$

iii. $\mathrm{BML} \vdash \Box\boxed{s} A \leftrightarrow \Box A$

$\leftrightarrow \boxed{s}\Box A$.

As usual, I leave the proof of the Lemma as an exercise to the reader.

We are now ready to discuss substitution. In addition to the usual substitution lemma of the propositional calculus, there are two variants of the modal substitution lemma.

1.9. FIRST SUBSTITUTION LEMMA. (FSL). Let $A(p)$ be given.

$$\mathrm{BML} \vdash \boxed{s}(B \leftrightarrow C) \to . A(B) \leftrightarrow A(C).$$

1.10. SECOND SUBSTITUTION LEMMA. (SSL). Let $A(p)$ be given.

$$\mathrm{BML} \vdash \Box(B \leftrightarrow C) \to \Box\big(A(B) \leftrightarrow A(C)\big).$$

Proof of 1.9: By induction on the complexity of $A(p)$. The basis consists of several cases:

i. Let $A(p) = p$. By Lemma 1.8.1,

$$\mathsf{BML} \vdash \boxed{s}(B \leftrightarrow C) \to .B \leftrightarrow C$$

i'. Let $A(p) = q$. Trivially,

$$\mathsf{BML} \vdash \boxed{s}(B \leftrightarrow C) \to .q \leftrightarrow q.$$

ii-iii. $A(p)$ is t or f. This is handled exactly as in case i'.

For the induction step there are also several cases:

iv-vii. $A(p)$ is $\sim D$, $D \wedge E$, $D \vee E$, or $D \to E$. For the case of negation note that

$$D(B) \leftrightarrow D(C). \to .\sim D(B) \leftrightarrow \sim D(C)$$

is a tautology. Together with the induction hypothesis,

$$\mathsf{BML} \vdash \boxed{s}(B \leftrightarrow C) \to .D(B) \leftrightarrow D(C)$$

this yields

$$\mathsf{BML} \vdash \boxed{s}(B \leftrightarrow C) \to .\sim D(B) \leftrightarrow \sim D(C).$$

The other propositional connectives are treated similarly.

viii. $A(p)$ is $\Box D(p)$. This is the interesting case. Observe

$$\mathsf{BML} \vdash \boxed{s}(B \leftrightarrow C) \to .D(B) \leftrightarrow D(C), \quad \text{by induction hypothesis}$$

$$\vdash \Box\boxed{s}(B \leftrightarrow C) \to \Box\big(D(B) \leftrightarrow D(C)\big), \quad \text{by 1.3.iii}$$

$$\vdash \boxed{s}(B \leftrightarrow C) \to \Box\big(D(B) \leftrightarrow D(C)\big), \quad \text{by 1.8.iii}$$

$$\vdash \boxed{s}(B \leftrightarrow C) \to .\Box D(B) \leftrightarrow \Box D(C), \quad \text{by 1.3.iv.}$$

QED

(REMARK: Following Definition 1.1 of BML, I remarked that the modal system customarily chosen as one's basic system is the theory K obtained by deleting axiom schema *A3*. I note the use, in the key step of the above proof, of Lemma 1.8.iii--a consequence of *A3*. Our need of a nice Substitution Lemma in section 3, below, thus dictates the choice of BML over K as our basic system.)

The Second Substitution Lemma can be proven similarly; or, it can be reduced quickly to the First.

Proof of 1.10 (SSL): Write D for $B \leftrightarrow C$, E for $A(B) \leftrightarrow A(C)$, and observe

$$\mathsf{BML} \vdash \boxed{s} D \to E, \quad \text{by the } FSL$$

$$\vdash \Box(\boxed{s} D \to E), \quad \text{by } R2$$

$\vdash \Box \boxed{s} D \to \Box E$

$\vdash \Box D \to \Box E$, by 1.8.iii,

i.e. $\quad$ $\mathsf{BML} \vdash \Box(B \leftrightarrow C) \to \Box\big(A(B) \leftrightarrow A(C)\big)$. QED

Modulo the ordinary propositional substitution lemma, the *FSL* also more-or-less follows from the *SSL*. For, from $\boxed{s}(B \leftrightarrow C)$ one concludes $B \leftrightarrow C$, whence substitutability in non-modal contexts, and $\Box(B \leftrightarrow C)$, whence (by *SSL*) substitutability inside modal contexts. There is, however, a subtler reason why *FSL* follows from the *SSL*: They are equivalent, though distinct, formalisations of the same closure principle for PRA.

Given a schematic closure property of PRA, say

$$\forall \phi \left(\mathsf{PRA} \vdash A(\phi) \Rightarrow \mathsf{PRA} \vdash B(\phi) \right),$$

there is one very natural modal simulation, namely

$$\Box A(p) \to \Box B(p),$$

and one not quite so natural one,

$$\boxed{s}\, A(p) \to B(p).$$

The two Substitution Lemmas are two such formalisations of the closure of PRA under a substitution rule. For some interesting modal logics ML, any two such formalisations are equivalent in the sense that, if an instance of one is provable in ML then the corresponding instance of the other is also provable in ML.

1.11. FORMALISATION LEMMA. For any A, B, the following are equivalent:

i. $\mathsf{BML} \vdash \Box A \to \Box B$

ii. $\mathsf{BML} \vdash \boxed{s}\, A \to B$.

As in the derivation of *SSL* from *FSL*, the proof that 1.11.ii $\Rightarrow$ 1.11.i is a fairly simple syntactic matter, depending only on closure under *R2*. The converse is not so simple a matter. Model theoretic proofs for BML and PRL will be given in Chapter 2, below.

The Formalisation Lemma is due to Dick de Jongh and is mildly interesting. For one thing, the closure rule 1.11.i $\Rightarrow$ 1.11.ii is one of the logic of provability but is not one of arithmetic. It will thus serve as a warning of what not to read into PRL as a complete analysis of $Pr(\cdot)$. This fact also suggests that, since

$\Box A \to \Box B$ is a natural simulation of a closure rule, $\boxed{s} A \to B$ is not. Nonetheless, in the sequel we will generally prefer the unnatural $\boxed{s} A \to B$. This not only gives us quickly the natural $\Box A \to \Box B$, but it also allows conclusions about sentences B and not only the sentences $\Box B$.

EXERCISES

1. Prove Lemma 1.3.
2. Give arithmetic counterexamples to the converses to 1.3.iii-v.
3. Prove Lemmas 1.6 and 1.8.
4. How should one modally represent the closure of PRA under Löb's Rule: PRA $\vdash Pr(\ulcorner\psi\urcorner) \to \psi \;\Rightarrow\;$ PRA $\vdash \psi$? Prove the equivalence over BML (using $R2$) of the two schemata.

2. PROVABILITY LOGICS

The system BML of Basic Modal Logic was not designed to capture all properties of $Pr(\cdot)$. Specifically, it was not intended to reflect the subtleties inherent in the arithmetic language's self-referential capability. Such is the purpose of the system PRL.

2.1. DEFINITION. PRL is the modal theory extending BML by accepting the axioms $A1$-$A3$ and rules $R1$-$R2$ of BML and the additional axiom schema

$$A4.\quad \Box(\Box A \to A) \to \Box A.$$

As we have already remarked, axiom $A4$ simulates the Formalised Löb's Theorem and, hence, PRL is valid under arithmetic interpretations. The converse, that PRL is complete with respect to these interpretations will not be proven until Chapter 3 and, although all the early workers in the modal logic of provability believed this early on, it is not at all obvious that PRL is sufficient. A more natural source for additional principles would seem to be diagonalisation itself. If we bear in mind (i) that, in an instance,

$$\text{PRA} \vdash \phi \leftrightarrow \psi(\ulcorner\phi\urcorner),$$

of diagonalisation, the fixed point ϕ occurs only as a code in the operator $\psi(\cdot)$ and (ii) that the modal source of such a code, i.e. the inverse image of such under an arithmetic interpretation of the modal language, is the scope of a box, we see that the condition on interpretability within PRA of fixed point assertions $p \leftrightarrow A(p)$ is that p lie only within the scopes of $\Box$'s in $A(p)$. More briefly stated, if every instance of p in $A(p)$ is within the scope of a box, then, under any arithmetic interpretation $*$, it is $\ulcorner p^* \urcorner$ and not p^* itself that occurs in $A(p)^*$. Thus, modulo the interpretations of the other sentential variables in $A(p)$, one can choose $p^* = \phi$ so that

$$\text{PRA} \vdash p^* \leftrightarrow A(p)^*.$$

The condition that p occur only within the scopes of boxes in $A(p)$ is the appropriate modal restriction on fixed points.

2.2. DEFINITION. Let p and $A(p)$ be given. We say p obeys the *Diagonalisation Restriction* (*DR*) with respect to $A(p)$, or, more simply, p is *boxed* in $A(p)$, if every occurrence of p in $A(p)$ lies within the scope of a box.

The terminology "p obeys the *DR* w.r.t. $A(p)$" is to be preferred to "p is boxed in $A(p)$" not only because of its more sophisticated tone, but also because it extends more straightforwardly in Chapter 4, below, when we introduce new operators. Nonetheless, I shall use "boxed" throughout most of Part I.

For now, our interest in the *DR* is simply in giving the condition for reasonably postulating the existence of fixed points. There are two simple ways of doing this. First, one can apply brute force:

2.3. DEFINITION. The system DOL of *D*iagonalisation *O*perator *L*ogic is the extension of BML-- its language, axioms, and rules-- by the adjunction, for each formula $A(p, q_1, \ldots, q_n)$ in which the variable p is boxed, of new operators $\delta_A(q_1, \ldots, q_n)$ in the free variables $q_1, \ldots, q_n$ of A other than p and the axiom schema,

$$\delta_A(B_1, \ldots, B_n) \leftrightarrow A(\delta_A(B_1, \ldots, B_n), B_1, \ldots, B_n).$$

A stronger version DOL$^+$ assumes also the functoriality of the δ-operators:

$$\boxed{s}\,(B_i \leftrightarrow C_i) \rightarrow .\, \delta_A(B_1, \ldots, B_n) \leftrightarrow \delta_A(C_1, \ldots, C_n).$$

While it is a *fairly* immediate corollary to the Diagonalisation Lemma that any arithmetic interpretation of the modal language extends to one of DOL-- i.e. a fixed point ϕ interpreting δ_A can be found-- it requires further thought to realise that the functoriality schema can also be satisfied. Moreover, this functoriality is *not* a general property of the arithmetic fixed points constructed by the proof of the Diagonalisation Lemma in PRA. Thus, the system DOL^+ is less natural as a modal explication of self-reference than DOL and we shall stress here the more natural DOL. (However, the main result of the next section will allow the present subtleties to be labelled "beside the point": Both systems have PRL as their modal fragment.)

The disadvantages to extending the language to simulate diagonalisation are manifold. As we have just seen, new problems-- choose between DOL and DOL+ (and some other DOL++?)-- arise. Also, the semantics is a bit more difficult to work with. To handle fixed points without assuming their existence (and thus without having to answer irrelevant questions about them and without being responsible for their semantic interpretations), we must resort to an elimination rule.

2.4. DEFINITION. The system DIL of *Di*agonalisation *L*ogic is the extension of BML-- in the language of BML-- by the addition to the axioms and rules of BML the new rule:

$$DiR: \quad \boxed{s}\,(p \leftrightarrow A(p)) \rightarrow B \;/\; B,$$

where (i) p is boxed in $A(p)$, and (ii) p has no occurrence in B.

I leave to the reader the verification that the rule is valid with respect to arithmetic interpretations.

The systems DOL and DIL are equivalent to PRL:

2.5. THEOREM. Let B be any sentence of the usual modal language (i.e. B has none of the diagonalisation operators δ_A). The following are equivalent:

i. $PRL \vdash B$

ii. $DIL \vdash B$

iii. $DOL \vdash B$.

Partial proof: The implications i $\Rightarrow$ ii $\Rightarrow$ iii are relatively easily established and will be done so here. The converses are a little less immediate. Implication iii $\Rightarrow$ ii is given in the exercises, thus leaving only implication ii $\Rightarrow$ i in need of proof in the next section. We will actually prove iii $\Rightarrow$ i by showing the fixed point to be explicitly definable in PRL. (In fact, the functoriality needed to prove the conservation of DOL^+ over PRL will also be established.)

Let us start with the most mundane step in the proof: ii $\Rightarrow$ iii. We wish to show DOL closed under the rule *DiR*. Suppose

$$\mathrm{DOL} \vdash \boxed{s}\,(p \leftrightarrow A(p)) \to B,$$

where p is boxed in $A(p)$ and has no occurrence in B. Now, the only failure of the axioms and rules of DOL to be schematic is the new axioms about the diagonalisation operators-- and then only if one regards $\delta_A(B_1,\ldots,B_n)$ as a new propositional constant rather than $\delta_A(\cdot,\ldots,\cdot)$ as a connective. Thus, p can be replaced throughout the derivation by δ_A and one has

$$\mathrm{DOL} \vdash \boxed{s}\,(\delta_A \leftrightarrow A(\delta_A)) \to B.$$

Applying *R2* to the axiom $\delta_A \leftrightarrow A(\delta_A)$ yields

$$\mathrm{DOL} \vdash \boxed{s}\,(\delta_A \leftrightarrow A(\delta_A)),$$

whence $\mathrm{DOL} \vdash B$.

i $\Rightarrow$ ii. The proof that DIL extends PRL reduces to showing that DIL proves the axiom *A4* of PRL:

$$\mathrm{DIL} \vdash \Box(\Box A \to A) \to \Box A.$$

In other words, the proof consists of deriving the Formalised Löb's Theorem by diagonalisation-- being careful not to use non-propositional reasoning other than that provided by the Derivability Conditions.

To avoid long formulas, we will use the Deduction Theorem-- which is valid so long as we do not apply *R2*.

Assume $\boxed{s}\,(p \leftrightarrow \Box(p \to A))$. From this and the instance,

$$\Box(p \to A) \to \Box\Box(p \to A),$$

of *A3* the *FSL* yields $p \to \Box p$. Thus

$$p \to \Box p \wedge \Box(p \to A),$$

whence $p \to \Box A$.

Conversely, $\mathrm{BML} \vdash \Box A \to \Box(p \to A)$ and the assumption on p yields $\Box A \to p$. Thus:

$$\mathrm{BML} \vdash \boxed{s}\,(p \leftrightarrow \Box(p \to A)) \to .p \leftrightarrow \Box A$$

$$\vdash \boxed{s}\,(p \leftrightarrow \Box(p \to A)) \to \boxed{s}\,(p \leftrightarrow \Box A), \quad \text{by 1.7.v}$$

$$\vdash \boxed{s}\,(p \leftrightarrow \Box(p \to A)) \to .\Box A \leftrightarrow \Box(\Box A \to A), \quad \text{by } FSL.$$

One application of *DiR* yields

$\mathrm{DIL} \vdash .\Box A \leftrightarrow \Box(\Box A \rightarrow A).$ QED

EXERCISES

1. The strong box $\boxed{s}$ has nicer properties than the ordinary box $\Box$. Why, therefore, does the Diagonalisation Restriction mention the box rather than the strong box? Give an example of $A(p)$ in which every occurrence of p lies within the scope of a strong box but for which $p \leftrightarrow A(p)$ is contradictory.

2. Let p be boxed in $A(p,q_1,\ldots,q_n)$. Show: For any interpretations $\psi_1 = q_1{}^*,\ldots,$ $\psi_n = q_n{}^*$, there is an interpretation $\phi = p^*$ such that $\mathrm{PRA} \vdash \phi \leftrightarrow A(\phi,\psi_1,\ldots,\psi_n)$. Conclude that DiR is a valid rule of inference: If

   $$\mathrm{PRA} \vdash (\boxed{s}(p \leftrightarrow A(p)) \rightarrow B)^*$$

   for all arithmetic interpretations $*$ and p obeys the rule's conditions, then $\mathrm{PRA} \vdash B^*$ for all arithmetic interpretations.

3. Derive the Incompleteness Theorems in DIL, i.e.

   i. Show: $\mathrm{BML} \vdash \boxed{s}(p \leftrightarrow \sim\Box p) \wedge \sim\Box f \rightarrow \sim\Box p$

   ii. Show: $\mathrm{BML} \vdash \boxed{s}(p \leftrightarrow \sim\Box p) \rightarrow .p \leftrightarrow \sim\Box f$

   iii. Show: $\mathrm{DIL} \vdash \sim\Box f \rightarrow \sim\Box\sim\Box f$

   iv. (Macintyre and Simmons). Show:

   $$\mathrm{BML} \vdash \boxed{s}(p \leftrightarrow .\Box p \rightarrow A) \rightarrow .p \leftrightarrow (\Box A \rightarrow A).$$

4. Prove the implication iii $\Rightarrow$ ii of Theorem 2.5: For B a purely modal sentence, $\mathrm{DOL} \vdash B \Rightarrow \mathrm{DIL} \vdash B$. (Hint: By remarks of section 4, below (Unfair!),

   $$\mathrm{DOL} \vdash B \quad \text{iff} \quad \mathrm{BML} \vdash \bigwedge \boxed{s}(\delta_{A_i} \leftrightarrow A_i(\delta_{A_i})) \rightarrow B$$

   for some $A_1,\ldots,A_n$.)

5. (Other formulations of PRL). i. Derive schema $A4$ over BML from the schema,

   $$\Box(A \leftrightarrow \Box A) \rightarrow \Box A,$$

   asserting the provability of Henkin's fixed point $p \leftrightarrow \Box p$.

   ii. Show that axiom $A3$ is redundant in PRL.

   iii. (Macintyre and Simmons). Show that PRL is equivalent to the extension of BML by Löb's Theorem formulated as a rule of inference:

   $$LR\text{:} \quad \Box A \rightarrow A \;/\; A.$$

   (Hints: i & ii. Consider $A \wedge \Box A$; iii. start with the instance

$\Box(\Box A \to A) \to \Box\Box(\Box A \to A)$ of $A3$ and show

$$\text{BML} \vdash \Box\left(\Box(\Box A \to A) \to \Box A\right) \to .\Box(\Box A \to A) \to \Box A.\ )$$

3. SELF-REFERENCE IN PRL

One of the goals of the present section is the completion of the proof of Theorem 2.5 by showing PRL to be closed under the Diagonalisation Rule *Dir*. Model theoretically this can be done fairly directly-- as we shall see in Chapter 2; syntactically, one must do much more. Short of performing a complete syntactic analysis of PRL via a sequent calculus or a tableau system (which analyses have been partially carried out in the literature), the obvious syntactic proof of closure under the Diagonalisation Rule consists of finding explicit fixed points within PRL-- i.e. actually interpreting DOL within PRL. In other words, the goal of the present section is to offer an analysis of self-reference within PRL and incidentally, as it were, complete the proof of Theorem 2.5.

There are two theorems on modal fixed points-- existence and uniqueness. Although existence, i.e. explicit definability, is the more immediately pressing result, syntactically uniqueness is easier to prove.

3.1. UNIQUENESS OF FIXED POINTS. Let p be boxed in $A(p)$ and let q be a new variable. Then:

$$\text{PRL} \vdash \boxed{s}\left(p \leftrightarrow A(p)\right) \wedge \boxed{s}\left(q \leftrightarrow A(q)\right) \to .p \leftrightarrow q.$$

The Uniqueness Theorem was proven independently by Claudio Bernardi, Dick de Jongh, and Giovanni Sambin. De Jongh's proof is model theoretic and will be given in the next chapter. Sambin's proof is rather syntactically involved; it shows directly that any possible fixed point is equivalent to the specific sentence constructed in his explicit definability proof. We will not encounter this proof in the present monograph. Bernardi's syntactic proof is the simplest and prettiest of the three proofs and will be presented here.

Proof of 3.1: Because of my preference for schemata of the form $\boxed{s}A \to B$ over schemata of the form $\Box A \to \Box B$, I will prove

$$\boxed{s}\left(p \leftrightarrow A(p)\right) \wedge \boxed{s}\left(q \leftrightarrow A(q)\right) \to .p \leftrightarrow q$$

instead of the more natural

$$\Box(p \leftrightarrow A(p)) \wedge \Box(q \leftrightarrow A(q)) \rightarrow \Box(p \leftrightarrow q).$$

This means a few extra steps in the proof and a slight obscuring of the main idea, which is to derive

$$\Box(p \leftrightarrow q) \rightarrow (p \leftrightarrow q)$$

from the fixed point hypotheses and then appeal to Löb's Theorem.

Write $A(p) = B(\Box C_1(p),\ldots,\Box C_n(p))$, with p not occurring in $B(q_1,\ldots,q_n)$ and observe,

$$\begin{aligned} \mathrm{PRL}\vdash \Box(p \leftrightarrow q) &\rightarrow \Box(C_i(p) \leftrightarrow C_i(q)), \quad \text{by } SSL \\ &\rightarrow .\Box C_i(p) \leftrightarrow \Box C_i(q) \\ &\rightarrow \boxdot(\Box C_i(p) \leftrightarrow \Box C_i(q)), \quad \text{by Lemma 1.4} \\ &\rightarrow .A(p) \leftrightarrow A(q), \quad \text{by } FSL. \end{aligned}$$

We now bring in the fixed point hypotheses to conclude

$$\begin{aligned} \mathrm{PRL}\vdash \boxdot(p \leftrightarrow A(p)) \wedge \boxdot(q \leftrightarrow A(q)) &\rightarrow .\Box(p \leftrightarrow q) \rightarrow (p \leftrightarrow q) \qquad (*) \\ &\rightarrow \Box(\Box(p \leftrightarrow q) \rightarrow (p \leftrightarrow q)), \quad \text{by 1.4} \\ &\rightarrow \Box(p \leftrightarrow q), \quad \text{by } A4 \\ &\rightarrow .p \leftrightarrow q, \quad \text{by } (*). \end{aligned}$$

QED

By the Formalisation Lemma, or by simplifying the proof just presented, we also have

$$\mathrm{PRL}\vdash \Box(p \leftrightarrow A(p)) \wedge \Box(q \leftrightarrow A(q)) \rightarrow \Box(p \leftrightarrow q).$$

In this form it has a metamathematical meaning: If $\psi(v)$ is a formula arising from a modal context, e.g. if $\psi(v)$ is $Pr(v) \rightarrow \theta$ or $\sim Pr(\ulcorner\theta\urcorner \dot{\rightarrow} v)$, then

$$\mathrm{PRA}\vdash \phi_1 \leftrightarrow \psi(\ulcorner\phi_1\urcorner) \quad \text{and} \quad \mathrm{PRA}\vdash \phi_2 \leftrightarrow \psi(\ulcorner\phi_2\urcorner)$$

yield

$$\mathrm{PRA}\vdash \phi_1 \leftrightarrow \phi_2.$$

For, letting $\phi_1 = p^*$, $\phi_2 = q^*$, $\psi = A^*$,

$$Pr(\ulcorner\phi_1 \leftrightarrow \psi(\ulcorner\phi_1\urcorner)\urcorner) \wedge Pr(\ulcorner\phi_2 \leftrightarrow \psi(\ulcorner\phi_2\urcorner)\urcorner) \rightarrow Pr(\ulcorner\phi_1 \leftrightarrow \phi_2\urcorner)$$

is provable-- whence true. If, in particular, $\psi(v)$ is $\sim Pr(v)$, we can conclude that any two Gödel sentences ϕ_1, ϕ_2, i.e. sentences ϕ_1, ϕ_2, satisfying

$$\mathrm{PRA}\vdash \phi_1 \leftrightarrow \sim Pr(\ulcorner\phi_1\urcorner) \quad \text{and} \quad \mathrm{PRA}\vdash \phi_2 \leftrightarrow \sim Pr(\ulcorner\phi_2\urcorner),$$

are equivalent in PRA, i.e. $\mathrm{PRA}\vdash \phi_1 \leftrightarrow \phi_2$. But we already know this-- and more:

PRA$\vdash \phi_i \leftrightarrow Con$.

The explicit definability of *the* Gödel sentence is basically the Second Incompleteness Theorem, that of the Henkin sentence $\phi \leftrightarrow Pr(\ulcorner\phi\urcorner)$ is Löb's Theorem, that of Kreisel's variant $\phi \leftrightarrow Pr(\ulcorner\phi \to \psi\urcorner)$ of Löb's sentence is the Formalised Löb's Theorem, and that of Löb's sentence $\phi \leftrightarrow .Pr(\ulcorner\phi\urcorner) \to \psi$ is a quick corollary to the explicit definability of Kreisel's fixed point (To see this, note that, if $\vdash \phi \leftrightarrow .Pr(\ulcorner\phi\urcorner) \to \psi$, then $\vdash Pr(\ulcorner\phi\urcorner) \leftrightarrow Pr(\ulcorner Pr(\ulcorner\phi\urcorner) \to \psi\urcorner)$ and $Pr(\ulcorner\phi\urcorner)$ is a Kreisel sentence.). These are fairly easy, but have seemingly diverse derivations. There is, however, a unifying method behind these.

Giovanni Sambin and C. Smoryński first conjectured the explicit definability in PRL of fixed points. After a special case had been proven by Bernardi and Smoryński (The latter's model theoretic algorithm will be given in the next chapter.), de Jongh and, a bit later, Sambin proved the full result. De Jongh's original proof was semantic and so complicated he never showed it to anyone; Sambin's original proof was syntactic and so complicated I never read it completely. However, de Jongh did share with me his easy syntactic proof of a special case:

<u>3.2. LEMMA</u>. PRL$\vdash \Box C(t) \leftrightarrow \Box C(\Box C(t))$.

Proof: To prove the left-to-right implication, note

$$\text{PRL}\vdash \Box C(t) \to .t \leftrightarrow \Box C(t)$$
$$\to \boxed{s}\,(t \leftrightarrow \Box C(t)) \qquad (*)$$
$$\to .\Box C(t) \leftrightarrow \Box C(\Box C(t)),$$

by the *FSL*, whence

$$\text{PRL}\vdash \Box C(t) \to \Box C(\Box C(t)).$$

For the converse implication, start with (*):

$$\text{PRL}\vdash \Box C(t) \to \boxed{s}\,(t \leftrightarrow \Box C(t))$$
$$\to .C\Box C(t) \leftrightarrow C(t), \quad \text{by } FSL$$
$$\to .C\Box C(t) \to C(t),$$

whence

$$\text{PRL}\vdash C\Box C(t) \to .\Box C(t) \to C(t)$$
$$\vdash \Box C(\Box C(t)) \to \Box(\Box C(t) \to C(t))$$

$\vdash \Box C(\Box C(t)) \to \Box C(t)$, by $A4$. QED

3.3. COROLLARY. Let $A(p) = B(\Box C(p))$. Then

$$\text{PRL} \vdash AB(t) \leftrightarrow A(AB(t)).$$

Proof: The Corollary follows from the Lemma by simple algebra:

$$\text{PRL} \vdash \Box CB(t) \leftrightarrow \Box CB(\Box CB(t))$$

$$\vdash B\Box CB(t) \leftrightarrow B\Box CB(\Box CB(t)), \quad \text{by } R2 \text{ and } FSL$$

$$\vdash AB(t) \leftrightarrow A(AB(t)),$$

since $A(p) = B\Box C(p)$. QED

Corollary 3.3 is already sufficiently strong to account for the "classical" fixed point calculations:

3.4. EXAMPLES. i. (Gödel sentences). $A(p) = \sim\Box p$: Here $B(q) = \sim q$, $C(p) = p$. The fixed point is $D = AB(t) = \sim\Box\sim t$, which is equivalent to $\sim\Box f$.

ii. (Henkin sentence). $A(p) = \Box p$: The Lemma applies as $A = \Box C(p)$ for the trivial $C(p) = p$. D is $\Box t$, i.e. t.

iii. (Löb's sentence). $A(p,q) = \Box p \to q$: Here $B(r,q) = r \to q$, $C(p) = p$, and $D = AB(t) = B\Box CB(t) = (\Box(t \to q)) \to q$, which is equivalent to $\Box q \to q$.

iv. (Kreisel's sentence). $A(p,q) = \Box(p \to q)$: Again, the Lemma applies: $D = \Box(t \to q)$, i.e. $\Box q$.

Corollary 3.3 is actually the basis and a lemma for the inductive proof of the full result and not just a "do-able" special case:

3.5. EXPLICIT DEFINABILITY THEOREM. Let p be boxed in $A(p)$. There is a sentence D possessing only those variables of $A(p)$ other than p, and such that

i. $\text{PRL} \vdash \boxed{s}(p \leftrightarrow A(p)) \to .p \leftrightarrow D$

ii. $\text{PRL} \vdash D \leftrightarrow A(D)$.

Since we already have the Uniqueness Theorem at hand, we need only prove 3.5.ii. As I just said, we will prove this by induction. To have something to induct on, we look at a decomposition of $A(p)$.

3.6. DEFINITION. Let p be boxed in $A(p)$. Then we can write $A(p)$ in the form,

$$A(p) = B(\Box C_1(p), \ldots, \Box C_n(p)),$$

where p does not occur in $B(q_1,\ldots,q_n)$. Such a representation of $A(p)$ is called a *decomposition*; the sentences $\Box C_i(p)$ are the *components* of the decomposition. The components are assumed distinct, non-overlapping, and to contain non-vacuous occurrences of p.

3.7. EXAMPLE-WARNING. A given sentence $A(p)$ may have more than one decomposition. Even the number of components need not be unique. Let

$$A(p) \;=\; \Box(\Box p \to q) \wedge \sim\Box p.$$

$A(p)$ has two decompositions:

i. $A(p) = B(\Box C(p))$; where

$$B(q_1,q) \;=\; \Box(q_1 \to q) \wedge \sim q_1, \quad C(p) \;=\; p.$$

ii. $A(p) = B(\Box C_1(p), \Box C_2(p))$; where

$$B(q_1,q_2) \;=\; q_1 \wedge \sim q_2, \quad C_1(p) \;=\; \Box p \to q, \quad C_2(p) \;=\; p.$$

Proof of 3.5: We prove the Theorem by induction on the number of components in a decomposition of a formula $A(p)$ in which p is boxed. Thus, let p be boxed in $A(p)$ and choose a fixed decomposition $A(p) \;=\; B(\Box C_1(p),\ldots,\Box C_n(p))$.

If $n = 1$, we know by Corollary 3.3 that $D = AB(t)$ is a fixed point of $A(p)$.

I will not first prove the case of $n > 1$ components directly by induction. Such will be done, but then we must state the induction hypothesis carefully and the actual calculation of the fixed points therefrom is slightly more complicated. Instead, I first give a slightly awkward inductive proof for each n.

We are given $A(p) \;=\; B(\Box C_1(p),\ldots,\Box C_n(p))$, $n > 1$. If we simply relabel the p's in the various components, we get

$$A_1(p_1,\ldots,p_n) \;=\; B_1(\Box C_1(p_1),\ldots,\Box C_n(p_n)),$$

where $B_1(q_1,\ldots,q_n) = B(q_1,\ldots,q_n)$. With respect to any p_i, A_1 has only one component. Thus, for $i = 1$, there is by the case $n = 1$ a fixed point $D_1 = D_1(p_2,\ldots,p_n)$ of $A_1(p_1)$:

$$\mathsf{PRL} \vdash D_1 \leftrightarrow A_1(D_1,p_2,\ldots,p_n).$$

Now, by the construction of D_1 ($D_1 = A_1(B_1(t),p_2,\ldots,p_n)$), there are no new components with respect to $p_2,\ldots,p_n$. Thus, if we write

$$A_2(p_2,\ldots,p_n) \;=\; D_1(p_2,\ldots,p_n),$$

we can also write

$$A_2(p_2,\ldots,p_n) \;=\; B_2(\Box C_2(p_2),\ldots,\Box C_n(p_n)).$$

Let $D_2(p_3,\ldots,p_n)$ diagonalise this with respect to p_2, write

$$A_3(p_3,\ldots,p_n) \;=\; D_2(p_3,\ldots,p_n)$$
$$\;=\; B_3(\Box C_3(p_3),\ldots,\Box C_n(p_n)),$$

and continue to define $D_3,\ldots,D_n$, $A_3,\ldots,A_n$, and $B_4,\ldots,B_n$ such that

$$\mathrm{PRL}\vdash D_k(p_{k+1},\ldots,p_n) \leftrightarrow A_k(D_k,p_{k+1},\ldots,p_n).$$

Finally, define $D \;=\; D_n$.

Claim. $\mathrm{PRL}\vdash D \leftrightarrow A(D)$.

Proof: We show by induction on $n - k$ that

$$\mathrm{PRL}\vdash D \leftrightarrow A_k(D,\ldots,D).$$

Basis. $k = n$. This is the definition of $D = D_n$.

Induction step. Assume

$$\mathrm{PRL}\vdash D \leftrightarrow A_{k+1}(D,\ldots,D).$$

Then

$$\mathrm{PRL}\vdash D \leftrightarrow D_k(D,\ldots,D), \text{ by the definition of } A_{k+1} \qquad (*)$$
$$\vdash D \leftrightarrow A_k(D_k(D,\ldots,D),D,\ldots,D), \text{ by definition of } D_k$$
$$\vdash D \leftrightarrow A_k(D,D,\ldots,D), \text{ by } (*) \text{ and the } FSL.$$

This completes the induction. We thus get,

$$\mathrm{PRL}\vdash D \leftrightarrow A_1(D,\ldots,D)$$
$$\vdash D \leftrightarrow B_1(\Box C_1(D),\ldots,\Box C_n(D)), \text{ by definition of } A_1$$
$$\vdash D \leftrightarrow A(D), \text{ by the decomposition of } A.$$

This completes the proof of the Claim and therewith the Theorem. QED

This proof is not the best. A slightly easier proof can, as noted in the proof given, be given: Proceed by a genuine induction. For the induction step, let $A(p)$ be decomposed as $B(\Box C_1(p),\ldots,\Box C_n(p))$. Replace only the occurrences of p in the occurrences of the last component by a new variable q. Call the result A^*:

$$A^*(p,q) \;=\; B(\Box C_1(p),\ldots,\Box C_{n-1}(p),\Box C_n(q)).$$

$A^*(p)$ has only $n - 1$ components, whence the induction hypothesis yields a fixed point $D^* = D^*(q)$:

$$\mathrm{PRL}\vdash D^*(q) \leftrightarrow B(\Box C_1(D^*),\ldots,\Box C_{n-1}(D^*),\Box C_n(q)).$$

Strengthening the induction hypothesis, all occurrences of q in D^* lie in occurrences of the component $\Box C_n(q)$. Thus, Corollary 3.3 applies and we can find D satisfying

$$\mathrm{PRL} \vdash D \leftrightarrow D^*(D).$$

Substituting D for $D^*(D)$ and for q in the fixed point equation for D^* yields

$$\begin{aligned}\mathrm{PRL} \vdash D &\leftrightarrow D^*(D)\\ &\leftrightarrow B(\Box C_1(D^*),\ldots,\Box C_{n-1}(D^*),\Box C_n(D))\\ &\leftrightarrow B(\Box C_1(D),\ldots,\Box C_{n-1}(D),\Box C_n(D))\\ &\leftrightarrow A(D).\end{aligned}$$

This finishes the "easy" proof.

The difficulty with the easy proof is in performing the computations. The first step in finding the fixed point in the n component case is to find the fixed point in the $n - 1$ component case; the first step in finding the fixed point in the $n - 1$ component case is to find the fixed point in the $n - 2$ component case; etc. The grubbier proof proceeds by successively finding fixed points in the single component case and is more direct in application.

3.8. EXAMPLE. Let $A(p) = \Box(p \to r) \vee \Box(p \to s) \vee \Box(p \to u)$. The decomposition $A(p) = B(\Box C_1(p),\Box C_2(p),\Box C_3(p))$ with

$$B(q_1,q_2,q_3) = q_1 \vee q_2 \vee q_3$$

$$C_1(p) = p \to r,\quad C_2(p) = p \to s,\quad C_3(p) = p \to u$$

is obvious. This yields

$$A_1(p_1,p_2,p_3) = \Box(p_1 \to r) \vee \Box(p_2 \to s) \vee \Box(p_3 \to u)$$

$$B_1(q_1) = q_1 \vee \Box(p_2 \to s) \vee \Box(p_3 \to u).$$

(N.B. I am not following the *notation* of the proof here.) Since A_1 has only one component with respect to p_1, its fixed point is $D_1 = A_1(B_1(t))$. But

$$\begin{aligned}B_1(t) &= t \vee \Box(p_2 \to s) \vee \Box(p_3 \to u)\\ &\leftrightarrow t,\end{aligned}$$

whence

$$\begin{aligned}D_1 &\leftrightarrow A_1(t)\\ &\leftrightarrow \Box(t \to r) \vee \Box(p_2 \to s) \vee \Box(p_3 \to u)\\ &\leftrightarrow \Box r \vee \Box(p_2 \to s) \vee \Box(p_3 \to u).\end{aligned}$$

Next we diagonalise on this with respect to p_2:

$$A_2(p_2,p_3) = \Box r \vee \Box(p_2 \to s) \vee \Box(p_3 \to u)$$
$$= B_2(\Box C_2(p_2)),$$

where

$$B_2(q) = \Box r \vee q \vee \Box(p_3 \to u).$$

The fixed point is $D_2 = A_2B_2(t)$. Again,

$$\mathrm{PRL} \vdash B_2(t) \leftrightarrow t$$
$$\vdash D_2 \leftrightarrow A_2(t)$$
$$\vdash D_2 \leftrightarrow \Box r \vee \Box(t \to s) \vee \Box(p_3 \to u)$$
$$\vdash D_2 \leftrightarrow \Box r \vee \Box s \vee \Box(p_3 \to u).$$

We are now left with only the third variable p_3. The calculation yields $D_3 \leftrightarrow \Box r \vee \Box s \vee \Box u$, whence we can take $D = \Box r \vee \Box s \vee \Box u$.

The tediousness of this calculation illustrates an important point: The smaller the number of components, the quicker the calculation. Short-cuts are to be welcomed. Note that the fixed point D to the formula $A(p)$ of Example 3.8 is simply $A(t)$-- as in Lemma 3.2. Several explanations for this can be given-- cf. Exercises 4 and 7, below, and Chapter 4, much farther below.

I have not been quite fair to Sambin. His original proof of Theorem 3.5 is similar in many respects to the one given here. The main difference is due to his not having fully exploited the *FSL*: He only made propositional substitutions, which required him to use *maximal* components in the decomposition of $A(p)$-- no $\Box C_i(p)$ could be embedded in a larger modal context. This meant he had to prove a stronger result than de Jongh's Lemma 3.2, namely:

3.9. LEMMA. Let $C(p)$ be given. Then:

$$\mathrm{PRL} \vdash \Box C(E) \to .E \leftrightarrow F \quad \Rightarrow \quad \mathrm{PRL} \vdash \Box C(E) \leftrightarrow \Box C(F).$$

Sambin's Lemma contains the key to a computational improvement (in some cases) of the fixed point calculation-- cf. Exercise 7, below.

As I've already noted, the Explicit Definability Theorem was first proven independently by de Jongh and Sambin. The present simple syntactic proof is essentially a later simplification due to de Jongh. There are now several proofs--

some of which will be commented on in the next chapter.

EXERCISES

1. Find the fixed point of the formula $A(p) = \Box(\Box p \to q) \wedge \sim\Box p$ of Example 3.7 using each of the two decompositions given.

2. Give two decompositions of the formula $A(p) = \Box(\Box\sim p \vee \Box p) \to \Box p$ and find the fixed point by each decomposition.

3. Find the fixed point D of $\Box(p \to \Box f) \to \Box\sim p$.

4. Let $A^i(p) = \Box C^i(p)$ for $i = 1,2$. Show: The fixed points of $A^1 \vee A^2$ and $A^1 \wedge A^2$ are $\Box C^1(t) \vee \Box C^2(t)$ and $\Box C^1(t) \wedge \Box C^2(t)$, respectively.

5. Let $A^1(p) = \sim\Box p$, $A^2(p) = \Box\sim p$, $A^3(p) = \Box p$. Considering $A^1(p) \vee A^2(p)$, $A^1(p) \wedge A^3(p)$, $\sim A^3(p)$, and $A^3(p) \to A^2(p)$, show that the boolean operations are generally not preserved by the process of finding fixed points.

6. (Sambin). Let *S*ambin's *L*ogic SAL extend BML by the adjunction of the rule of inference:

   SaR: $\Box C(E) \to .E \leftrightarrow F \;/\; \Box C(E) \leftrightarrow \Box C(F)$.

   i. Show directly: $\mathsf{SAL} \vdash \Box A \to A \;\Rightarrow\; \mathsf{SAL} \vdash A$

   ii. Show directly: $\mathsf{SAL} \vdash \Box(\Box A \to A) \to \Box A$

   iii. Show: PRL is closed under SaR.

7. Define C to be *almost boxed* if $\mathsf{PRL} \vdash C \to \Box C$.

   i. Prove the following generalisation of Sambin's Lemma: If C is almost boxed, then $\mathsf{PRL} \vdash C(E) \to .E \leftrightarrow F \;\Rightarrow\; \mathsf{PRL} \vdash C(E) \leftrightarrow C(F)$.

   ii. Let $C(p)$ be almost boxed and p boxed in $C(p)$. Show: $\mathsf{PRL} \vdash C(t) \leftrightarrow C(C(t))$.

   iii. Apply ii to Example 3.8. Generalise Exercise 4.

   iv. Define $A(p) = B(C_1(p),\ldots,C_n(p))$ to be a decomposition of $A(p)$ if (i) each $C_i(p)$ is almost boxed, (ii) p is boxed in each $C_i(p)$, and (iii) p has no occurrence in $B(q_1,\ldots,q_n)$. Verify that the proof of Theorem 3.5 remains valid using this kind of decomposition.

8. Define UFPL to be the extension of BML by the axiom schema expressing the

uniqueness of fixed points. Show: UFPL = PRL.

9. Let p_1, p_2 be boxed in A_1 and A_2. Show: There are formulae D_1, D_2 such that PRL$\vdash D_1 \leftrightarrow A_1(D_1,D_2)$ and PRL$\vdash D_2 \leftrightarrow A_2(D_1,D_2)$.

4. AVOIDING *R2*

Although the necessitation rule *R2* is a very natural one-- particularly in the arithmetic context under which it simulates the first Derivability Condition, not every system of modal logic we will be interested in will be closed under it. In Chapter 3, for example, we will be interested in those modal sentences translating into *true* (not merely provable) schemata of arithmetic. The resulting modal system is not closed under *R2*, but obviously extends PRL. Moreover, in the next chapter, in proving completeness with respect to the given model theory, we will need to consider many theories extending BML but not closed under *R2*. This requires a reformulation of BML without *R2*. This does not offer a difficult problem-- the solution is quite simple (and quite inelegant).

4.1. CONVENTION. The only rules of inference allowed are *R1* and *R2*. Given a set Γ of axioms, we write

i. $\Gamma \vdash A$ if A is derivable from Γ by means of *R1* only; and

ii. $\Gamma \vdash_2 A$ if A is derivable from Γ by means of *R1* and *R2*.

4.2. DEFINITION. For any set Γ of sentences, define $\Gamma^2 = \Gamma \cup \{\Box B: B \in \Gamma\}$.

Modulo the axioms of BML, the passage from Γ to Γ^2 allows us to avoid *R2*:

4.3. LEMMA. Let Γ include all instances of *A1*-*A3*. Then:

$$\Gamma \vdash_2 A \quad \text{iff} \quad \Gamma^2 \vdash A.$$

Proof: The right-to-left implication is trivial. To prove the converse, it suffices to prove the closure of the set of theorems of Γ^2 with respect to *R1* under *R2*, i.e. it suffices to show: $\Gamma^2 \vdash A \Rightarrow \Gamma^2 \vdash \Box A$.

Let $\Gamma^2 \vdash A$. Then there is a sequence $A_1,\ldots,A_n = A$ such that each A_i is either (i) an axiom $A_i \in \Gamma$, (ii) the necessitation $\Box B_i$ of an axiom $B_i \in \Gamma$, or (iii) a consequence by *R1* of A_j, $A_k = A_j \to A_i$, where $j,k < i$. We show by induction on the length i of an initial segment $A_1,\ldots,A_i$ of this derivation that $\Gamma^2 \vdash \Box A_i$.

If $A_i \in \Gamma$, then $\Box A_i \in \Gamma^2$ by definition and $\Gamma^2 \vdash \Box A_i$.

If $A_i = \Box B_i$ for $B_i \in \Gamma$, then we derive $\Box A_i$ as follows:

$$\Gamma^2 \vdash \Box B_i, \quad \text{since } \Box B_i \in \Gamma^2$$

$$\vdash \Box B_i \to \Box\Box B_i, \quad \text{by } A3$$

$$\vdash \Box\Box B_i \ (= \Box A_i), \quad \text{by } R1.$$

Finally, if A_i follows from A_j, $A_k = A_j \to A_i$ by $R1$, with $j,k < i$, we have $\Gamma^2 \vdash \Box A_j$ and $\Gamma^2 \vdash \Box(A_j \to A_i)$ by induction hypothesis. By $A2$, we also have

$$\Gamma^2 \vdash \Box A_j \wedge \Box(A_j \to A_i) \to \Box A_i.$$

Simple propositional logic ($A1$ and $R1$) yields $\Gamma^2 \vdash \Box A_i$. QED

It follows that we get $R2$-free axiomatisations BML^2 and PRL^2 of BML and PRL, respectively. For convenience, we still denote the newly presented theories BML and PRL.

Since we are only interested in theories extending BML, we allow mention of BML to be dropped from $\vdash$, $\vdash_2$:

$$\Gamma \vdash A \text{ means } \mathsf{BML} \cup \Gamma \vdash A$$

$$\Gamma \vdash_2 A \text{ means } \mathsf{BML} \cup \Gamma \vdash_2 A.$$

Finally, notice that the Deduction Theorem for the propositional calculus yields:

4.4. LEMMA. $\Gamma \vdash A$ iff there is a finite set $\{C_0, \ldots, C_{n-1}\} \subseteq \Gamma$ such that

$$\mathsf{BML} \vdash \bigwedge C_i \to A.$$

I should warn the reader: These conventions will be invoked without warning whenever they are needed.

EXERCISE

1. Define a modal theory S to be *regular* if S is closed under $R2$. Let A, B be sentences, S a regular theory. Prove:

   i. $\mathsf{S} \vdash A \to B$ iff $\forall \mathsf{T} \supseteq \mathsf{S}(\mathsf{T} \vdash A \Rightarrow \mathsf{T} \vdash B)$

   ii. $\mathsf{S} \vdash \Box A \to \Box B$ iff $\forall$ *regular* $\mathsf{T} \supseteq \mathsf{S}(\mathsf{T} \vdash A \Rightarrow \mathsf{T} \vdash \Box B)$

   iii. $\mathsf{S} \vdash \boxed{s} A \to B$ iff $\forall$ *regular* $\mathsf{T} \supseteq \mathsf{S}(\mathsf{T} \vdash A \Rightarrow \mathsf{T} \vdash B)$

Chapter 2
Modal Model Theory

As we saw in the last chapter, one can get pretty far in studying self-reference in arithmetic by purely syntactic means. Unfortunately, syntax tends to be forgettable as well as limited and somewhat involved. Fortunately, there are various semantic interpretations with suggestive terminologies to aid the memory or, at least, rediscovery.

The original set theoretic semantics for modal logic was algebraic: The collection of equivalence classes of sentences of a modal theory (e.g. BML or PRL) under provable equivalence constitutes a boolean algebra with an additional operator; this algebra is quickly recognised to be the "free algebra in the category of modal algebras" satisfying the laws of the given modal theory. The homomorphisms from this free algebra into these others constitute a form of interpretation for which the modal theory in question is complete-- whence one can use the algebraic semantics to analyse the theory. Unfortunately, such algebraic semantics have never been particularly useful by themselves. What was needed was a representation, or duality, theory for such modal algebras. This duality theory was introduced independently by Paul R. Halmos, who payed no attention to modal logic, and Saul Kripke, who ignored the algebra, in the 1960s. In Chapter 5, below, we will take an algebraic view of PRL and extensions; for now, we follow Kripke's lead.

Kripke's models for modal logic are based on the notion of a *possible* world. Instead of having a single truth valuation for the world at hand, one has a collection of such valuations for a collection of such worlds. The usual propositional connectives are given, in each world, their usual interpretations; the modal connective refers to all possible worlds-- with one special proviso: Which worlds are possible depends on which world one is in. In the present chapter, we interpret this to mean that, among possible worlds $\alpha,\beta,\ldots$, there is an *accessibility relation* R which determines those worlds considered possible by others. ($\alpha R \beta$ means β is possible as far as α is concerned.) Once one has this idea, the notion of a model is clear. However, the appropriate constraints on R are not clear.

Mathematically, there are two games one can play with Kripke models: i. Given a modal theory, look for a decent type of relation R (equivalence relation, partial ordering, etc.) giving a complete semantics for the theory; and ii. given a type of relation, find a modal theory for which it offers a complete semantics. Neither game is guaranteed a good solution (Game ii always has a solution, but the relation R need not be sufficiently closely related to the outcome for it to be a *good* solution.) nor are they proper goals for us here; nonetheless, it is good to keep such in mind: The first game is, if not the ultimate, at least the immediate goal of the present chapter; moreover, the newcomer to modal logic will find most of the modern literature on the subject devoted to the playing of these games and he will be better prepared for them if he bears them in mind in the sequel.

The goals of the next two sections are proofs of completeness of BML and PRL for their respective classes of Kripke models. These basic results are due to Kripke and Krister Segerberg, respectively. The third section applies the model theory to the study of fixed points in PRL. This includes a "soft" model theoretic rehash of the Explicit Definability Theorem and some additional results. Finally, in section 4 we introduce a second provability logic that will be important in Chapter 3.

1. MODEL THEORY FOR BML

We might as well begin with a definition:

1.1. DEFINITION. A *frame* is a triple (K,R,α_0), where K is a non-empty set of *nodes* $\alpha,\beta,\gamma,\ldots$ (including α_0), R is a transitive binary relation on K (i.e. for $\alpha,\beta,\gamma \in K$, $\alpha R \beta$ and $\beta R \gamma$ imply $\alpha R \gamma$), and α_0 is a minimum element of K with respect to R (i.e. for any $\beta \in K$ other than α_0, $\alpha_0 R \beta$).

A frame, or *Kripke* frame, is to be thought of as the set of possible worlds underlying one's modal metaphysics together with the corresponding accessibility relation. The minimum node can be taken to represent the "actual" world. The assumption on the existence of a minimum node is not essential and is merely a convenient convention. There is no uniform terminology or notation in the literature and the use of lower case Greek letters to denote "nodes" is merely what I am used to.

Another definition:

1.2. DEFINITION. A *Kripke model* is a quadruple $\underline{K} = (K, R, \alpha_0, \Vdash)$, where (K, R, α_0) is a frame and $\Vdash$ is a satisfaction relation between nodes α and modal sentences. The assertion "$\alpha \Vdash A$" is read either "α forces A" (whence $\Vdash$ is a "forcing relation") or "A is true at α" and is assumed to satisfy the following conditions:

i. nothing special is assumed for atomic formulae

ii-iii. $\alpha \Vdash t$; $\alpha \nVdash f$

iv. $\alpha \Vdash \sim A$ iff $\alpha \nVdash A$

v-vii. $\alpha \Vdash A \circ B$ iff $(\alpha \Vdash A) \circ (\alpha \Vdash B)$, for $\circ \in \{\wedge, \vee, \rightarrow\}$

viii. $\alpha \Vdash \Box A$ iff $\forall \beta(\alpha R \beta \Rightarrow \beta \Vdash A)$.

A few immediate lemmas are collected in the following

1.3. REMARKS. i. A forcing relation $\Vdash$ on a frame (K, R, α_0) is completely and freely determined by its decisions on the atoms. That is, any decision on the truth or falsity of atomic formulae at nodes (i.e. the decision for each α and p whether or not $\alpha \Vdash p$) extends uniquely to a forcing relation $\Vdash$ making those same decisions; for, clauses ii-viii of Definition 1.2 are merely the inductive clauses of an inductive definition. In practical terms, this means that to describe a model $(K, R, \alpha_0, \Vdash)$ on a given frame (K, R, α_0) we need only specify the choices $\alpha \Vdash p$ or $\alpha \nVdash p$ for atoms p.

ii. The relation $\alpha \Vdash A$ depends only on α and those β such that $\alpha R \beta$. A technical expression of this is the following: Given $\underline{K} = (K, R, \alpha_0, \Vdash)$, define $\underline{K}_\alpha = (K_\alpha, R_\alpha, \alpha, \Vdash)$ by

a. $K_\alpha = \{\alpha\} \cup \{\beta \in K: \alpha R \beta\}$

b. $R_\alpha = R \restriction K_\alpha \times K_\alpha$, the restriction of R to K_α

c. $\Vdash_\alpha$: For $\beta \in K_\alpha$, $\beta \Vdash_\alpha p$ iff $\beta \Vdash p$.

$\underline{K}_\alpha$ is a Kripke model and, for all $\beta \in K_\alpha$ and all sentences A, $\beta \Vdash_\alpha A$ iff $\beta \Vdash A$. (Exercise.)

iii. $\Box$ is persistent with respect to R: If $\alpha \Vdash \Box A$ and $\alpha R \beta$, then $\beta \Vdash \Box A$. For, let $\alpha R \beta$ and note that

$$\begin{aligned} \alpha \Vdash \Box A \quad &\Rightarrow \quad \forall\gamma(\alpha R \gamma \Rightarrow \gamma \Vdash A), \text{ by definition} \\ &\Rightarrow \quad \forall\gamma(\beta R \gamma \Rightarrow \gamma \Vdash A), \text{ by transitivity} \\ &\Rightarrow \quad \beta \Vdash \Box A, \text{ by definition.} \end{aligned}$$

The points to Remarks 1.3.i and 1.3.iii are fairly obvious and these Remarks will be applied often in the sequel. The point to Remark 1.3.ii is less immediate. Basically, it justifies directly our restriction to models with minimum nodes: The behaviour of any node α in any model $\underline{K}$ coincides with that of the minimum node in $\underline{K}_\alpha$. Exactly what we are looking for in such behaviour has yet to be explained.

<u>1.4. DEFINITIONS</u>. Let $\underline{K} = (K, R, \alpha_0, \Vdash)$ be a Kripke model.

i. A sentence A is *true* in $\underline{K}$, written $\underline{K} \models A$, iff it is forced at α_0: $\underline{K} \models A$ iff $\alpha_0 \Vdash A$.

ii. A sentence A is *valid* in $\underline{K}$, written $\underline{K} \models_2 A$, iff it is forced at all nodes $\alpha \in K$: $\underline{K} \models_2 A$ iff $\forall\alpha \in K(\alpha \Vdash A)$.

iii. A set Γ of sentences is *true* (*valid*) in $\underline{K}$, written $\underline{K} \models \Gamma$ (respectively: $\underline{K} \models_2 \Gamma$) iff every sentence $A \in \Gamma$ is true (respectively: valid) in $\underline{K}$.

But for the bifurcation of the concept of truth in a model, this definition is fairly unremarkable. That we now have two notions-- truth and validity-- is reasonable enough: In a modal context there should be ordinary truth and a more stringent notion thereof. Since we have the modal operator, we can avoid the notion of validity: $\underline{K} \models_2 A$ iff $\underline{K} \models A \wedge \Box A$. (Exercise.) It is, however, of moderate interest to consider both notions.

Generalising notions of truth and validity are notions of semantic entailment:

<u>1.5. DEFINITIONS</u>. The semantic consequence relations $\models$ and $\models_2$ are defined by:

i. $\Gamma \models A$ iff, for all $\underline{K}$, $\underline{K} \models \Gamma \Rightarrow \underline{K} \models A$

ii. $\Gamma \models_2 A$ iff, for all $\underline{K}$, $\underline{K} \models_2 \Gamma \Rightarrow \underline{K} \models_2 A$.

With syntactic and semantic consequence relations, the customary thing to do is to prove their coincidence:

1.6. STRONG COMPLETENESS THEOREM. For all Γ, A,

i. $\Gamma \vdash A$ iff $\Gamma \models A$

ii. $\Gamma \vdash_2 A$ iff $\Gamma \models_2 A$.

Proof: Recall from section 4 that $\Gamma \vdash A$ means A is derivable from Γ and the augmented system of BML by means only of the rule *R1* of modus ponens, and that $\Gamma \vdash_2 A$ means A is derivable from Γ and the axioms of BML by means of both *R1* and the rule *R2* of necessitation.

To prove the left-to-right implications of the Theorem one:

i. shows by inspection all axioms of BML to be valid,

ii. observes both the sets of true and of valid sentences to be closed under *R1*,

and iii. notes validity to be preserved under *R2*.

I leave these routine matters to the reader.

The converses are established as follows: First, we note that part ii of the Theorem reduces quickly to part i:

$$\begin{aligned} \Gamma \models_2 A &\Rightarrow \Gamma \cup \{\Box B: B \in \Gamma\} \models A, \quad \text{(why?)} \\ &\Rightarrow \Gamma \cup \{\Box B: B \in \Gamma\} \vdash A, \quad \text{by part i} \\ &\Rightarrow \Gamma \vdash_2 A, \quad \text{by } R2. \end{aligned}$$

Next, of course, we prove the right-to-left implication of part i. This is done contrapositively: We assume $\Gamma \nvdash A$ and fairly canonically construct a model $\underline{K}$ in which Γ is true and A is not. The only non-canonical part of the construction is dictated by our insistence on having a minimum node in K: Among the many possibilities, we must single one out.

The plan of the proof is quite simple: We economically identify nodes with sets of sentences true at them. Since we do not know which sets to take, we take all of them. Moreover, we make one world accessible to another if it is at all possible to do so-- thus maximising the amount of information available to falsify untrue necessitations. Finally, the identification of a node with the set of sentences true at it tells us how to define the forcing relation. To state all of this formally, we first need our minimum node. For this and later applications as well, we need a lemma.

1.7. LEMMA. Let $\Delta \nvdash B$. There is a *completion* $\hat{\Delta} \supseteq \Delta$ such that $\hat{\Delta} \nvdash B$.

Proof: Let $C_1, C_2, \ldots$ enumerate all sentences of the modal language. Define Δ_n by induction on $n \geq 1$ as follows: First, $\Delta_1 = \Delta$. To define Δ_{n+1}, observe that, if

$$\Delta_n \cup \{C_n\} \vdash B \quad \text{and} \quad \Delta_n \cup \{\sim C_n\} \vdash B,$$

then $\Delta_n \vdash B$ by appeal to the appropriate tautology and modus ponens. Hence, one of $\Delta_n \cup \{C_n\}$ and $\Delta_n \cup \{\sim C_n\}$ does not derive B. Let Δ_{n+1} be such:

$$\Delta_{n+1} = \begin{cases} \Delta_n \cup \{C_n\}, & \text{if } \Delta_n \cup \{C_n\} \nvdash B \\ \Delta_n \cup \{\sim C_n\}, & \text{if } \Delta_n \cup \{\sim C_n\} \nvdash B. \end{cases}$$

The completion $\hat{\Delta}$ desired is $\hat{\Delta} = \bigcup_n \Delta$. $\hat{\Delta} \nvdash B$ since any proof would use only axioms from some Δ_n, thereby yielding $\Delta_n \vdash B$, contrary to construction. $\hat{\Delta}$ is complete since, for every C, either C or $\sim C$ is in $\hat{\Delta}$.

QED

A small remark before completing the proof of the Completeness Theorem: A complete consistent theory Δ is deductively closed: $\Delta \vdash C$ iff $C \in \Delta$. For, $C \notin \Delta$ implies $\sim C \in \Delta$ by completeness; but $\sim C \in \Delta$ together with $\Delta \vdash C$ makes Δ inconsistent.

Continuation of the proof of 1.6: We must show that $\Gamma \models A$ implies $\Gamma \vdash A$. Assume, by way of contraposition, that $\Gamma \nvdash A$. Fix $\hat{\Gamma}$ to be an arbitrary completion of Γ for which $\hat{\Gamma} \nvdash A$. We will construct a Kripke model $\underline{K}$ from completions of BML. But for our insistence on having a minimum element, which will be $\hat{\Gamma}$, we would choose K to consist of all such completions; instead, we must choose only those completions accessible to $\hat{\Gamma}$, i.e. we first define R and then define K.

In what follows, Δ and its variants denote completions of BML.

Given Δ, define

$$\Delta_{\Box} = \{C: \Box C \in \Delta\}; \text{ and: } \Delta R \Delta' \text{ iff } \Delta_{\Box} \subseteq \Delta'.$$

(Observe that, if Δ is precisely what is forced at (the node identified with) Δ, then $\Delta_{\Box} \subseteq \Delta'$ is the minimum requirement Δ' must satisfy to be accessible to Δ.)

Given R, we can easily define $\underline{K}$:

$$K = \{\hat{\Gamma}\} \cup \{\Delta: \hat{\Gamma} R \Delta \text{ and } \Delta \text{ is a completion of BML}\}$$

$R \;=\;$ the restriction of the relation R just defined to K

$\alpha_0 \;=\; \hat{\Gamma}$

$\Vdash$: $\Delta \Vdash p$ iff $p \in \Delta$, for p atomic.

First note that $(K, R, \hat{\Gamma})$ is a frame: K is non-empty, $\hat{\Gamma}$ is an R-minimum node by definition, and R is transitive by $A3$: If $\Delta R \Delta'$ and $\Delta' R \Delta''$, then

$$\begin{aligned} C \in \Delta \;&\Rightarrow\; \Box\Box C \in \Delta, \quad \text{by } A3 \\ &\Rightarrow\; \Box C \in \Delta', \quad \text{since } \Delta R \Delta' \\ &\Rightarrow\; C \in \Delta'', \quad \text{since } \Delta' R \Delta''. \end{aligned}$$

We thus conclude $\Delta_{\Box} \subseteq \Delta''$, i.e. $\Delta R \Delta''$. Observe that this argument does not assume $\Delta, \Delta', \Delta'' \in K$. Thus: if $\Delta \in K$ and $\Delta R \Delta'$, then $\Delta' \in K$.

We conclude $\underline{K}$ is a Kripke model. It only remains to check that it is the model we want:

Claim. For any $\Delta \in K$ and any modal sentence B,

$$\Delta \Vdash B \quad \text{iff} \quad B \in \Delta.$$

Proof of the Claim: By induction on the complexity of B.

i. $B = p$ is atomic. By the definition of $\Vdash$.

ii-iii. $B = t$ or f. By the consistency and completeness of Δ and the definition of a forcing relation.

iv-vii. B is a propositional combination of C, D. The result in this case follows by the consistency and completeness of Δ and the definition of a forcing relation.

viii. $B = \Box C$. This is the interesting case. One implication is trivial:

$$\begin{aligned} \Box C \in \Delta \;&\Rightarrow\; \forall \Delta' (\Delta R \Delta' \;\Rightarrow\; C \in \Delta'), \quad \text{by definition of } R \\ &\Rightarrow\; \forall \Delta' (\Delta R \Delta' \;\Rightarrow\; \Delta' \Vdash C), \quad \text{by induction hypothesis} \\ &\Rightarrow\; \Delta \Vdash \Box C. \end{aligned}$$

To prove the converse, we argue contrapositively:

$$\Box C \notin \Delta \;\Rightarrow\; C \notin \Delta_{\Box}.$$

The claim now is that $\Delta_{\Box} \nvdash C$. For, otherwise there are $C_1, \ldots, C_n \in \Delta_{\Box}$ such that

$$\begin{aligned} \text{BML} &\vdash \textstyle\bigwedge C_i \to C \\ &\vdash \Box \textstyle\bigwedge C_i \to \Box C \end{aligned}$$

$$\vdash \bigwedge \Box C_i \to \Box C$$

whence $\quad \Delta \vdash \bigwedge \Box C_i \to \Box C$

$$\vdash \Box C,$$

since $\Delta \vdash \Box C_i$ for each i. But this last conclusion is contrary to hypothesis: $\Box C \notin \Delta$. Thus $\Delta_\Box \nvdash C$.

By Lemma 1.7, there is a completion Δ' of Δ such that $\Delta \nvdash C$. But $\Delta_\Box \subseteq \Delta'$ means $\Delta R \Delta'$. Thus $\Delta' \in K$ and we conclude:

$$\Box C \notin \Delta \;\Rightarrow\; \forall \Delta'(\Delta R \Delta' \text{ and } C \notin \Delta')$$

$$\Rightarrow\; \forall \Delta'(\Delta R \Delta' \text{ and } \Delta' \Vdash\!\!\!\!/\; C), \text{ by induction hypothesis}$$

$$\Rightarrow\; \Delta \Vdash\!\!\!\!/\; C.$$

This completes the proof of the Claim. QED

Completion of the proof of Theorem 1.6: We are done. For, from the fact that $\hat{\Gamma} \nvdash A$ we can conclude via the Claim that, in $\underline{K}$, $\hat{\Gamma} \Vdash\!\!\!\!/\; A$, i.e. we have constructed a model of Γ in which A is not true. QED

Our main interest in proving the Strong Completeness Theorem for BML is in having a model theory for PRL:

1.8. COROLLARY. For any sentence A,

$$\text{PRL} \vdash A \quad \text{iff} \quad \text{PRL} \models A \quad \text{iff} \quad \text{PRL} \models_2 A.$$

This follows immediately from 1.6 and the closure of PRL under *R2*.

Corollary 1.8, as a completeness theorem for PRL, is not of much use. What we will need will be completeness with respect to a manageable class of models. We will find this in the immediately following section; the section immediately following that will apply such a model theory to self-reference. For now, let us pause to consider some easy consequences of Theorem 1.6.

1.9. COMPACTNESS THEOREM. $\Gamma \models A$ iff, for some finite $\Gamma_0 \subseteq \Gamma$, $\Gamma_0 \vdash A$.

This is immediate: The right-to-left implication is trivial and the left-to-right implication follows from the obvious analogue for derivability.

Compactness can also be established by appeal to the compactness theorem for first-order logic as Kripke models for BML are first-order definable structures.

With such an independently established compactness theorem, one can reduce strong completeness ($\Gamma \vdash A$ iff $\Gamma \models A$) to simple completeness ($\vdash A$ iff $\models A$) and, in fact, Kripke's original completeness theorem for BML was simple completeness with respect to finite models. This finiteness is rather useful and can be derived from Theorem 1.6:

<u>1.10. COMPLETENESS THEOREM WITH RESPECT TO FINITE MODELS</u>. Let A be a modal sentence. The following are equivalent:

i. $\mathsf{BML} \vdash A$

ii. A is true in all finite Kripke models

iii. A is valid in all finite Kripke models.

Proof: That i implies each of ii and iii follows from that half of Theorem 1.6 left to the reader to prove, i.e. the soundness of BML with respect to all, whence all finite, Kripke models. That iii implies ii is trivial and that ii implies iii follows from Remark 1.3.ii. (If A is true in all finite Kripke models and $\alpha \in K$ is any node in a finite Kripke model $\underline{K}$, then A is true in $\underline{K}_\alpha$, whence $\alpha \Vdash A$ in $\underline{K}$.) Thus, we need only show that one of ii, iii implies i. We show that ii implies i by contraposition.

Suppose $\mathsf{BML} \nvdash A$. By the Strong Completeness Theorem, there is a model $\underline{K} = (K, R, \alpha_0, \Vdash)$ in which $\alpha_0 \nVdash A$. We will show that a finite quotient of $\underline{K}$ is a countermodel to A. To construct such, let S denote the set of all subformulae of A (including A) and define, for all $\alpha, \beta \in K$,

i. $S(\alpha) = \{B \in S: \alpha \Vdash B\}$

ii. $S(\alpha) R_S S(\beta)$ iff $\forall C(\Box C \in S(\alpha) \Rightarrow C, \Box C \in S(\beta))$.

(Remark: The definition is analogous to that for the construction in the proof of Theorem 1.6. The $S(\alpha)$'s are the restrictions to S of complete theories and R_S is very similar to the relation R defined on such complete theories. If we define $S(\alpha)_\Box = \{C: \Box C \in S(\alpha)\}$, we almost have

$$S(\alpha) R_S S(\beta) \text{ iff } S(\alpha)_\Box \subseteq S(\beta).$$

We do not quite have this because we also want-- for the sake of transitivity-- to insist that $\Box C \in S(\alpha)$ implies $\Box C \in S(\beta)$. In the previous construction, this was automatic; in the present case, since S is not closed under $\Box$, we must add the assumption.)

We now define our candidate $\underline{K}_S = (K_S, R_S, S(\alpha_0), \Vdash_S)$ for the finite countermodel as follows:

$K_S = \{S(\alpha): \alpha \in K\}$

R_S is as just defined

$S(\alpha_0)$ is as just defined

$\Vdash_S$: $S(\alpha) \Vdash_S p$ iff $p \in S(\alpha)$, for p atomic.

Claim 1. $(K_S, R_S, S(\alpha_0))$ is a finite frame, i.e. K_S is a finite nonempty set, R_S is transitive and $S(\alpha_0)$ is an R_S-minimum element.

K_S is clearly nonempty. Finiteness is also fairly easy: K_S consists of subsets of a fixed finite set S.

The transitivity of R_S is not too subtle: Let $S(\alpha)\, R_S\, S(\beta)\, R_S\, S(\gamma)$. Let $\Box C \in S(\alpha)$ be arbitrary. By the definition of R_S, $\Box C \in S(\beta)$, whence $C, \Box C \in S(\gamma)$. Since $\Box C$ was arbitrary, $S(\alpha)\, R_S\, S(\gamma)$.

Finally, if $S(\alpha)$ is any element of K_S other than $S(\alpha_0)$, we have $\alpha_0\, R_S\, \alpha$, whence, for any $\Box C \in S$,

$$\begin{aligned}\Box C \in S(\alpha_0) &\Rightarrow \alpha_0 \Vdash \Box C\\ &\Rightarrow \alpha \Vdash C, \Box C\\ &\Rightarrow C, \Box C \in S(\alpha).\end{aligned}$$

Thus, $S(\alpha_0)\, R_S\, S(\alpha)$.

We next claim that α and $S(\alpha)$ agree on S:

Claim 2. For any $B \in S$,

$S(\alpha) \Vdash_S B$ iff $\alpha \Vdash B$.

The proof of Claim 2, like that of the analogous claim in the proof of Theorem 1.6, is an induction on the complexity of B. As before, the only interesting case is that in which $B = \Box C$:

$$\begin{aligned}\alpha \Vdash \Box C &\Rightarrow \Box C \in S(\alpha), \text{ by definition}\\ &\Rightarrow \forall \beta\big(S(\alpha)\, R_S\, S(\beta) \Rightarrow C \in S(\beta)\big)\\ &\Rightarrow \forall \beta\big(S(\alpha)\, R_S\, S(\beta) \Rightarrow \beta \Vdash C\big)\end{aligned}$$

$$\Rightarrow \quad \forall \beta \big(S(\alpha)\, R_S\, S(\beta) \;\Rightarrow\; S(\beta) \Vdash_S C \big),$$

by induction hypothesis. But this last conclusion yields $S(\alpha) \Vdash_S \Box C$. To prove the converse, we prove the inverse:

$$\alpha \nVdash \Box C \quad \Rightarrow \quad \exists \beta \big(\alpha R \beta \;\&\; \beta \nVdash C \big)$$

$$\Rightarrow \quad \exists \beta \big(S(\alpha)\, R_S\, S(\beta) \;\&\; \beta \nVdash C \big),$$

since $\alpha R \beta \Rightarrow S(\alpha)\, R_S\, S(\beta)$ (Why?). But now the induction hypothesis yields

$$\Rightarrow \quad \exists \beta \big(S(\alpha)\, R_S\, S(\beta) \;\&\; S(\beta) \nVdash_S C \big)$$

$$\Rightarrow \quad S(\alpha) \nVdash_S \Box C.$$

The proof of the Theorem is completed by recalling $A \in S$ and $\alpha_0 \nVdash A$; for, then $S(\alpha_0) \nVdash_S A$. QED

The immediate corollary to the *Finite Model Property*, i.e. completeness with respect to finite models, is decidability.

1.11. COROLLARY. BML is decidable.

For, to test if BML $\vdash A$, one merely needs to check if A is true in all models having at most 2^n nodes, where n is the number of subformulae of A. (This is not much of a decision procedure. As in the case with the propositional calculus, no *feasible* decision procedure for BML is known. In actual practice, however, one rarely wants to test derivability; but, for these rare occasions, more efficient tableau methods are available.)

We thus have compactness, the finite model property, and decidability as applications of a complete semantics. A good model theory has other mundane applications, e.g. proofs of closure under rules of inference. The following such result was promised in Chapter 1 (1.1.11).

1.12. FORMALISATION LEMMA. For any A,B, the following are equivalent:

i. BML $\vdash \Box A \to \Box B$

ii. BML $\vdash \boxed{s}\, A \to B$.

Proof: We have already proven implication ii $\Rightarrow$ i syntactically by observing simply that it followed from closure under *R2*. A semantic proof is also possible. (Exercise.)

i $\Rightarrow$ ii. We prove the contrapositive. Suppose

$$\mathrm{BML} \nvdash \boxed{s}\, A \to B.$$

Let $\underline{K} = (K, R, \alpha_0, \Vdash) \nvDash \boxed{s}\, A \to B$ be a countermodel, i.e.

$$\alpha_0 \Vdash \boxed{s}\, A, \qquad \alpha_0 \nVdash B.$$

Let α_{-1} be a new node and define $\underline{K}_{-1}$ by:

$$K_{-1} = K \cup \{\alpha_{-1}\}$$

$$R_{-1} = R \cup \{(\alpha_{-1}, \alpha): \ \alpha \in K\}$$

α_{-1} is the minimum of $\underline{K}_{-1}$

$\Vdash_{-1}$: for α K, $\alpha \Vdash_{-1} p$ iff $\alpha \Vdash p$ for p atomic; α_{-1} forces no (or all, some, ...) atoms.

In pictorial terms, $\underline{K}$ is some model with a minimum α_0 and $\underline{K}_{-1}$ is obtained by putting a new minimum below $\underline{K}$:

$\underline{K}$: α_0 $\qquad\qquad$ $\underline{K}_{-1}$: α_0, α_{-1}.

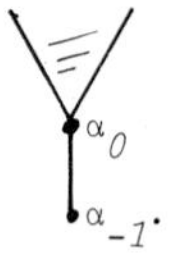

We can apply Remark 1.3.ii to conclude that

$$\alpha \Vdash_{-1} C \quad \text{iff} \quad \alpha \Vdash C$$

for all C and all $\alpha \in K$: for, $(\underline{K}_{-1})_\alpha = \underline{K}_\alpha$ for $\alpha \in K$. Thus, $\alpha_0 \Vdash_{-1} \boxed{s}\, A$ and it follows that $\alpha_{-1} \Vdash_{-1} \Box A$. If $\mathrm{BML} \vdash \Box A \to \Box B$, it follows that $\alpha_{-1} \Vdash_{-1} \Box B$ and, thus, $\alpha_0 \Vdash_{-1} B$, i.e. $\alpha_0 \Vdash B$. But this contradicts our assumption. QED

<u>1.13. REMARK</u>. If we assume $\underline{K} \models \mathrm{PRL}$, a routine check will reveal that $\underline{K}_{-1} \models \mathrm{PRL}$ as well, whence the Formalisation Lemma also holds for PRL. Alternatively, we can simply wait until the next section, where we get an adequate class of models for PRL that is visibly closed under the construction just given.

EXERCISES

1. Prove the assertion of Remark 1.3.ii: If $\underline{K} = (K, R, \alpha_0, \Vdash)$ is a Kripke model and $\alpha \in K$, then, for $\underline{K}_\alpha$ defined as in 1.3.ii and all $\beta \in K_\alpha$, one has $\beta \Vdash_\alpha A$ iff $\beta \Vdash A$, for all sentences A.

2. Prove the soundness of BML left unproven in the text:

$$\Gamma \vdash A \Rightarrow \Gamma \models A\ ; \quad \Gamma \vdash_2 A \Rightarrow \Gamma \models_2 A.$$

3. Let BML^- be the system obtained from BML by deleting the axiom schema *A3* (and its adjoined boxed instances $\Box(\Box A \to \Box\Box A)$) while adding $\Box^n A$ for all other instances A of axioms. (Without *A3*, the trick of Chapter 1, section 4 of avoiding *R2* by adding boxed instances of axioms does not work. One must iterate the boxing.) Prove the Strong Completeness Theorem for BML^- with respect to those models $\underline{K} = (K, R, \alpha_0, \Vdash)$ resulting by dropping the requirement of the transitivity of R.

(Remark: In the last chapter I announced that BML was the obvious choice of a base system over which to work. For our meta-arithmetic purposes this was true. In model theory, however, BML^- (most commonly known as K) is the obvious choice: It is the minimum system for which the proof given works nicely. Moreover, with respect to the two games, cited at the beginning of the present chapter, of matching axioms of modal logic with conditions on R, our base system BML then becomes the prototype: Schema *A3* corresponds to transitivity. The next exercise offers another example.)

4. The modal system $S4$ extends BML by the schemata

$$\Box A \to A \quad \text{and} \quad \Box(\Box A \to A).$$

Prove the completeness of $S4$ with respect to models whose frames are (transitive and) reflexive. (Suggestion: Rather than repeat the construction in the proof of Theorem 1.6, modify that of Theorem 1.10.)

5. Prove the following: For any sentences A, B,

   i. $\mathsf{BML} \vdash A$ iff $\mathsf{BML} \vdash \Box A$

   ii. $\mathsf{BML} \vdash \Box A \vee \Box B$ iff $\mathsf{BML} \vdash \Box A$ or $\mathsf{BML} \vdash \Box B$.

6. Construct Kripke models showing the underivability in BML of the following:

   i. $\Box(p \vee q) \to \Box p \vee \Box q$

   ii. $(\Box p \to \Box q) \to \Box(p \to q)$

   iii. $(\Box p \leftrightarrow \Box q) \to \Box(p \leftrightarrow q)$.

7. Show that $S4$ (Exercise 4) is the logic of the strong box. I.e., for any modal sentence A, let A^s result from A by repalcing each occurrence of $\Box$ by $\boxed{s}$. Show: $\mathsf{BML} \vdash A^s$ iff $S4 \vdash A$.

2. MODEL THEORY FOR PRL

With Corollary 1.8, we have a completeness theorem for PRL with respect to those Kripke models in which PRL is true (respectively, valid). The proof yields, in fact, a strong completeness theorem therefor. Such a result is useful, but it does have its limitations. What one needs is completeness with respect to readily recognisable models, i.e. models we can recognise as models of PRL without having to pay much attention to the actual forcing relation. Not paying much attention boils down to ignoring outright the forcing relation.

2.1. DEFINITION. A sentence A is *valid* in a frame (K,R,α_0) if $\alpha \Vdash A$ for all $\alpha \in K$ and all forcing relations $\Vdash$ on the frame, i.e. if A is valid in all models $\underline{K} = (K,R,\alpha_0,\Vdash)$ on the frame. A set Γ of sentences is *valid* in a given frame if every sentence in Γ is valid in the frame.

The reader who has faithfully worked the exercises of the preceding section should already have an idea of what we are after: S4 is complete with respect to models on reflexive frames; hence S4 is valid in reflexive frames. The converse also holds: S4 is valid in a frame iff it is reflexive. (Exercise.) Our immediate goal is to find a characterisation for the frames in which PRL is valid.

Which frames is PRL valid in? To see which, we simply write down the condition for validity of the instance,

$$\Box(\Box p \to p) \to \Box p, \quad p \text{ a fixed atom,}$$

of $A4$ in a given frame (K,R,α_0). For notational convenience in doing this, we let $\check{R}$ denote the relation converse to R and $X \subseteq K$ the set of nodes at which p is to be forced. (With respect to p, the forcing relations are determined by these sets.) In terms of X,

$$\alpha \Vdash \Box(\Box p \to p) \to \Box p$$

iff (after some unravelling)

$$\forall \beta \check{R} \alpha\big(\forall \gamma \check{R} \beta(\gamma \in X) \Rightarrow \beta \in X\big) \Rightarrow \forall \beta \check{R} \alpha(\beta \in X).$$

In words: $A4$ is valid in (K,R,α_0) iff transfinite induction on $\check{R}$ holds.

2.2. DEFINITION. A frame (K,R,α_0) is *reverse well-founded* if it has no ascending sequences of length ω, i.e. if there is no infinite sequence $\alpha_0 R \alpha_1 R \ldots$.

2.3. CHARACTERISATION THEOREM. The frames in which PRL is valid are precisely the reverse well-founded frames.

Proof: Actually, the unravelling of the meaning of "$\alpha \Vdash \Box(\Box p \to p) \to \Box p$" already proves the Theorem. Nonetheless, I reproduce a more pedestrian version of the proof here.

First, we must show PRL to be valid in reverse well-founded frames. To this end, let (K,R,α_0) be such a frame. Since every model on a reverse well-founded frame is a Kripke model, we have by results of section 1 the validity of all theorems of BML as well as the closure under modus ponens *and* the necessitation rule *R2*. (Reminder: We are dealing with validity and not merely truth.) Thus, it suffices to verify the validity of the schema *A4*,

$$\Box(\Box A \to A) \to \Box A;$$

the validity of the extra schema,

$$\big(\Box(\Box A \to A) \to \Box A\big)$$

(cf. section 1.4) follows by the preservation of validity by *R2*.

Suppose, by way of contradiction,

$$\alpha_1 \nVdash \Box(\Box A \to A) \to \Box A,$$

for some $\alpha_1 \in K$ and some forcing relation $\Vdash$ on (K,R,α_0). This means

$$\alpha_1 \Vdash \Box(\Box A \to A) \qquad (*)$$

$$\alpha_1 \nVdash \Box A. \qquad (**)$$

By (*) there is an $\alpha_2 \in K$ such that $\alpha_1 R \alpha_2$ and $\alpha_2 \nVdash A$. By (*), $\alpha_2 \Vdash \Box A \to A$, whence $\alpha_2 \nVdash \Box A$. This means there is an $\alpha_3 \in K$ such that $\alpha_2 R \alpha_3$ and $\alpha_3 \nVdash A$. By (*), ... We get an infinite sequence $\alpha_1 R \alpha_2 R \ldots$, contrary to the assumption of reverse well-foundedness. Hence, *A4* is valid in (K,R,α_0).

(Variation: If $\Box(\Box A \to A) \to \Box A$ is not valid in a model $(K,R,\alpha_0,\Vdash)$ on the given frame, there is an R-maximal node α with properties (*) and (**). As above, we get $\alpha R \beta$ with β also enjoying these properties, a contradiction to the R-maximality of α.)

The converse is also established contrapositively: Let $\alpha_0 R \alpha_1 R \ldots$. We construct a model $(K,R,\alpha_0,\Vdash)$ on the frame so that $\Box(\Box p \to p) \to \Box p$ is not valid.

This construction is quite simple: Define $\Vdash$ by,

$$\beta \Vdash p \quad \text{iff} \quad \beta \notin \{\alpha_0, \alpha_1, \ldots\}$$

for any $\beta \in K$ and any atom p. Observe:

(1) $\alpha_i \nVdash \Box p$ for any i, since $\alpha_i R \alpha_{i+1} \nVdash p$

(2) $\beta \Vdash \Box p \to p$ for all β since

$\beta \Vdash p$ for $\beta \notin \{\alpha_0, \alpha_1, \ldots\}$

$\beta \nVdash \Box p$ for $\beta \in \{\alpha_0, \alpha_1, \ldots\}$

(3) $\beta \Vdash \Box(\Box p \to p)$ for all β by (2)

(4) $\alpha_0 \Vdash \Box(\Box p \to p) \to \Box p$, by (1) and (3). QED

We have, as mentioned earlier, a strong completeness theorem for PRL with respect to *models* and we now also have a characterisation of those *frames* in which PRL is valid. These two results do not tautologically yield the desired conclusion. However, it is a weakly valid conclusion: Completeness, albeit not strong completeness, holds with respect to (finite) reverse well-founded frames:

<u>2.4. COMPLETENESS THEOREM</u>. For any modal sentence A, the following are equivalent:

i. $\mathsf{PRL} \vdash A$

ii. A is true in all models on (finite) reverse well-founded frames

iii. A is valid in all (finite) reverse well-founded frames.

We could prove this by mimicking the proof of Theorem 1.10 on the completeness of BML with respect to finite models. There are a few differences in the proofs, however, chiefly caused by the desire to make the finite frame irreflexive. (For, a finite transitive frame is reverse well-founded iff it is irreflexive.) The induced relation R_S must be defined slightly differently, whence the set K_S of nodes must be defined slightly differently. I leave this proof to the reader (cf. the Exercises) and present instead an alternate proof of a useful strengthened version of the Theorem.

First, a definition:

<u>2.5. DEFINITION</u>. By a *tree* is meant a frame $(K, <, \alpha_0)$ in which

i. $<$ is a strict partial ordering, i.e. $<$ is transitive and asymmetric

ii. the set of predecessors of any element is finite and linearly ordered by $<$, i.e. for $\alpha \in K$, $\{\beta \in K: \ \beta < \alpha\}$ consists of $\beta_1 < \beta_2 < \ldots < \beta_k$ for some k.

2.6. FINITE TREE THEOREM. Let A be a modal sentence. The following are equivalent:

i. $\mathsf{PRL} \vdash A$

ii. A is true in all models on finite trees

iii. A is valid in all models on finite trees.

Proof: Since finite trees are reverse well-founded frames, the implications i $\Rightarrow$ ii and i $\Rightarrow$ iii follow from the Characterisation Theorem. The equivalence ii $\Leftrightarrow$ iii follows as in the proof of Theorem 1.10. Thus, we need only show ii $\Rightarrow$ i. As usual, we do this contrapositively.

Let $\mathsf{PRL} \nvdash A$ and let $\underline{K} = (K, R, \alpha_0, \Vdash)$ be a countermodel to A, i.e. $\alpha_0 \nVdash A$. As in the proof of Theorem 1.10, let

$$S \ = \ \{B: \ B \text{ is a subformula of } A\}.$$

We construct a finite tree model $\underline{K}_T \ = \ (K_T, <_T, (\alpha_0), \Vdash_T)$ by stages. K_T will consist of carefully chosen finite R-increasing sequences in K.

Stage 0. Put the sequence (α_0) into K_T.

Stage n+1. For each sequence $(\alpha_0, \ldots, \alpha_n) \in K_T$, look at $\{\Box B \in S: \ \alpha_n \nVdash \Box B\}$. If this set is empty, do not extend $(\alpha_0, \ldots, \alpha_n)$. Otherwise, for each such $\Box B$, choose by *A4* a node $\beta \in K$ such that $\alpha_n R \beta$ and

$$\beta \Vdash \Box B, \quad \beta \nVdash B. \qquad (*)$$

Add $(\alpha_0, \ldots, \alpha_n, \beta)$ to K_T.

The rest of $\underline{K}_T$ is readily determined:

$<_T$ is the usual strict ordering by extension of finite sequences

(α_0) is as above

$\Vdash_T$: $(\alpha_0, \ldots, \alpha_n) \Vdash_T p$ iff $\alpha_n \Vdash p$.

Claim 1. $(K_T, <_T, (\alpha_0))$ is a finite tree with origin (α_0).

That this is a tree with origin (α_0) is obvious. Finiteness follows from König's Lemma: The tree is finitely branching because branches are correlated with elements of the finite set S and there are no infinite paths because the succession from $(\alpha_0, \ldots, \alpha_n)$ to $(\alpha_0, \ldots, \alpha_n, \alpha_{n+1})$ results in at least one additional sentence

$\Box B \in S$ being forced by α_{n+1}-- after one has gone through all such sentences, the process stops. (Actually, we can bound the cardinality of K_T explicitly: It is at most $(m+1)!$, where S has m elements of the form $\Box B$. (Exercise.))

Claim 2. For all $B \in S$ and all $(\alpha_0,\ldots,\alpha_n) \in K_T$,

$$(\alpha_0,\ldots,\alpha_n) \Vdash_T B \quad \text{iff} \quad \alpha_n \Vdash B.$$

The proof is again by induction on the complexity of B and, again, the only interesting case is that in which $B = \Box C$:

$$\begin{aligned} \alpha_n \Vdash \Box C \;&\Rightarrow\; \forall \beta(\alpha_n R \beta \;\Rightarrow\; \beta \Vdash C) \\ &\Rightarrow\; \forall \beta((\alpha_0,\ldots,\alpha_n,\beta) \in K_T \;\Rightarrow\; \beta \Vdash C) \\ &\Rightarrow\; \forall \beta((\alpha_0,\ldots,\alpha_n,\beta) \in K_T \;\Rightarrow\; (\alpha_0,\ldots,\alpha_n,\beta) \Vdash_T C), \end{aligned}$$

by induction hypothesis. But this last readily implies $(\alpha_0,\ldots,\alpha_n) \Vdash_T \Box C$. Inversely,

$$\begin{aligned} \alpha_n \nVdash \Box C \;&\Rightarrow\; \exists \beta(\alpha_n R \beta \;\&\; \beta \nVdash C) \\ &\Rightarrow\; \exists \beta((\alpha_0,\ldots,\alpha_n,\beta) \in K_T \;\&\; \beta \nVdash C) \\ &\Rightarrow\; \exists \beta((\alpha_0,\ldots,\alpha_n,\beta) \in K_T \;\&\; (\alpha_0,\ldots,\alpha_n,\beta) \nVdash_T C), \end{aligned}$$

by induction hypothesis. But, again, this last readily yields the desired conclusion: $(\alpha_0,\ldots,\alpha_n) \nVdash_T \Box C$.

The Theorem follows immediately:

$$\alpha_0 \nVdash A \;\Rightarrow\; (\alpha_0) \nVdash_T A.$$

QED

Before discussing any applications of Theorem 2.6 and its corollary, Theorem 2.4, let me make a few quick remarks.

First, there are remarks on the proof given:

2.7. REMARKS. i. The existence of a node β accessible to α_n and satisfying (*) is the key to the proof and it should not go unnoticed. It is precisely this application of *A4* that guarantees the finiteness of the resulting tree.

ii. As parenthetically noted in the proof, we get an explicit bound on the size of a tree countermodel $\underline{K}$ to A whenever PRL $\nvdash A$. This overall bound is, of course, relevant to a discussion of completeness, but another bound is also important-- namely, one on the *height* of the tree. If A has only m boxed subformulae, i.e. if S contains only m formulae of the form $\Box B$, then the height of the tree, defined to be the maximum length of a strictly increasing sequence in the tree, can be weakly bounded by $m + 1$.

iii. The proof can be modified to yield a related result: If $\underline{K}$ is a model on a reverse well-founded frame and $\underline{K}_T$ is the equivalent tree model constructed by taking all strictly R-increasing finite sequences in K, then $(K_T, <_T, (\alpha_0))$ is a reverse well-founded tree. (Cf. the Exercises.) Modulo this remark, we could reverse our order and deduce 2.6 as a corollary to 2.4.

Remark 2.7.ii will prove useful to us in the next section.

Next, a practical, notational remark:

2.8. NOTATIONAL CONVENTION. A reverse well-founded frame is, among other things, a strict partial ordering. Thus, we can use the usual "<" in place of "R" in denoting the accessibility relation of a frame for a model of PRL. We shall do this in the sequel.

Finally, it is worth mentioning the existence of alternate proofs of the Completeness Theorem (2.4). As we have already noted, the method of proof of Theorem 1.10 applies without much trouble. Another proof proceeds by mimicking the proof of the Strong Completeness Theorem for BML. These variations are explored in the Exercises.

As for applications, the most interesting ones will be given in the next section; a few will be given in the Exercises; and I cite a few immediate ones here.

2.9. COROLLARY. PRL is decidable.

2.10. COROLLARY. For all sentences A, B,

i. PRL $\vdash A$ iff PRL $\vdash \Box A$

ii. PRL $\vdash \Box A \vee \Box B$ iff PRL $\vdash \Box A$ or PRL $\vdash \Box B$

iii. PRL $\vdash \Box A \rightarrow \Box B$ iff PRL $\vdash \boxed{s} A \rightarrow B$.

2.11. COROLLARY. PRL does not prove any of the following:

i. $\Box(p \vee q) \rightarrow \Box p \vee \Box q$

ii. $(\Box p \rightarrow \Box q) \rightarrow \Box(p \rightarrow q)$

iii. $(\Box p \leftrightarrow \Box q) \rightarrow \Box(p \leftrightarrow q)$.

I leave the proofs of 2.9, 2.10, and 2.11 as exercises to the reader.

Speaking of exercises:

EXERCISES

1. Show that the Second Incompleteness Theorem, $\sim\Box f \rightarrow \sim\Box\sim\Box f$, does not axiomatise PRL over BML (with *R2*). (Hint: Consider the frame: .)

2. Show: For any A there is an n_0 such that, for all $n \geq n_0$,

   PRL $\vdash A$ iff PRL $+ \Box^n f \vdash A$.

3. Prove Theorem 2.4 by the method of proof of Theorem 1.10. (Hint: $S, S(\alpha)$ are defined as before, but R_S is defined by: $S(\alpha) R_S S(\beta)$ iff

   i. $\forall C(\Box C \in S(\alpha) \Rightarrow C, \Box C \in S(\beta))$, as in 1.10

   and ii. $\exists C(\Box C \in S(\beta)\ \&\ \Box C \notin S(\alpha))$.

   Then define K_S to be $S(\alpha_0)$ together with those $S(\alpha)$'s accessible to it.)

4. Let A be a modal sentence and S the set of its subformulae. A set $X \subseteq S \cup \{\sim B: B \in S\}$ is an *S-completion* iff i. PRL $+ X$ is consistent, and ii. for $B \in S$, one of B and $\sim B$ is in X. Using S-completions, prove Theorem 2.4 by mimicking the proof of the Strong Completeness Theorem (1.6) for BML.

(So many proofs of completeness! There are several such techniques in modal logic. Some are more readily applicable in certain circumstances than others and it does pay to be aware of the variants. The proof of Exercise 4, for example, generalises fairly directly to yield the Interpolation Lemma, as the reader will now see.)

5. (Interpolation Lemma for PRL). For each sentence A, let L_A be the language of A, i.e. the set of modal sentences all the atoms of which occur in A. Define $S(A)$ to be the set of subformulae of A and $S^+(A) = S(A) \cup \{\sim C: C \in S(A)\}$ to be the closure of $S(A)$ under negation. Now: Fix A, B. Define $X \subseteq L_A$, $Y \subseteq L_B$ to be *separable* if there is a $C \in L_A \cap L_B$ such that

   a. PRL $+ X \vdash C$

b. $\mathsf{PRL} + Y \vdash \sim C$;

X,Y are *inseparable* otherwise. Finally, define $X \subseteq S^+(A)$, $Y \subseteq S^+(B)$ to be *S-complete* iff

a. X,Y are inseparable

b. $\forall D \in S(A)(D \in X$ or $\sim D \in X)$

c. $\forall D \in S(B)(D \in Y$ or $\sim D \in Y)$

i. Show: If $X_0 \subseteq S(A)$, $Y_0 \subseteq S(B)$ are inseparable, there is an S-complete pair X,Y such that $X_0 \subseteq X$, $Y_0 \subseteq Y$.

ii. Show: If X,Y is S-complete and $\Box E \notin X \cup Y$, there is another S-complete pair X',Y' such that

a. $\forall D(\Box D \in X \cup Y \Rightarrow D, \Box D \in X' \cup Y')$

b. $\Box E, \sim E \in X' \cup Y'$.

iii. Show: If $\{A\},\{\sim B\}$ are inseparable, there is a finite reverse well-founded model $\underline{K} \models A \wedge \sim B$.

iv. Prove the Interpolation Lemma: If $\mathsf{PRL} \vdash A \rightarrow B$, there is a sentence $C \in L_A \cap L_B$ such that $\mathsf{PRL} \vdash A \rightarrow C$ and $\mathsf{PRL} \vdash C \rightarrow B$.

v. Prove Beth's Theorem: Let $A(r)$ contain neither the atoms p nor q. If $\mathsf{PRL} \vdash A(p) \wedge A(q) \rightarrow .p \leftrightarrow q$, then, for some C containing only atoms of $A(r)$ other than r, $\mathsf{PRL} \vdash A(p) \rightarrow .p \leftrightarrow C$.

6. (Finite Linear Models). The theory $\mathsf{PRL} + Lin$ is axiomatised by adding to PRL the schema,

Lin: $\Box(\Box A \rightarrow B) \vee \Box(\boxed{s} B \rightarrow A)$.

i. Show: $\mathsf{PRL} + Lin \vdash \Box(\Box(\Box A \rightarrow B) \vee \Box(\boxed{s} B \rightarrow A))$. Conclude that $\mathsf{PRL} + Lin$ is closed under $R2$.

ii. Show: $\mathsf{PRL} + Lin$ is valid in precisely the reverse well-founded linear frames, i.e. those linear orderings whose converse relations are well-orderings.

iii. Prove the Lemma: Let $\underline{K} = (K, R, \alpha_0, \Vdash)$ be a model of $\mathsf{PRL} + Lin$. Let S be a finite set of modal sentences and define, for $\alpha \in K$,

$S(\alpha) = \{B \in S: \alpha \Vdash B\}$

$S(\alpha)_\Box = \{B: \Box B \in S(\alpha)\}$.

Then: For all $\alpha \in K$ and all $\beta,\gamma \in K$ such that $\alpha R \beta$ and $\alpha R \gamma$, $S(\beta)_{\Box} \subseteq S(\gamma)$ or $S(\gamma)_{\Box} \subseteq S(\beta)$

iv. Prove the Completeness Theorem: For all modal sentences A, PRL + $Lin \vdash A$ iff A is valid in all finite strict linear orderings. (Hint: If PRL + $Lin \nvdash A$, construct, via iii, a linear subtree of (K,R,α_0) and an induced forcing relation.)

(PRL + Lin serves several purposes: It has a minor rôle to play in provability logic, as we will see in the next section; it serves here as an additional exercise in modal technique; and it serves next as an easy counterexample. With respect to this second purpose, the extension of Exercise 1.7 characterising S4 as the logic of the strong box in BML to a characterisation of a logic named S4GRZ as the logic thereof in PRL is a rather more popular example. Cf. Boolos' book for details.)

7. (Characterisation, Completeness, and Strong Completeness). i. Show: Let T be a modal theory complete with respect to its characteristic frames, i.e. those in which T is valid. If this set of frames is definable in first-order logic, then T is strongly complete with respect to models on these frames.

   ii. Show: PRL + Lin is *not* strongly complete with respect to linear reverse well-founded frames. (Hint: Let $\Gamma = \{\Box^n p_n \wedge \sim\Box^n p_{n+1}: n \in \omega\}$, where $p_0, p_1, \ldots$ is an enumeration of atoms and $\Box^n$ indicates a string of n $\Box$'s.)

8. (Trees). Let $\underline{K} = (K,R,\alpha_0,\Vdash)$ be a Kripke model. Define $\underline{K}_T$ by

   K_T consists of all non-empty R-increasing finite sequences starting at α_0, i.e. $K_T = \{(\alpha_0,\ldots,\alpha_n): \alpha_0 R \ldots R \alpha_n\}$

   $<_T$ is the usual ordering-by-extension of finite sequences

   (α_0) is the one-element sequence

   $\Vdash_T$: $(\alpha_0,\ldots,\alpha_n) \Vdash_T p$ iff $\alpha_n \Vdash p$, for any atom p.

   i. Show: For all sentences A,

   $$(\alpha_0,\ldots,\alpha_n) \Vdash_T A \quad \text{iff} \quad \alpha_n \Vdash A.$$

   ii. Show: If (K,R,α_0) is reverse well-founded, then $(K_T,<_T,(\alpha_0))$ is a reverse well-founded tree (usually called a "well-founded" tree).

(Our final exercise is not so interesting in itself. It proves PRL is not strongly complete with respect to its characteristic frames. In view of the fairly easy example afforded by 7.ii, this exercise is hardly of raw pedagogic interest. However, it does illustrate the rôle of well-foundedness in reverse well-founded frames.)

9. The theory PRL + *Lin* + Γ of the hint of Exercise 7.ii is obviously consistent, whence it has a model $\underline{K} = (K, R, \alpha_0, \Vdash)$.

   i. Considering the instance

$$\Box(\Box^{k+1} f \to p_n) \vee \Box(\boxdot p_n \to \Box^k f),$$

   of *Lin*, show

$$\alpha_0 \Vdash \Box(\Box^{k+1} f \to p_n)$$

   for all n, k.

   ii. Considering the instance,

$$\Box(\Box p_{n+1} \to p_n) \vee \Box(\boxdot p_n \to p_{n+1}),$$

   of *Lin*, show

$$\alpha_0 \Vdash \Box(\Box p_{n+1} \to p_n),$$

   for all n.

   iii. Show: (K, R, α_0) cannot be reverse well-founded. (Hint: Choose α maximal such that, for some n, $\alpha \nVdash p_n$.)

3. MODELS AND SELF-REFERENCE

A completeness theorem only weakly establishes the equivalence of syntax and semantics. Though essentially the same, the two remain intensionally distinct and offer different approaches to logic and, of particular interest here, self-reference. One difference in approach is heuristic: Whereas syntactic proofs are often rather involved, model theoretic ones tend to have some motivation underlying them. With respect to PRL, the relevant models of which are reverse well-founded, this motivation is transfinite induction. A nice illustration of this is given by the following result of Dick de Jongh.

3.1. IMPLICIT DEFINABILITY OF FIXED POINTS. Let $\underline{K} = (K, <, \alpha_0, \Vdash)$ be a reverse well-founded model. Suppose p obeys the Diagonalisation Restriction with respect to $A(p)$, i.e. p is boxed in $A(p)$. Then:

   i. for all sentences D, E,

$$\alpha_0 \Vdash \boxed{s}(D \leftrightarrow A(D)) \wedge \boxed{s}(E \leftrightarrow A(E)) \rightarrow .D \leftrightarrow E$$

ii. if p_0 is an atom new to the language of PRL, there is an extension, say $\Vdash_0$, of $\Vdash$ to include p_0 and such that

$$\alpha_0 \Vdash \boxed{s}(p_0 \leftrightarrow A(p_0)).$$

Before proving this, which we do without appeal to the syntactic results of section 1, let us note a couple of immediate consequences. First, there is the closure of PRL under the Diagonalisation Rule *DiR* of 1.2.4:

3.2. COROLLARY. PRL is closed under the rule

$$DiR: \quad \boxed{s}(p \leftrightarrow A(p)) \rightarrow B \;/\; B,$$

where (i) p is boxed in $A(p)$ and (ii) p has no occurrence in B.

Proof: By contraposition. Let $\underline{K} = (K, <, \alpha_0, \Vdash)$ be a reverse well-founded countermodel to B, i.e. $\alpha_0 \nVdash B$. Alter the interpretation of $\Vdash$ on p to $\Vdash'$ so that

$$\alpha_0 \Vdash \boxed{s}(p \leftrightarrow A(p)).$$

Since p does not occur in B, the forcing relation thereon is not affected, i.e. $\alpha_0 \nVdash' B$. (Exercise.) Thus, we have a countermodel to $\boxed{s}(p \leftrightarrow A(p)) \rightarrow B$. QED

This is de Jongh's original proof of the closure of PRL under *DiR*. Modulo 3.1, by which the modification invoked in the proof is possible, this is certainly simpler and quicker than the proof by appeal to the *explicit* definability of fixed points. It happens that this result follows from the implicit definability result on general logical grounds:

3.3. COROLLARY. (Explicit Definability of Fixed Points). Let p be boxed in $A(p)$. There is a sentence D containing only those variables of $A(p)$ other than p and such that

i. $\mathrm{PRL} \vdash \boxed{s}(p \leftrightarrow A(p)) \rightarrow .p \leftrightarrow D$

ii. $\mathrm{PRL} \vdash D \leftrightarrow A(D)$.

Proof: i follows directly from the Implicit Definability of Fixed Points by Beth's Theorem (Exercise 2.5, above); ii follows from i by the First Substitution Lemma and Corollary 3.2. QED

Despite the ease with which the Explicit Definability of Fixed Points follows from their Implicit Definability, this is not the original proof of 3.3: At the time, no-one knew Beth's Theorem to hold for PRL. Besides, once 3.1 was known, the ability to *find* the prototypical fixed points (of Chapter 0, section 2) suggested looking for a method of finding the fixed points and the issue of Beth's Theorem was forgotten for a few years. One method is based on the proof of Theorem 3.1, to which proof we now turn.

Proof of Theorem 3.1: Let $(K,<,\alpha_0,\Vdash)$ be a given model on a reverse well-founded frame. Both parts of the Theorem are proven by induction on the well-founded converse to the frame ordering. To see how this is done, let $A(p) = B(\Box C_1(p),\ldots,\Box C_n(p))$, where p has no occurrence in $B(q_1,\ldots,q_n)$. The truth value of $A(p)$ at a node α (i.e. the decision whether or not to force $A(p)$), insofar as it depends on that of p, depends only on the truth values of p at nodes $\beta > \alpha$.

To simplify the details, assume each $\Box C_i(p)$ to be maximal so that $B(q_1,\ldots,q_n)$ is propositional in $q_1,\ldots,q_n$.

i. Suppose

$$\alpha_0 \Vdash \boxed{s}(D \leftrightarrow A(D)),\ \boxed{s}(E \leftrightarrow A(E)).$$

We show $\alpha \Vdash D \leftrightarrow E$ for all $\alpha \in K$ (whence $\alpha_0 \Vdash \boxed{s}(D \leftrightarrow E)$) by reverse induction. Note:

$$\begin{aligned} \forall\beta > \alpha(\beta \Vdash D \leftrightarrow E) &\Rightarrow \forall\beta > \alpha(\beta \Vdash \boxed{s}(D \leftrightarrow E)) \\ &\Rightarrow \forall\beta > \alpha(\beta \Vdash C_i(D) \leftrightarrow C_i(E)), \quad \text{by } FSL \\ &\Rightarrow \alpha \Vdash \Box C_i(D) \leftrightarrow \Box C_i(E) \qquad (*) \\ &\Rightarrow \alpha \Vdash A(D) \leftrightarrow A(E) \qquad (**) \\ &\Rightarrow \alpha \Vdash D \leftrightarrow E, \end{aligned}$$

where (**) follows from (*) by the propositional assumption on B.

(Remark: To make this proof totally semantic, the purist should give a semantic proof of the *FSL*.)

ii. A similar induction allows us to define $\Vdash_0$. Since $\Vdash_0$ is to extend $\Vdash$, we first define

$$\alpha \Vdash_0 q \quad \text{iff} \quad \alpha \Vdash q$$

for all old atoms q. A quick induction shows

$$\alpha \Vdash_0 C \quad \text{iff} \quad \alpha \Vdash C$$

for all sentences C of the original language. The crucial thing is the inductive determination of the truth value of p_0.

Suppose $\Vdash_0$ has been defined for all $\beta > \alpha$ and that

$$\forall \beta > \alpha \big(\beta \Vdash_0 p_0 \leftrightarrow A(p_0)\big).$$

Because $\Vdash_0$ is defined for $\beta > \alpha$, the statements

$$\alpha \Vdash_0 \Box C_i(p_0) \qquad \big(\forall \beta > \alpha \ \beta \Vdash_0 C_i(p_0)\big)$$

are already decided; hence so is their propositional combination,

$$\alpha \Vdash_0 A(p_0).$$

Thus, *define*

$$\alpha \Vdash_0 p_0 \quad \text{iff} \quad \alpha \Vdash_0 A(p_0).$$

By transfinite induction, this last equivalence holds for all $\alpha \in K$, whence

$$\alpha_0 \Vdash_0 \boxed{s}\,\big(p_0 \leftrightarrow A(p_0)\big).$$

QED

A marvellous thing about the proof of Theorem 3.1 is that it lends itself well to illustration: Let $\underline{K}$ be the model pictured and let

$$A(p_0) \;=\; \Box p_0 \to q:$$

(I.e. $\gamma \Vdash q$ and no other node does.)

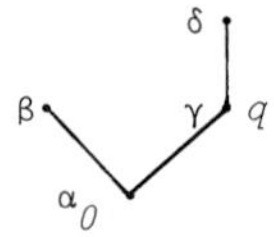

How do we define $\Vdash_0$? Observe $\beta, \delta \Vdash_0 \Box p_0$ vacuously, but do not force q. Thus, $\beta, \delta \nVdash_0 \Box p_0 \to q$ and we must decide $\beta, \delta \nVdash_0 p_0$:

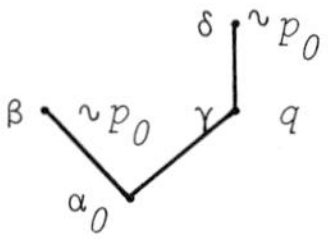

Now look at γ. $\gamma \nVdash_0 \Box p_0$ (since $\delta \nVdash_0 p_0$) and $\gamma \Vdash_0 q$. So $\gamma \Vdash_0 \Box p_0 \to q$, whence $\gamma \Vdash_0 p_0$:

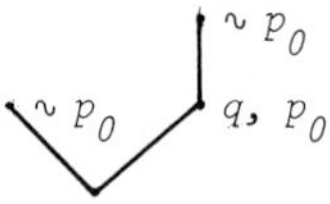

Finally, consider α_0. $\alpha_0 \nVdash_0 \Box p_0, q$, so $\alpha_0 \Vdash_0 \Box p_0 \to q$ and we have $\alpha_0 \Vdash_0 p_0$:

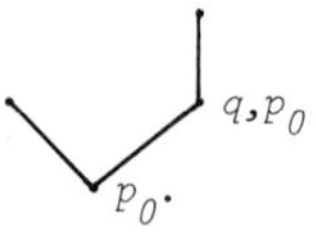

To see how the construction of p_0 depends on q, let δ, instead of γ, force q and see what happens:

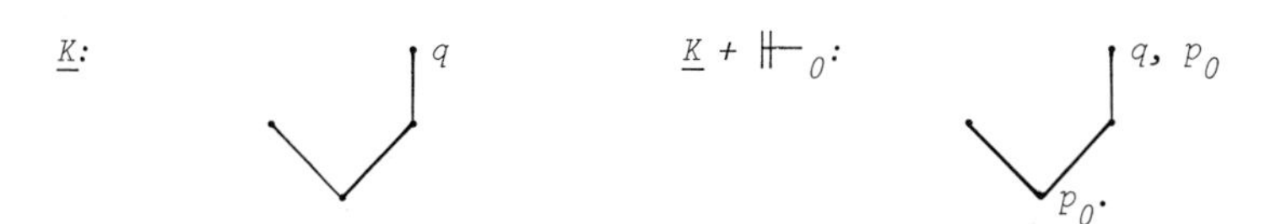

Observe that the endpoints β,δ, which originally behaved alike, now no longer do. This cannot happen if there are no side variables q. For, in such a case the endpoints will have the same information on which to base their respective decisions on the truth value of p_0 and they will, thus, come to the same decision. Those nodes having only terminal nodes as successors will thus base their respective decisions on common information and will, thus, make a common decision... In short, if there are no side variables, the truth value of p_0 at α is determined solely by the depth of α in the frame and we can determine this truth value by the construction of the proof of Theorem 3.1.

3.4. EXAMPLE. Let $A(p) = \Box(p \to \Box f) \to \Box\sim p$. ($A(p)$ appeared in Exercise 1.3.3 and will appear again in the Exercises at the end of the present section.) If, as I've very briefly suggested, the truth values of p depend only on the depths of nodes in models, we need only consider linear models. Let $(K,<,\alpha_0)$ be given by $K = \{\alpha_0\} \cup \{\gamma_0 > \gamma_1 > \gamma_2 > \ldots\}$. We determine

$$\gamma_n \Vdash p \quad \text{or} \quad \gamma_n \nVdash p$$

by induction on n:

$n = 0$. $\gamma_0 \Vdash \Box B$ for all B; in particular, $\gamma_0 \Vdash \Box\sim p$, whence $\gamma_0 \Vdash A(p)$ and we must define $\gamma_0 \Vdash p$.

$n = 1$. Since $\gamma_0 \Vdash p$, $\gamma_1 \nVdash \Box\sim p$. But $\gamma_1 \Vdash \Box(p \to \Box f)$ since, for any γ,

$$\begin{aligned} \gamma_1 < \gamma \quad &\Rightarrow \quad \gamma = \gamma_0 \\ &\Rightarrow \quad \gamma \Vdash \Box f \\ &\Rightarrow \quad \gamma \Vdash (p \to \Box f). \end{aligned}$$

Thus $\gamma_1 \nVdash A(p)$ and we define $\gamma_1 \nVdash p$.

$n = 2$. Again, $\gamma_2 \nVdash \Box\sim p$ and $\gamma_2 \Vdash \Box(p \to \Box f)$ since the only extension of γ_2 which forces p is γ_0 which also forces $\Box f$. Thus $\gamma_2 \nVdash A(p)$ and we define $\gamma_2 \nVdash p$.

$n = 3, 4, \ldots$. $\gamma_n \nVdash p$ by similar reasoning (or an induction-- exercise).

Thus,

$$\gamma_n \Vdash p \quad \text{iff} \quad n = 0;$$

but we can also easily see that

$$\gamma_n \Vdash \Box f \quad \text{iff} \quad n = 0.$$

Thus, $\alpha_0 \Vdash \boxed{s}\,(p \leftrightarrow A(p)) \rightarrow .p \leftrightarrow \Box f,$

(given the proper decision on $\alpha_0 \Vdash p$ or $\alpha_0 \nVdash p$ --but let's ignore this for now). Since linear models are supposed to be typical in this case and since we have the finite model property, I claim it will follow that

$$\mathrm{PRL} \vdash \boxed{s}\,(p \leftrightarrow A(p)) \rightarrow .p \leftrightarrow \Box f,$$

whence

$$\mathrm{PRL} \vdash \Box f \leftrightarrow A(\Box f),$$

i.e. $\mathrm{PRL} \vdash \Box f \leftrightarrow .\Box(\Box f \rightarrow \Box f) \rightarrow \Box \sim \Box f.$

(Exercise: Prove this directly.)

Our next goal is to clarify and justify all that has just been done. That is, we will show generally how to find the fixed point $p \leftrightarrow A(p)$, when $A(p)$ has no parameters, by the procedure just illustrated. To prove that this procedure works and is workable we must show:

i. we can restrict ourselves to finite linear models

ii. we can effectively tell when to stop

and iii. we can always read the fixed point off our model.

In performing these three tasks, we will make use of the following

3.5. TEMPORARY NOTATIONAL CONVENTIONS. i. $A(p)$ is a fixed sentence containing only the variable p and p is boxed in $A(p)$.

ii. We relax the requirement that a frame have a minimum node. By Remark 1.3.ii this has no great effect on our work. In any event, the finite models we end with will have minimum nodes.

iii. All frames $(K,<)$ are assumed reverse well-founded.

3.6. DEFINITIONS. Let $(K,<)$ be given. The *ordinal*, $o(\alpha)$, of a node $\alpha \in K$ is inductively defined by

$$o(\alpha) = sup\{\, o(\beta) + 1 : \ \alpha < \beta \,\}.$$

The *ordinal* of the frame $(K,<)$ is

$$o(K,<) = sup\{\, o(\alpha) : \ \alpha \in K \,\}.$$

(If $(K,<)$ has a minimum element α_0, then $o(K,<) = o(\alpha_0)$.) We also call this the *ordinal* of a model $\underline{K} = (K,<,\Vdash)$ and write $o(\underline{K})$.

In a finite frame $(K,<)$, we can define $o(\alpha)$ as the largest n for which there is a path

$$\alpha < \alpha_1 < \ldots < \alpha_n$$

in the frame. Thus, e.g., $o(\alpha) = 0$ iff α is terminal, i.e. there is no $\beta > \alpha$.

Preceding Example 3.4 I declared the behaviour of a node relative to formulae mentioning only p (assumed equivalent to $A(p)$) to be determined entirely by the depth of the node in the frame of a given model. We now know exactly what "depth" means-- it is the ordinal of a node. Our next definition isolates the appropriate meaning of "behaviour".

3.7. DEFINITIONS. Given p, $\underline{K}$, and $\alpha \in K$, define

$$S_p = \{B\colon\ B \text{ contains no atoms other than } p\}$$
$$S_p(\alpha) = \{B \in S_p\colon\ \alpha \Vdash B\}.$$

(It is *not* assumed in the definition of S_p that p *does* occur in B-- merely that no other variable does.)

We are now in position to prove our claim:

3.8. LEMMA. Let $p \leftrightarrow A(p)$ be valid in $\underline{K}$. Then: For all $\alpha,\beta \in K$,

i. $o(\alpha) = o(\beta) \Rightarrow S_p(\alpha) = S_p(\beta)$

ii. if $o(\alpha), o(\beta)$ are finite,

$$S_p(\alpha) = S_p(\beta) \Rightarrow o(\alpha) = o(\beta).$$

Proof: There is really no harm if the reader wishes to consider only the case in which all nodes have finite ordinal. This is really what we will need to know; by allowing transfinite ordinals, we can conclude all nodes of infinite depth behave alike, but in our applications we will not have such nodes.

i. We prove i by induction on $o(\alpha) = o(\beta)$. Assume as induction hypothesis: For all $B \in S_p$,

$$\forall\alpha'\beta'\big(o(\alpha') = o(\beta') < o(\alpha) \Rightarrow \alpha' \Vdash B \text{ iff } \beta' \Vdash B\big). \qquad (*)$$

We wish to conclude: For all $B \in S_p$,

$$\alpha \Vdash B \text{ iff } \beta \Vdash B. \qquad (**)$$

We shall require a tiny sublemma:

3.9. SUBLEMMA. Let $(K,<)$, $\alpha \in K$, and an ordinal $\kappa < o(\alpha)$ be given. There is a node $\beta > \alpha$ with $o(\beta) = \kappa$.

Proof: *Finite case*. Choose a chain $\alpha < \alpha_1 < \ldots < \alpha_n$ where $n = o(\alpha)$ and let $\beta = \alpha_{n-\kappa}$. That $o(\beta) = \kappa$ is not hard to see.

General case. By the reverse well-foundedness of $<$, we can assume α maximal so that $o(\alpha) > \kappa$ and $\forall \beta > \alpha\ (o(\beta) \neq \kappa)$. It follows that, for any $\beta > \alpha$, $o(\beta) < \kappa$. (Why?) But then $o(\beta) + 1 \leq \kappa$ and

$$o(\alpha) = sup\{o(\beta) + 1: \beta > \alpha\} \leq \kappa,$$

contrary to assumption. QED

Proof of 3.8 continued: Every $B \in S_p$ is a propositional combination of sentences of the form $\Box C$ and the atom p. We establish (**) first for $B = \Box C$, second for $B = p$, and finally for arbitrary B. (Note the similarity with the proof of Theorem 3.1. We are doing the same thing-- determining the value of p by induction from the top down.)

Step 1. $B = \Box C$. We have

$$\begin{aligned} \alpha \Vdash \Box C \quad &\text{iff} \quad \forall \alpha'(\alpha < \alpha' \Rightarrow \alpha' \Vdash C) && (1)\\ &\text{iff} \quad \forall \alpha'(o(\alpha) > o(\alpha') \Rightarrow \alpha' \Vdash C) && (2)\\ &\text{iff} \quad \forall \beta'(o(\beta) > o(\beta') \Rightarrow \beta' \Vdash C) && (3)\\ &\text{iff} \quad \forall \beta'(\beta < \beta' \Rightarrow \beta' \Vdash C) && (4)\\ &\text{iff} \quad \beta \Vdash \Box C, \end{aligned}$$

where the equivalences between (1) and (2) and between (3) and (4) follow from the induction hypothesis and Lemma 3.9. (Exercise: Explain this.)

Step 2. Write $A(p) = P(\Box C_1,\ldots,\Box C_n)$, where $P(q_1,\ldots,q_n)$ is purely propositional and has no occurrence of p. (Thus, as before, all occurrences of p in $A(p)$ are in subformulae $\Box C_i$; unlike our earlier decompositions, however, we do not assume p occurs in each-- or any-- of the $\Box C_i$'s.) By ordinary propositional logic, we have

$$\alpha \Vdash P(\Box C_1,\ldots,\Box C_n) \quad \text{iff} \quad \beta \Vdash P(\Box C_1,\ldots,\Box C_n).$$

In other words,

$$\alpha \Vdash A(p) \text{ iff } \beta \Vdash A(p),$$

i.e. $\alpha \Vdash p$ iff $\beta \Vdash p$.

Step 3. B is arbitrary. Write $B = P(p, \Box C_1, \ldots, \Box C_n)$, where $P(p, q_1, \ldots, q_n)$ is propositional and repeat the argument of Step 2.

This completes the proof of i.

ii. We wish to show that among nodes of finite depth there is no collapsing. To see this, define

$$\Box^0 f = f, \quad \Box^{n+1} f = \Box\Box^n f,$$

and notice

$$\alpha \Vdash \Box^{n+1} f \text{ iff } o(\alpha) \leq n.$$

QED

Recall the three tasks we have set ourselves to perform: We must show

i. we can restrict ourselves to finite linear models

ii. we can effectively tell when to stop

and iii. we can always read the fixed points from our model.

With Lemma 3.8, we can now quickly perform the major part of task i. First, a bit of notation:

3.10. DEFINITION. By $(L,<)$, we mean any infinite linearly ordered set of order type ω^*. For concreteness we let $L = \{\gamma_n : n \in \omega\}$ and define

$$\gamma_n < \gamma_m \text{ iff } m < n.$$

By $\underline{L}$ we mean *the* model on $(L,<)$ in which $p \leftrightarrow A(p)$ is valid (and all atoms other than p are ignored).

With this notation, we can state the conclusion of linearity:

3.11. COROLLARY. Let $\underline{K}$ be a finite model in which $p \leftrightarrow A(p)$ is valid. For any $\alpha \in K$, $B \in S_p$,

$$\alpha \Vdash B \text{ in } \underline{K} \text{ iff } \gamma_n \Vdash B \text{ in } \underline{L},$$

where $n = o(\alpha)$.

Proof: We could either note that the proof of Lemma 3.8 never required α, β to be in the same model, or we can combine $\underline{K}$ and $\underline{L}$ into one model, $\underline{K} + \underline{L}$, by taking their disjoint union:

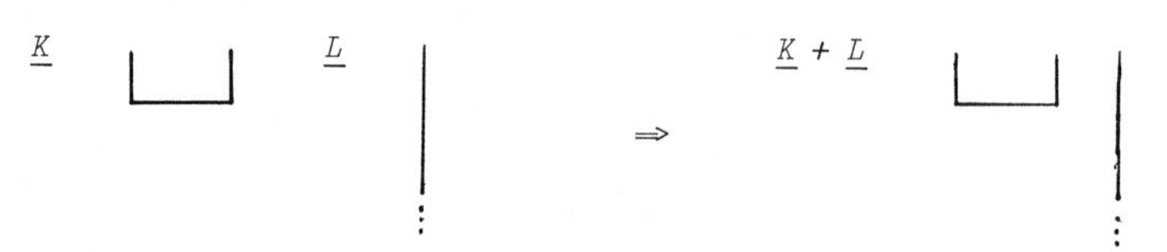

Because we dropped the requirement that models have minimum elements, $\underline{K} + \underline{L}$ is a model; had we not done so, we would form $(\underline{K} + \underline{L})'$ by tacking on a new minimum node α_0 and defining $\Vdash$ at α_0 in the unique way preserving the validity of $p \leftrightarrow A(p)$.

QED

This almost completes the first task. Our ultimate goal is to determine D such that $\mathsf{PRL} \vdash D \leftrightarrow A(D)$. If we find D such that $D \leftrightarrow A(D)$ is valid in $\underline{L}$, it will follow from the Corollary that $D \leftrightarrow A(D)$ is valid in all finite models, whence $\mathsf{PRL} \vdash D \leftrightarrow A(D)$. Since the Corollary also tells us that $D \leftrightarrow A(D)$ is valid in all finite linear orderings iff it is valid in $\underline{L}$, we have completed our first task.

Our second task is to determine which finite models we have to consider. This means we have to determine when the procedure "stops" as it did in Example 3.4, when it was clear $\gamma_n \not\Vdash p$ for all $n \geq 2$.

3.12. DEFINITION. Let $B \in S_p$. B is *eventually constant of order* n_0 iff one of the following holds in $\underline{L}$:

i. $\forall n \geq n_0 (\gamma_n \Vdash B)$

ii. $\forall n \geq n_0 (\gamma_n \Vdash \sim B)$.

3.13. LEMMA. Every $B \in S_p$ is eventually constant.

Proof: By replacing any unboxed occurrence of p by $A(p)$, we can assume without loss of generality that p is boxed in B and write $B = P(\Box C_1, \ldots, \Box C_k)$, where $P(q_1, \ldots, q_k)$ is purely propositional and contains no occurrence of p. If each $\Box C_i$ is eventually constant of order n_i, then B must be eventually constant of order $n = max\{n_1, \ldots, n_k\}$. Thus, it suffices to consider the case $B = \Box C$.

Case 1. $\mathsf{PRL} \vdash \boxed{s} (p \leftrightarrow A(p)) \to \Box C$. Then $\Box C$ is eventually constant of order 0.

Case 2. $\mathsf{PRL} \nvdash \boxed{s} (p \leftrightarrow A(p)) \to \Box C$. By the completeness theorem for PRL, there is a finite model

$$\underline{K} = (K, <, \alpha_0, \Vdash) \not\Vdash \boxed{s} (p \leftrightarrow A(p)) \to \Box C.$$

If $n_0 = o(\alpha_0)$, Corollary 3.11 yields $\quad \gamma_{n_0} \not\Vdash \Box C,$

whence, for all $n \geq n_0$, $\gamma_n \not\Vdash \Box C$. QED

In particular, p is eventually constant of some order. This order is readily determined.

3.14. REMARK. Let $h = card\{\Box C\colon \Box C$ is a subformula of $A\}$. Then p is eventually constant of order h.

Proof: Essentially, this is just a double repetition of the above proof and that of Remark 2.7.ii: As in the proof of 3.13, write $A(p)$ in the form $P(\Box C_1, \ldots, \Box C_k)$, where P is purely propositional and note that each sentence

$$\boxed{s}\,(p \leftrightarrow A(p)) \rightarrow \Box C_i \qquad (*)$$

has, aside from $\Box(p \leftrightarrow A(p))$, only the h boxed subformulae of $A(p)$ as boxed subformulae.

By Remark 2.7.ii, if (*) can be falsified, it can be so in a model of ordinal at most $h + 1$. In fact, we can reduce this by 1 since the estimate is based on the slowest possible succession of falsified boxed subformulae and $\Box(p \leftrightarrow A(p))$ is not falsified in a model falsifying (*). (Exercise: Verify this.)

By the proof of 3.13, this estimate shows $\Box C_i$ to be eventually constant of order at most h and $A(p)$ to be eventually constant of order at most the maximum of the orders of the individual $\Box C_i$'s, hence of order at most h. QED

With Remark 3.14, we have completed the second of our three tasks. The third is now immediate.

3.15. THEOREM. (Parameter-Free Fixed Point Calculation). The fixed point D of $A(p)$ is defined by

$$D = \begin{cases} \bigvee\{\Box^{k+1}f \wedge \sim\Box^k f\colon\ k \in Q\}, & n_0 \notin Q \\ \bigvee\{\Box^{k+1}f \wedge \sim\Box^k f\colon\ k \in Q\} \vee \sim\Box^{n_0}f, & n_0 \in Q, \end{cases}$$

where i. p is eventually constant of order n_0

ii. $Q = \{k \leq n_0\colon\ \gamma_k \Vdash p \text{ in } \underline{L}\}$

Proof: Note the obvious facts:

i. $\gamma_n \Vdash \Box^k f$ iff $n < k$

ii. $\gamma_n \Vdash \sim\Box^k f$ iff $k \leq n$

iii. $\gamma_n \Vdash \Box^{k+1}f \wedge \sim\Box^k f$ iff $n = k$.

Thus, for D as described,

$$\gamma_n \Vdash D \quad \text{iff} \quad \begin{cases} n \in Q, & n_0 \notin Q \\ n \in Q \text{ or } n \geq n_0, & n_0 \in Q \end{cases}$$

$$\text{iff} \quad \gamma_n \Vdash p.$$

Thus, $\quad \underline{L} \models_2 p \leftrightarrow D,$

whence $\quad \underline{L} \models_2 \boxed{s}\,(p \leftrightarrow A(p)) \rightarrow . p \leftrightarrow D \qquad (1)$

$\quad \underline{L} \models_2 D \leftrightarrow A(D). \qquad (2)$

But, by Corollary 3.11, it follows that (1) and (2) are valid in *all* finite models, whence $\quad \mathsf{PRL} \vdash \boxed{s}\,(p \leftrightarrow A(p)) \rightarrow . p \leftrightarrow D$

$\quad \mathsf{PRL} \vdash D \leftrightarrow A(D).$ QED

Theorem 3.15 and the algorithm behind it are due to the author, who marvels at their sheer beauty. A lovely algorithm like this demands application. I shall, nevertheless, not illustrate it here. Instead, I suggest the reader take a second look at Example 3.4 and I also direct him to the Exercises at the end of this section.

The key issue to discuss is the calculation of fixed points in the more general case. Before doing this, however, we digress to discuss an interesting corollary to the Parameter-Free Fixed Point Calculation, namely a Normal Form Theorem for modal sentences with no propositional variables. Such a sentence B is vacuously the fixed point D to $A(p) = B$. The calculation of 3.15 yields

$$\mathsf{PRL} \vdash B \leftrightarrow \bigvee\{\Box^{k+1} f \wedge \sim\Box^k f : \ k \in Q\}$$

or

$$\mathsf{PRL} \vdash B \leftrightarrow \bigvee\{\Box^{k+1} f \wedge \sim\Box^k f : \ k \in Q\} \vee \sim\Box^{n_0} f,$$

for some n_0, Q.

The "normal form" just cited is not quite a normal form in that it is not unique. The sentence

$$B: \quad \Box^2 f \wedge \sim\Box f \vee \sim\Box^3 f$$

has, for example, the additional normal forms

$$\Box^2 f \wedge \sim\Box f \vee \Box^4 f \wedge \sim\Box^3 f \vee \sim\Box^4 f$$

and

$$\Box^2 f \wedge \sim\Box f \vee \Box^4 f \wedge \sim\Box^3 f \vee \Box^5 f \wedge \sim\Box^4 f \vee \sim\Box^5 f,$$

etc. Unicity, as well as some simplification, can be imposed if we base the calculation of D on the *maximal intervals* rather than on the individual nodes at which p is forced.

3.16. NORMAL FORM THEOREM. Let B contain no atoms. Then B can be written in the form

$$\Box^{k_0}f \vee \bigvee_{i=1}^{n}(\Box^{k_i}f \wedge \sim\Box^{m_i}f) \vee \sim\Box^{m_{n+1}}f, \qquad (*)$$

where

$$0 \leq k_0 < m_1 < k_1 < \ldots < m_n < k_n < m_{n+1} \leq \omega$$

and $\Box^{\omega}f = t$. Moreover, except for the degenerate case in which B is provable, the representation (*) is unique.

The proof is immediate: Let $A(p) = B$ and calculate the fixed point using maximal intervals on which p is forced and observing that, for $0 \leq m \leq k \leq \omega$ and all finite n,

$$\gamma_n \Vdash \Box^{k+1}f \text{ iff } n \leq k$$

$$\gamma_n \Vdash \Box^{k+1}f \wedge \sim\Box^{m}f \text{ iff } m \leq n \leq k$$

$$\gamma_n \Vdash \sim\Box^{m}f \text{ iff } m \leq n.$$

I leave the details to the reader.

The Normal Form Theorem was multiply discovered; those discoverers I know of are (in alphabetical order): Johann van Bentham, George Boolos, and Roberto Magari. Magari was the first to publish it; Boolos reports he knew the result earliest, but delayed in publishing.

The multiplicity of a discovery can be interpreted as an indication of its importance. That the Normal Form Theorem is of some importance is borne out by its applications. I defer these to the Exercises and make only one brief remark here: As I noted above, Theorem 3.15 and the algorithm it provides for finding a fixed point D (in normal form) to $A(p)$ is due to the author. However, the *existence* of such fixed points had, unbeknownst to the author, already been proven by Claudio Bernardi, who used the Normal Form Theorem to show D was the limit, in a suitable sense, of $A^n(t)$ as $n \to \infty$. Unfortunately, as Giovanni Sambin showed, again by appeal to the Normal Form Theorem, this limit is not always attained in finitely many steps: The formula $A(p)$ of Exercise 1.3.3 and Example 3.4, above, has the property that $\mathrm{PRL} \vdash D \leftrightarrow A(B)$ iff $\mathrm{PRL} \vdash D \leftrightarrow B$. (Exercise.)

The history of the de Jongh-Sambin Theorem on the Explicit Definability of Fixed Points, although not too involved, is complex enough for me not to go into detail about it. Suffice it to say that, after the author had proven Theorem 3.15 during his stay in Amsterdam, Dick de Jongh proved the full result-- by a completely different method. It wasn't until a few years later that George Boolos showed that the proof in the parameter-free case can extend to the general one. To see this, one must make a shift in perspective. As presented, the algorithm for finding $D \leftrightarrow A(D)$ proceeds inductively down a linear model. At some point the procedure starts spewing out a constant value and we can stop; the proof of the Completeness Theorem tells us in advance when this will happen. With parameters, this might not happen: Even in $\underline{L}$, a continual change in the truth value of a parameter q could result in a continual change in the truth value of the fixed point variable p. Boolos' insight is that one is not working one's way down a single model $\underline{L}$, stopping when the Completeness Theorem tells us to, and then reading the fixed points off the nodes which force it in the one model, but rather one is searching through all models of a height determined by the proof of the Completeness Theorem and reading the fixed point off the minimum nodes thereof.

3.17. EXAMPLE. Let $A(p) = \Box p \rightarrow q$ be the Löb sentence. By the Completeness Theorem, we need only look at models of ordinal ≤ 1, since there is only one boxed formula in $A(p)$. Now, through reasoning analogous to that behind Lemma 3.8, we need only consider the following models:

$\underline{K}_1$ • q $\qquad$ $\underline{K}_2$ •

$\underline{K}_3$ q, q $\qquad$ $\underline{K}_4$ q $\qquad$ $\underline{K}_5$ q $\qquad$ $\underline{K}_6$

$\underline{K}_7$ q, q $\qquad$ $\underline{K}_8$ q

Here, the presence or absence of the letter "q" indicates whether or not the atom q is to be forced at the given node. With such information, from the top down, we determine those nodes at which p is forced:

$\underline{K}_1$ • q,p $\qquad$ $\underline{K}_2$ •

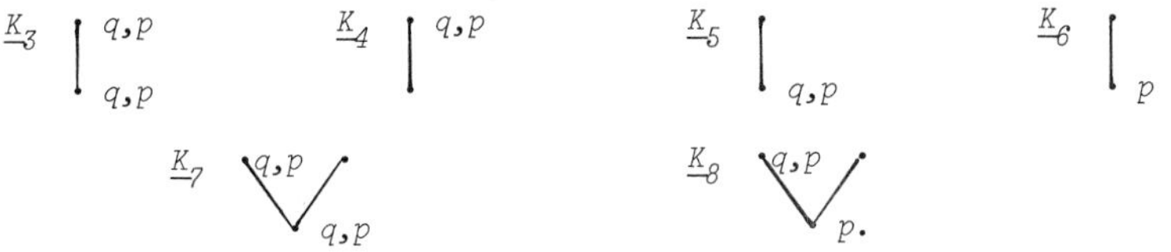

Now, the behaviour of the origins of these models is readily described:

$\underline{K}_1$: $D_1 = q \wedge \Box q \wedge \Box \sim q$ $\qquad$ $\underline{K}_2$: $D_2 = \sim q \wedge \Box q \wedge \Box \sim q$

$\underline{K}_3$: $D_3 = q \wedge \Box q \wedge \sim\Box\sim q$ $\qquad$ $\underline{K}_4$: $D_4 = \sim q \wedge \Box q \wedge \sim\Box\sim q$

$\underline{K}_5$: $D_5 = q \wedge \sim\Box q \wedge \Box\sim q$ $\qquad$ $\underline{K}_6$: $D_6 = \sim q \wedge \sim\Box q \wedge \Box\sim q$

$\underline{K}_7$: $D_7 = q \wedge \sim\Box q \wedge \sim\Box\sim q$ $\qquad$ $\underline{K}_8$: $D_8 = \sim q \wedge \sim\Box q \wedge \sim\Box\sim q.$

Since p is forced at the origins of $\underline{K}_1, \underline{K}_3, \underline{K}_5, \underline{K}_6, \underline{K}_7$, and $\underline{K}_8$, we have

$$p \leftrightarrow .D_1 \vee D_3 \vee D_5 \vee D_6 \vee D_7 \vee D_8,$$

i.e. $$p \leftrightarrow .(q \wedge \Box q \wedge \Box\sim q) \vee (q \wedge \Box q \wedge \sim\Box\sim q) \vee (q \wedge \sim\Box q \wedge \Box\sim q) \vee$$
$$\vee (\sim q \wedge \sim\Box q \wedge \Box\sim q) \vee (q \wedge \sim\Box q \wedge \sim\Box\sim q) \vee (\sim q \wedge \sim\Box q \wedge \sim\Box\sim q),$$

which simplifies to

$$p \leftrightarrow .(q \wedge \Box q) \vee (\sim q \wedge \Box\sim q) \vee (\sim\Box q \wedge \sim\Box\sim q),$$

which in turn simplifies to

$$p \leftrightarrow .q \wedge \Box q \vee \sim\Box q.$$

This can be rewritten as

$$p \leftrightarrow .\Box q \rightarrow q \wedge \Box q,$$

or, more familiarly, as

$$p \leftrightarrow .\Box q \rightarrow q.$$

Boolos' method is a nice generalisation of the algorithm of Theorem 3.15 and the more energetic reader might like to supply a complete formal description of the process, together with a proof that it works. However, I would not recommend relying on it for actually finding the fixed points to very complicated formulae $A(p)$: If m denotes the number of parameters and n the ordinal of a tree, then the function $F(n,m)$ telling how many distinct trees of height n one must look at in the case in which $A(p)$ has at least n boxed subformulae satisfies the recursion,

$$F(0,m) = 2^m$$
$$F(n+1,m) = F(0,m)\cdot(2^{F(n,m)} - 1).$$

The first few values of F are tabulated below:

$n \backslash m$	1	2	3	4
0	2	4	8	16
1	6	60	2040	1,048,560
2	126	$4\cdot 2^{60} - 4$	$8\cdot 2^{2040} - 8$	+
3	$2\cdot 2^{126} - 2$	+	+	+

From the Table we see that the procedure rapidly becomes impractical: With two variables and only one box we have 64 trees to look at in all; with one variable and two boxes, there are 134 such trees. One can use a computer for three variables and one box, and one can forget anything else.

The working of the recursion, not really well illustrated by Example 3.16, is this: Trees of ordinal $n + 1$ are obtained by choosing a tree of ordinal 0 as origin (possible in $F(0,m)$ ways) and a nonempty set of trees of ordinal n (possible in $2^{F(n,m)} - 1$ distinct ways) to place above the origin. Beginning with ordinal height 2, the formulae associated with the nodes become rather ugly: Let $\underline{K}$ result from $\underline{K}_1, \underline{K}_4$, and $\underline{K}_7$:

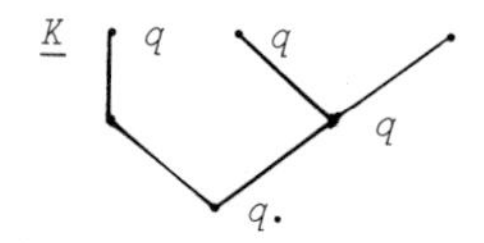

The formula E describing $\underline{K}$ is obtained from $D_1,\ldots,D_8$ describing $\underline{K}_1,\ldots,\underline{K}_8$ as follows: One has a huge conjunction of

q, since $\underline{K} \models q$

$\square \bigvee_{i \in X} D_i$, if $X \supseteq \{1,2,4,7\}$,

$\sim\square \bigvee_{i \in X} D_i$, if $X \not\supseteq \{1,2,4,7\}$ is nonempty.

Since $\square$ does not distribute over disjunction, these formulae cannot be much simplified. One can put $\square D_i$ in conjunctive normal form and pull the conjunctions outside the $\square$'s; but one is still left with boxed disjunctions. This ugliness is unavoidable; as one might guess from the extremely rapid growth of $F(n,m)$, there cannot be much of a simplification: There is, demonstrably, no nice Normal Form Theorem for formulae in $m > 0$ variables (cf. the Exercises).

What does all this tell us about self-reference? Well, for logically complex $A(p)$ (i.e. for $A(p)$ with several parameters or several boxes), it tells us we must go back to the syntactic determination, given in Chapter 1, of the fixed point. And here model theory can, at least, offer a tiny twist on the argument. In place of a clever application of Löb's Theorem, an obvious appeal to maximality provides a slightly easier proof.

3.18. LEMMA. $\mathrm{PRL} \vdash \Box C(t) \leftrightarrow \Box C(\Box C(t))$.

Proof: $\Rightarrow$ The proof of the implication is quite similar to the syntactic one in Lemma 1.3.2:

$$\begin{aligned} \alpha_0 \Vdash \Box C(t) &\Rightarrow \alpha_0 \Vdash \boxed{s}(\Box C(t) \leftrightarrow t) \\ &\Rightarrow \alpha_0 \Vdash \Box C(\Box C(t)) \leftrightarrow \Box C(t) \\ &\Rightarrow \alpha_0 \Vdash \Box C(\Box C(t)). \end{aligned}$$

$\Leftarrow$. A similar argument fails to show

$$\alpha_0 \Vdash \Box C(\Box C(t)) \rightarrow \Box C(t)$$

Thus, assume

$$\alpha_0 \Vdash \Box C(\Box C(t)), \quad \alpha_0 \nVdash \Box C(t),$$

and choose $\alpha \geq \alpha_0$ maximal with this property:

$$\alpha \Vdash \Box C(\Box C(t)) \qquad \alpha \nVdash \Box C(t)$$

but $\quad \forall \beta > \alpha\ (\beta \Vdash \Box C(t))$.

From this last,

$$\alpha \Vdash \Box(\Box C(t) \leftrightarrow t)$$

and the Second Substitution Lemma yields

$$\alpha \Vdash \Box(C(\Box C(t)) \leftrightarrow C(t)),$$

whence $\quad \alpha \Vdash \Box C(\Box C(t)) \leftrightarrow \Box C(t)$,

contrary to assumptions on α. QED

From Lemma 3.18, the derivation of the full Explicit Definability Theorem is identical to the syntactic derivation: It is, after all, a mere matter of calculation at this point.

EXERCISES

1. Let $(K,<)$ be a reverse well-founded frame and let $\alpha, \beta \in K$. Show: For o, S_p

defined as in 3.6, 3.7,

$$o(\alpha), o(\beta) \geq \omega \quad \Rightarrow \quad S_p(\alpha) = S_p(\beta).$$

2. Use the algorithm of 3.15 to find the fixed points D to

   i. $A(p) = \Box p$

   ii. $A(p) = \sim\Box p$

   iii. $A(p) = \Box p \rightarrow \sim\Box^3 f$

   iv. $A(p) = \Box^k p \rightarrow \Box^n f$.

(Remark: The formula in iv shows the bound of Remark 3.14 to be best possible.)

3. (Simultaneous Diagonalisation). Let p_1, p_2 be boxed in $A_1(p_1,p_2), A_2(p_1,p_2)$ and suppose A_1, A_2 have no other variables.

   i. Show that the method of 3.15 extends to an algorithm for finding D_1, D_2 such that

$$\text{PRL} \vdash D_1 \leftrightarrow A_1(D_1,D_2)$$
$$\text{PRL} \vdash D_2 \leftrightarrow A_2(D_1,D_2).$$

What is the estimate on the orders of which p_1, p_2 are eventually constant?

   ii. Find D_1, D_2 for

$$A_1(p_1,p_2) = \Box^2 p_1 \rightarrow \Box p_2$$
$$A_2(p_1,p_2) = \Box p_2 \rightarrow \Box\sim p_1.$$

4. Apply Boolos' algorithm (3.17) to find the fixed point of Kreisel's sentence $A(p) = \Box(p \rightarrow q)$.

5. Let $A(p) = \Box(p \rightarrow \Box f) \rightarrow \Box\sim p$ be the formula of Example 3.4. Show: For D, B in normal form, PRL $\vdash D \leftrightarrow A(B)$ iff $D = B$. Conclude that $D \not\leftrightarrow A^n(t)$ for any finite n.

6. (Linear Models). Recall the schema

$$Lin\text{:} \quad \Box(\Box A \rightarrow B) \vee \Box(\boxed{s}\, B \rightarrow A)$$

of Exercise 2.6, above.

   i. Prove: For any sentence A with no propositional variables,

$$\text{PRL} \vdash A \quad \text{iff} \quad \text{PRL} + Lin \vdash A.$$

   ii. Prove: Let $A(q_1,\ldots,q_n)$ have only the variables shown. The following are equivalent:

a. For all variable-free $B_1,\ldots,B_n$

$\mathsf{PRL} \vdash A(B_1,\ldots,B_n)$

b. $\mathsf{PRL} + Lin \vdash A(q_1,\ldots,q_n)$.

(The next Exercise is Boolos' proof of Solovay's result on the non-extendability of the Normal Form Theorem. For convenience, it is expressed in terms of a dual modal operator, $\Diamond$, of *possibility*, defined by $\Diamond A = \sim\Box\sim A$.

Thus, $\alpha \Vdash \Diamond A$ iff $\exists \beta > \alpha(\beta \Vdash A)$.

As $\Box$ distributes over $\wedge$ but not $\vee$, $\Diamond$ distributes over $\vee$ but not $\wedge$.)

7. (Anti-Normal Form Theorem). Let the variable p be given and let S_p be the set of all sentences containing no atoms other than p. Define a hierarchy $\{H_n\}_{n\in\omega}$ on S_p by:

H_0 is the set of all $B \in S_p$ equivalent in PRL to one of $p, \sim p, t$, or f

H_{n+1} is the set of all $B \in S_p$ equivalent in PRL to a propositional combination of sentences $\Diamond^k C$ for $C \in H_n$ and $k \in \omega$.

Further, define the sequence of sentences

$A_1 = \Diamond p$

$A_{n+1} = \Diamond(p \wedge A_n)$.

Observe that, for $n > 0$, $A_n \in H_n$. Finally, define the model $\underline{K}$:

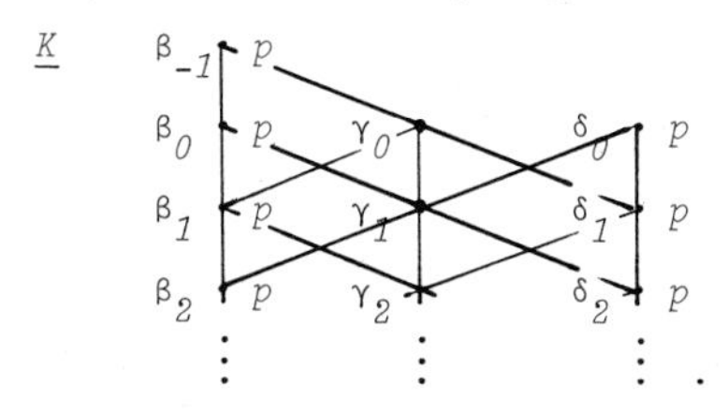

i. Show: For $B \in H_n$, $\beta_n \Vdash B$ iff $\delta_n \Vdash B$.

ii. Show: $\beta_n \Vdash A_{n+1}$, but $\delta_n \nVdash A_{n+1}$. Conclude $A_{n+1} \notin H_n$.

8. (Arithmetic Completeness; Elementary Case). The arithmetic interpretation B^* of a variable-free sentence B is defined inductively as follows:

t^*: $\overline{0} = \overline{0}$; f^*: $\overline{0} = \overline{1}$; $(\sim B)^*$: $\sim(B^*)$;

$(B \circ C)^*$: $B^* \circ C^*$, for $\circ \in \{\wedge, \vee, \rightarrow\}$; $(\Box B)^*$: $Pr(\ulcorner B^* \urcorner)$.

Show:

i. $\mathsf{PRA} \vdash B^*$ iff $\mathsf{PRL} \vdash B$

iff the normal form of B is t

ii. B^* is true iff $\mathsf{PRL} + \{\sim\Box^n f:\ n \in \omega\} \vdash B$

iff the normal form of B is t or $\sim\Box^n f$ for some *finite* n

iii. $\mathsf{PRL} \vdash B$ iff $\mathsf{PRL} + \{\sim\Box^n f:\ n \in \omega\} \vdash \Box B$

iv. i-iii with "PRL" replaced by "PRL + *Lin*".

(In i, use the fact that PRA proves no false Σ_1-sentences.)

4. ANOTHER PROVABILITY LOGIC

PRL is, as I announced in Chapters 0 and 1 and will show in Chapter 3, the Logic of Provability. That is, it axiomatises those schemata provable in PRA. What it does not axiomatise are the *true* schemata. These are yielded by the adjunction of the axiom schema of Reflexion:

$$Refl:\quad \Box A \to A.$$

That PRL + *Refl* axiomatises the true schemata is a result slightly less interesting, but more useful, than the axiomatisation of provable schemata by PRL and, like this latter, will be proven in the next chapter. For now, I wish merely to acquaint the reader with the elements of the model theory of PRL + *Refl*.

4.1. DEFINITION. PRL^{ω} is the extension of PRL (in its *R2*-free formulation) by the axiom schema of Reflexion,

$$Refl:\quad \Box A \to A.$$

The main theorem to prove is the following.

4.2. THEOREM. Let A be a modal sentence, $S_{\Box}$ the set of boxed subformulae of A, i.e. $S_{\Box} = \{\Box B:\ \Box B$ is a subformula of $A\}$. Then: The following are equivalent:

i. $\mathsf{PRL}^{\omega} \vdash A$

ii. $\mathsf{PRL} \vdash \bigwedge_{\Box B \in S_{\Box}} (\Box B \to B) \to A.$

Observe that this immediately reduces the decidability of PRL^{ω} to that of PRL. It also describes a sort of model theory for PRL^{ω}:

4.3. DEFINITION. A model $\underline{K} = (K, <, \alpha_0, \Vdash)$ is *A-sound* if, for every subformula $\Box B$ of A, i.e. every $\Box B \in S_{\Box}$, $\alpha_0 \Vdash \Box B \to B$.

Theorem 4.2 reduces quickly to the following:

4.4. THEOREM. Let A be a modal sentence.

i. $PRL^{\omega} \vdash A$ iff A is *true* in all A-sound models of PRL

ii. $PRL \vdash A$ iff A is *valid* in all A-sound models of PRL.

Proof that 4.4 ⇒ 4.2: Assume 4.4.i: $PRL^{\omega} \vdash A$ iff A is true in all A-sound models of PRL, i.e. all models of $PRL + \bigwedge_{\Box B \in S_{\Box}} (\Box B \to B)$. Thus, $PRL^{\omega} \vdash A$ iff $PRL + \bigwedge_{\Box B \in S_{\Box}} (\Box B \to B) \vdash A$. QED

From 4.4.ii, we derive another immediate corollary:

4.5. COROLLARY. Let A be a modal sentence.

$$PRL \vdash A \quad \text{iff} \quad PRL^{\omega} \vdash \Box A.$$

Proof: ⇒. This is easy:

$$PRL \vdash A \Rightarrow PRL \vdash \Box A, \text{ by closure under } R2$$
$$\Rightarrow PRL^{\omega} \vdash \Box A.$$

⇐. Observe,

$$PRL^{\omega} \vdash \Box A \Rightarrow \Box A \text{ is valid in all } \Box A\text{-sound models of PRL}$$
$$\Rightarrow PRL \vdash \Box A$$
$$\Rightarrow PRL \vdash A,$$

by the closure of PRL under the converse to $R2$ (Corollary 2.10). QED

Now, about the proof of Theorem 4.4: It is rather a simple matter and we could jump right into it. However, it relies on a rather useful construction which ought at some point to be singled out as such. We could have done this in sections 1 and 2 where we used it already in some minor matters, but the present place seems more suited to its display. Thus, before proving Theorem 4.4, we have the following definition.

4.6. DEFINITION. Let $\underline{K} = (K, <, \alpha_0, \Vdash)$ be a Kripke model. The *derived model*, $\underline{K}'$, is defined by

$K' = K \cup \{\alpha'\}$ (α' new)

$<'$: $\alpha <' \beta$ iff $\alpha < \beta$ for $\alpha, \beta \in K$; α' is the new minimum

$\Vdash'$: $\alpha \Vdash' p$ iff $\alpha \Vdash p$ for $\alpha \in K$, p any atom

$$\alpha' \Vdash' p \quad \text{iff} \quad \alpha_0 \Vdash p, \ p \text{ any atom.}$$

A sequence of successive derivations is denoted $\underline{K}^{(1)}, \underline{K}^{(2)}, \ldots$ with respective minima $\alpha_1, \alpha_2, \ldots$; i.e. define $\underline{K}^{(0)} = \underline{K}$, $\underline{K}^{(n+1)} = (\underline{K}^{(n)})'$ and let α_n denote the minimum node of $\underline{K}^{(n)}$.

Graphically, $\underline{K}^{(n)}$ is obtained from $\underline{K}$ by adding a short *tail* of n nodes $\alpha_1 > \alpha_2 > \ldots > \alpha_n$ below α_0:

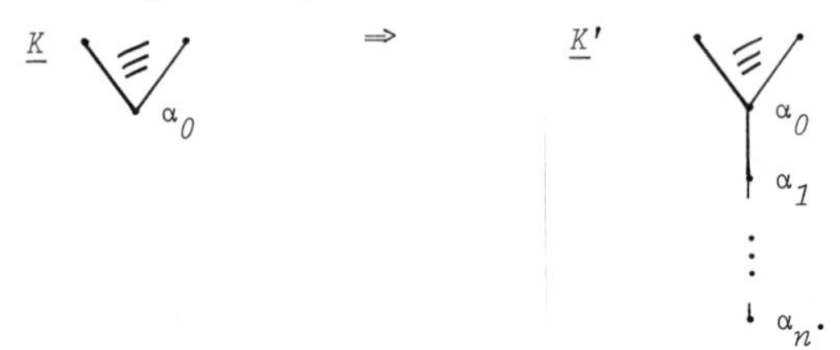

The forcing relation is standardised so each α_i forces the same atoms as α_0 in $\underline{K}$.

Before saying anything else, I should observe that, for $\alpha \in K$ and any sentence C,

$$\alpha \Vdash C \text{ in } \underline{K} \quad \text{iff} \quad \alpha \Vdash^{(n)} C \text{ in } \underline{K}^{(n)}.$$

This follows from the oft-cited Remark 1.3.ii. Since, moreover, for $\alpha, \beta \in K$,

$$\alpha < \beta \text{ in } \underline{K} \quad \text{iff} \quad \alpha <^{(n)} \beta \text{ in } \underline{K}^{(n)},$$

it is safe to omit all the nuisance-superscripts. We do so without further notice.

In our previous encounter with such a construction, in proving 1.12, we simply let the new node force no atoms (or, parenthetically, whatever atoms we chose). The following lemma explains the present choice.

4.7. LEMMA. Let $\underline{K}$ be an A-sound model, S the set of subformulae of A, and $\underline{K}^{(n)}$ the n-th derived model. Then: For all $B \in S$,

$$\alpha_0 \Vdash B \quad \text{iff} \quad \alpha_n \Vdash B.$$

Proof: By induction on n: It clearly suffices to prove the case $n = 1$. This is done by induction on the complexity of B:

i. B is an atom p. This follows by definition of the derived forcing relation.

ii-iii. B is t or f. These cases follow by the general definition of forcing.

iv-vii. B is $\sim C$, $C \wedge D$, $C \vee D$, or $C \rightarrow D$. These cases follow by propositional considerations.

viii. B is $\Box C \in S$. This is, as usual, the interesting case. Observe, for $\Box C \in S$,

$$\alpha_0 \Vdash \Box C \;\Rightarrow\; \alpha_0 \Vdash C, \text{ by } A\text{-soundness.}$$

Thus, $\forall \beta > \alpha_1 (\beta \Vdash C)$, whence $\alpha_1 \Vdash \Box C$.

Conversely,

$$\begin{aligned} \alpha_1 \Vdash \Box C &\Rightarrow \forall \beta > \alpha_1 (\beta \Vdash C) \\ &\Rightarrow \forall \beta > \alpha_0 (\beta \Vdash C) \\ &\Rightarrow \alpha_0 \Vdash \Box C. \end{aligned}$$

QED

We can now prove Theorem 4.4:

Proof of 4.4.i: Half of this is easy: If A is true in all A-sound models of PRL, then

$$\mathrm{PRL} + \bigwedge_{\Box B \in S_\Box} (\Box B \to B) \vdash A,$$

whence $\mathrm{PRL}^\omega \vdash A$.

The converse is proven contrapositively. Suppose A is false in an A-sound model $\underline{K} = (K, <, \alpha_0, \Vdash)$:

$$\alpha \Vdash \bigwedge_{\Box B \in S_\Box} (\Box B \to B), \quad \alpha_0 \nVdash A.$$

We show $\mathrm{PRL}^\omega \nvdash A$ by showing, for any finite set X,

$$\mathrm{PRL} + \bigwedge_{C \in X} (\Box C \to C) \nvdash A.$$

Note that, by the Lemma, each model in the sequence $\underline{K}^{(0)}, \underline{K}^{(1)}, \ldots$ is an A-sound model in which A is false. It suffices to show, for any given set X, there to be some number n_0 such that

$$\alpha_{n_0} \Vdash \bigwedge_{C \in X} (\Box C \to C).$$

To do this, it suffices to show, for any formula C, there is an n_0 such that for all $n \geq n_0$, $\quad \alpha_n \Vdash \Box C \to C$.

But this is easy: Either $\alpha_n \Vdash \Box C$ for all n, whence $\alpha_n \Vdash C$ for all n (since $\alpha_n > \alpha_{n+1} \Vdash \Box C$) or $\alpha_{n_0} \nVdash \Box C$ for some n_0, whence $\alpha_n \nVdash \Box C$ *and* $\alpha_n \Vdash \Box C \to C$ for all $n \geq n_0$.

QED

The proof of 4.4.ii is left as an Exercise for the reader.

EXERCISES

1. Prove Theorem 4.4.ii.

2. Compare Theorems 4.2 and 4.4 with Exercise 3.8. Show: PRL^{ω} is *not* axiomatised over PRL by $\{\sim\Box^{n}f: n \in \omega\}$.

3. Show that PRL^{ω} is closed under the rule,

   $\Box A/A$,

   but is not closed under the rule *R2*,

   $A/\Box A$.

Chapter 3
Arithmetic Interpretations of PRL

PRL has, as we have seen, moderately interesting syntactic properties and a similarly interesting model theory. Moreover, as I have intimated on occasion, it has much to say about the syntax of arithmetic: The de Jongh-Sambin Theorem, for example, applies directly to self-referential sentences of PRA-- those arising from modal contexts have explicit, non-self-referential explications. The relation between PRL and PRA, however, is deeper than this. PRL is the modal logic of provability within PRA and we now have the tools to prove this.

The system BML was originally studied by modal logicians without regard to its arithmetic significance, i.e. its axiomatisation via the Derivability Conditions. It was the model theory-- BML is the "logic of transitive structures"-- that provided the interest in the theory. I am not sure of the situation with PRL: I haven't read the literature fully. The system was studied by modal logicians in a purely modal context and later by others (proof theorists and universal algebraists) with arithmetic in mind. Whether the original modal logicians studying PRL were motivated by, interested in, or even aware of the arithmetical significance is what I cannot say. The earliest query about an axiomatisation of the logic of provability that I've seen is in an unpublished (?) paper of Michael Moss from the mid-1960s. In any event, the conjecture that PRL was *the* logic of provability was fairly widespread by the time the system was being arithmetically studied.

The first positive result was obtained in 1975 independently by George Boolos and by Claudio Bernardi and Franco Montagna (unpublished), who announced PRL to axiomatise the atom-free schemata provable in PRA (cf. Exercise 2.3.8, above). In response to Boolos' abstract, Robert M. Solovay proved the full result-- or, rather, two full results: PRL yields all modal schemata provable in PRA and PRL^{ω} yields all modal schemata true in arithmetic. Following Solovay's work, a number of researchers offered refinements.

The purpose of the present Chapter is to prove these results: In section 1, we will encounter Solovay's First Completeness Theorem, by which PRL is *the* modal logic of provability-- according to PRA; in section 2, we will see Solovay's Second Completeness Theorem, by which PRL^{ω} is *the* modal logic of provability-- in truth; and, in section 3, we will discuss some of the refinements.

Of course, Solovay's results are important in that they tell us that PRL and PRL^{ω} are *the* logics of provability. But they offer more than mere doctrinaire interest. Solovay's Second Completeness Theorem, for example, offers a very strong *in*completeness theorem, one much more general than Gödel's and Rosser's Theorems. I would like to explain this applicability by saying that Solovay's results are completeness theorems with respect to a metamathematical model theory (as opposed to an algebraic or a set theoretic model theory) and that, as with any good model theories, applications are to be expected. In truth, however, something more like the reverse holds: We use PRL and PRL^{ω}-- or, rather, their Kripke model theories-- to study the present interpretations. Solovay's completeness proof shows how to simulate Kripke models within PRA. Some incompleteness phenomena, easily exhibited modally, then translate immediately to arithmetic. We shall see this in section 2.

One last point: Solovay's proof can be viewed as a vast generalisation of that of Rosser's Theorem and the construction of Rosser sentences. These last will not be introduced until Chapter 6 where we present David Guaspari's and Solovay's modal analysis of them. The reader might bear this in mind later when he reads Chapter 6. For now, however, we avoid such subtleties.

1. SOLOVAY'S FIRST COMPLETENESS THEOREM

We begin with a formal definition.

1.1. DEFINITION. An *arithmetic interpretation* * is an assignment of arithmetic sentences A^* to modal sentences A that satisfies the following: For all A,B,p,

i. for atomic p, p^* is a sentence of the language of arithmetic ii-iii. t^* is $\overline{0} = \overline{0}$; f^* is $\overline{0} = \overline{1}$

iv. $(\sim A)^*$ is $\sim(A^*)$

v-vii. $(A \circ B)^*$ is $A^* \circ B^*$ for $\circ \in \{\wedge, \vee, \rightarrow\}$

viii. $(\Box A)^*$ is $Pr(\ulcorner A^* \urcorner)$.

Note that, by conditions ii-viii, * is freely determined by any assignment of arithmetic sentences p^* to atoms p.

The basic fact, which we are already familiar with and have already used in applying results on PRL to PRA, is the soundness of arithmetic interpretations.

1.2. SOUNDNESS LEMMA. For all modal sentences A,

$$\mathsf{PRL} \vdash A \quad \Rightarrow \quad \forall^* (\mathsf{PRA} \vdash A^*).$$

The proof is a routine induction on the length of a derivation in PRL and will not be presented here. (It is, after all, the motivation for, hence the prerequisite to, this monograph-- cf. Chapter 0.)

The converse is a bit deeper.

1.3. SOLOVAY'S FIRST COMPLETENESS THEOREM. For all modal sentences A,

$$\forall^* (\mathsf{PRA} \vdash A^*) \quad \Rightarrow \quad \mathsf{PRL} \vdash A.$$

The proof of this is by contraposition: If $\mathsf{PRL} \nvdash A$, then $\mathsf{PRL} \nvDash A$ and there is a finite Kripke model $\underline{K} = (K, <, \alpha_0, \Vdash)$ in which $\alpha_0 \nVdash A$. We will use self-reference, in the form of the Recursion Theorem, to simulate $\underline{K}$ and obtain * such that $\mathsf{PRA} \nvdash A^*$. To this end, fix, until the completion of the proof of Theorem 1.3, a sentence A such that $\mathsf{PRL} \nvdash A$. Let $\underline{K} = (K, <, \alpha_0, \Vdash)$ be such that $\alpha_0 \nVdash A$.

Without loss of generality, we can assume $K = \{1, \ldots, n\}$ for some finite n and that $\alpha_0 = 1$. Also, to avoid confusion with the ordinary ordering of the natural numbers, we can denote the accessibility relation of $\underline{K}$ by R. Thus, $\underline{K} = (\{1, \ldots, n\}, R, 1, \Vdash)$. For convenience, we extend R by setting $0 R i$ for each $1 \leq i \leq n$, i.e. each $i \in K$. (In the next section, we shall go one step further and make 0 a full-fledged member of a Kripke model by defining when 0 forces an atom. For now we don't need to. In fact, we really don't need to relate 0 to all elements of K; but it makes the ensuing definitions more uniform.)

We next define a function F as follows: We begin by setting $F(0) = 0$. To define $F(x+1)$, we look at $x+1$ and $F(x)$. If $x+1$ codes a proof that $\lim_{k\to\infty} F(k) \neq z$ for some z accessible to $F(x)$, we set $F(x+1) = z$; otherwise $F(x+1) = F(x)$. In other words, F is reluctantly attempting to climb its way through R by starting at 0 and only moving to an accessible node when it is proven that F will not stay there. Since K is finite, it will follow that F never leaves 0; however, we cannot prove this in PRA.

The formal definition of F is given by the Recursion Theorem (Theorem 0.6.12). Towards presenting such a definition, we introduce some notation. Let G be a partial recursive function with Σ_1 graph (cf. Chapter 0, section 6) $\psi v_0 v_1$. From ψ we can effectively obtain the formula,

$$\exists v_0 \forall v_1 > v_0\ \psi v_1 v,$$

asserting G to have the limit v. We abbreviate this as

$$L_\psi = v, \text{ or even } L = v.$$

Observe that the formula $L = v$ is not Σ_1, but Σ_2.

The function F is then defined in terms of its limit L (almost) formally by:

$$F(v_0) = v_1 \quad\leftrightarrow\quad \begin{cases} (v_0 = \overline{0} \wedge v_1 = \overline{0}) \vee \\ \vee\ (v_0 > \overline{0} \wedge Prov(v_0, \ulcorner L \neq \dot{v}_1 \urcorner) \wedge F(v_0 \dot{-} \overline{1})\, \overline{R}\, v_1) \vee \\ \vee\ (v_0 > \overline{0} \wedge \forall v_2 \leq v_0 \sim\!\big(Prov(v_0, \ulcorner L \neq \dot{v}_2 \urcorner) \wedge \overline{F}(v_0 \dot{-} \overline{1})\, \overline{R}\, v_2\big) \wedge \\ \qquad\qquad \wedge\ v_1 = \overline{F}(v_0 \dot{-} 1)). \end{cases}$$

For a fully formal definition, we actually define the graph $\psi v_0 v_1 = \exists v_3 \chi v_3 v_0 v_1$ of the partial recursive function F as discussed in Chapter 0, section 6, by applying a Σ_1-truth definition, the Selection Theorem (0.6.9), and diagonalisation to the disjunction of the following three Σ_1-formulae:

$$v_0 = \overline{0} \wedge v_1 = \overline{0}$$

$$v_0 > \overline{0} \wedge Prov(v_0, \ulcorner L_\psi \neq \dot{v}_1 \urcorner) \wedge \exists v_4 \big(\psi(v_0 \dot{-} \overline{1}, v_4) \wedge v_4 \overline{R} v_1\big)$$

$$v_0 > \overline{0} \wedge \exists v_3 v_4 \forall v_2 \leq v_0 \sim\!\big(Prov(v_0, \ulcorner L_\psi \neq \dot{v}_2 \urcorner) \wedge \chi(v_3, v_0 \dot{-} \overline{1}, v_4) \wedge v_4 \overline{R} v_1\big) \wedge \psi(v_0 \dot{-} \overline{1}, v_1).$$

Finally, of course, one must explain, beyond the obvious remark that it is to be a representation of R, what $\overline{R}$ is, i.e. how the representation is chosen. Since R is a finite set, we can either choose an index x to the set and let

$$v_0 \overline{R} v_1:\quad \langle v_0, v_1 \rangle \in D_{\overline{x}},$$

or we can just list all pairs $<x,y>$ R and write

$$v_0 \overline{R} v_1: \quad \bigvee_{<x,y> \in R} (v_0 = \overline{x} \wedge v_1 = \overline{y}).$$

The function F, so defined, does all I said it would do:

i. $F(0) = 0$

ii. if $x + 1$ proves $L \neq \overline{z}$ and $F(x) R z$, then $F(x + 1) = z$

iii. otherwise $F(x + 1) = F(x)$.

Now, the Recursion Theorem only guarantees the existence of partial functions. The function F, however, is total-- for it is defined by recursion: $F(0) = 0$ and $F(x + 1)$ is defined by cases from $F(x)$. Moreover, this fact can be proven in PRA:

1.4. LEMMA. Let ψ be the Σ_1-formula defining the graph of F. Then:

$$\text{PRA} \vdash \forall v_0 \exists ! v_1\ \psi v_0 v_1$$

i.e. PRA proves that F is a total function.

Sketch of the proof: The unicity of v_1 is automatic on assumption of uniformisation in the definition of F. The existence of a value is proven by induction on v_0 in the Σ_1-formula $\exists v_1\ \psi v_0 v_1$ and this proof is readily formalised in PRA. QED

Because ψ is provably the graph of a total function, we can henceforth use functional notation: Augment the language by a new function constant $\overline{F}$ and its defining axiom,

$$\overline{F} v_0 = v_1 \leftrightarrow \psi v_0 v_1.$$

Armed with such notation, we can now state and prove a few lemmas. The first two merely document some obvious facts about F and their provability within PRA.

1.5. LEMMA. i. $\text{PRA} \vdash \forall v_0 (\overline{F} v_0 \leq \overline{n})$

ii. for all $x \in \omega$,

$$\text{PRA} \vdash \forall v_0 \big(\overline{F} v_0 = \overline{x} \rightarrow \forall v_1 > v_0 (\overline{F} v_1 = \overline{x} \vee \overline{x} \overline{R} \overline{F} v_1)\big)$$

iii. $\text{PRA} \vdash \exists v_0 v_1 \forall v_2 > v_0 (\overline{F} v_2 = v_1)$, i.e. $\text{PRA} \vdash \exists v_1 (L = v_1)$.

Note that in assertion ii it is not assumed $x \leq n$. Of course, for $x > n$, the sentence being proved is vacuously true (by part i).

Sketch of the proof: i. The first assertion is a simple induction: $F0 = 0 \leq n$ and $F(x + 1)$ is either in the range of R-- whence $\leq n$-- or equal to $F(x)$--

whence $\leq n$.

ii. To prove ii it is convenient first to rewrite the formula in the equivalent form,

$$\forall v_1 \forall v_0 \big(\overline{F}v_0 = \overline{x} \rightarrow .F(v_0 + v_1 + \overline{1}) = \overline{x} \vee \overline{x}\overline{R}\overline{F}(v_0 + v_1 + \overline{1})\big)$$

and then to induct on v_1, using the transitivity of R. (Remark: The induction is on a Π_1-formula, not a Σ_1-formula as explicitly allowed in PRA. However, by Facts 0.6.16.ii and Theorem 0.6.17, such induction is available in PRA.)

iii. One proves iii by first proving

$$\forall v_0 \big(\exists v_1 (\overline{F}v_1 = v_0) \rightarrow \exists v_1 (L = v_1)\big). \qquad (*)$$

This is vacuously true for $v_0 > n$. For $v_0 \leq n$, one can induct *informally* on $\check{R}$, the converse to R: By ii, the assertion holds for maximal nodes $y \in K$. If x is not maximal and $\exists v_1 (\overline{F}v_1 = \overline{x})$ then, by ii again, either $L = \overline{x}$ or, for some y such that xRy, $\exists v_1 (\overline{F}v_1 = \overline{y})$ and the induction hypothesis yields $\exists v_1 (L = v_1)$. Ultimately,

$$\text{PRA} \vdash \exists v_1 (\overline{F}v_1 = \overline{0}) \rightarrow \exists v_1 (L = v_1).$$

but PRA $\vdash \overline{F}\overline{0} = \overline{0}$, whence PRA $\vdash \exists v_1 (L = v_1)$. QED

1.6. REMARK. The induction of this last proof was an *in*formal, metamathematical one performed outside PRA. Ostensibly, the reason for performing the induction informally is the Σ_2-complexity of the formula $\exists v_1 (L = v_1)$. In fact, this complexity is a red herring: By the finiteness of K, the assertion that the limit exists reduces to a disjunction,

$$L = \overline{0} \vee L = \overline{1} \vee \ldots \vee L = \overline{n}.$$

Moreover, this finiteness again reduces each $L = \overline{x}$ to

$$\exists v_1 (\overline{F}v_1 = \overline{x}) \wedge \bigwedge \forall v_1 (\overline{F}v_1 \neq \overline{y}),$$

the conjunction being over all y's accessible to x. Hence, the implication

$$\exists v_1 (\overline{F}v_1 = \overline{x}) \rightarrow \exists v_1 (L = v_1)$$

is equivalent to a boolean combination of Σ_1-sentences and, by Theorem 0.6.17, we can use induction on this formula to prove (*) in PRA. In a later refinement (3.4, below), we will have an infinite K and will need more induction at this step in the proof.

Let's get back on track. First, a corollary:

1.7. COROLLARY. PRA $\vdash L \leq \overline{n}$, i.e. PRA $\vdash \bigvee_{x \leq n} L = \overline{x}$.

This follows from 1.5.i, 1.5.iii, and the implication,

$$\text{PRA}\vdash v \le \overline{n} \to \bigvee_{x \le n} v = \overline{x},$$

by pure logic.

1.8. LEMMA. For $x, y \le n$,

i. $\text{PRA}\vdash L = \overline{x} \wedge \overline{x}R\overline{y} \to Con_{PRA+L=\overline{y}}$

ii. $\text{PRA}\vdash L = \overline{x} \wedge \overline{x} \ne \overline{y} \wedge \sim\overline{x}R\overline{y} \to \sim Con_{PRA+L=\overline{y}}$

iii. $\text{PRA}\vdash L = \overline{x} \wedge \overline{x} > \overline{0} \to Pr(\ulcorner L \ne \overline{x}\urcorner)$.

The first two assertions can be restated as follows:

i. $\text{PRA}\vdash L = \overline{x} \wedge \overline{x}R\overline{y} \to \sim Pr(\ulcorner L \ne \overline{y}\urcorner)$

ii. $\text{PRA}\vdash L = \overline{x} \wedge \overline{x} \ne \overline{y} \wedge \sim\overline{x}R\overline{y} \to Pr(\ulcorner L \ne \overline{y}\urcorner)$.

Proof: i. Let xRy and suppose, by way of contradiction, $L = \overline{x} \wedge Pr(\ulcorner L \ne \overline{y}\urcorner)$. From $L = \overline{x}$ we can choose v_0 so that $\forall v_2 \big(v_2 > v_0 \to \overline{F}v_2 = \overline{x}\big)$. We can also choose $v_1 + \overline{1} > v_0$ so that $Prov(v_1+\overline{1}, \ulcorner L \ne \overline{y}\urcorner)$. (For, any derivation can be arbitrarily extended by the addition of redundancies.) But,

$$\text{PRA}\vdash Prov(v_1+\overline{1}, \ulcorner L \ne \overline{y}\urcorner) \wedge \overline{F}v_1 = \overline{x} \wedge \overline{x}R\overline{y} \to \overline{F}(v_1 + \overline{1}) = \overline{y},$$

which contradicts the assumption $\forall v_2 > v_0(\overline{F}v_0 = \overline{x})$, i.e. the assumption $L = \overline{x}$. Thus, $\text{PRA}\vdash L = \overline{x} \wedge \overline{x}R\overline{y} \to \sim Pr(\ulcorner L \ne \overline{y}\urcorner)$.

ii. First, observe

$$\text{PRA}\vdash L = \overline{x} \to \exists v_0(\overline{F}v_0 = \overline{x})$$
$$\to Pr(\ulcorner \exists v_0(\overline{F}v_0 = \overline{x})\urcorner), \qquad (1)$$

by Demonstrable Σ_1-Completeness (0.6.). But, by 1.5.ii (using the obvious abbreviation),

$$\text{PRA}\vdash \forall v_0\big(\overline{F}v_0 = \overline{x} \to (L = \overline{x} \vee \overline{x}RL)\big)$$
$$\vdash Pr(\ulcorner \forall v_0\big(\overline{F}v_0 = \overline{x} \to (L = \overline{x} \vee \overline{x}RL)\big)\urcorner), \qquad (2)$$

by *D1*. Now (1) and (2) yield

$$\text{PRA}\vdash L = \overline{x} \to Pr(\ulcorner L = \overline{x} \vee \overline{x}RL\urcorner). \qquad (3)$$

As with (1),

$$\text{PRA}\vdash \overline{x} \ne \overline{y} \wedge \sim\overline{x}R\overline{y} \to Pr(\ulcorner \overline{x} \ne \overline{y} \wedge \sim\overline{x}R\overline{y}\urcorner).$$

With (3) this yields

$$\text{PRA}\vdash L = \overline{x} \wedge \overline{x} \ne \overline{y} \wedge \sim\overline{x}R\overline{y} \to Pr(\ulcorner L = \overline{x} \vee \overline{x}RL\urcorner) \wedge Pr(\ulcorner \overline{x} \ne \overline{y} \wedge \sim\overline{x}R\overline{y}\urcorner)$$

$$\vdash L = \overline{x} \wedge \overline{x} \neq \overline{y} \wedge \sim\overline{x} R \overline{y} \rightarrow Pr(\ulcorner L \neq \overline{y} \urcorner).$$

iii. By the Least Number Principle,

$$\mathrm{PRA} \vdash L = \overline{x} \wedge \overline{x} > \overline{0} \rightarrow \exists v \left(\overline{F}(v + \overline{1}) = \overline{x} \wedge \overline{F}v \neq \overline{x}\right).$$

For such v, by the definition of F,

$$\mathrm{PRA} \vdash \overline{F}(v + \overline{1}) = \overline{x} \wedge \overline{F}v \neq \overline{x} \rightarrow Prov(v + \overline{1}, \ulcorner L \neq \overline{x} \urcorner).$$

Hence $\quad \mathrm{PRA} \vdash L = \overline{x} \wedge \overline{x} > \overline{0} \rightarrow Pr(\ulcorner L \neq \overline{x} \urcorner).$ QED

Lemmas 1.5 and 1.8 offer a few basic facts about F and L provable in PRA. As background, we should also be aware of the following true results which are not provable in PRA:

1.9. LEMMA. The following are true, though unprovable in PRA:

i. $L = \overline{0}$

ii. for $0 \leq x \leq n$, PRA + $L = \overline{x}$ is consistent.

Proof: i. By Lemma 1.5.iii, L exists (provably in PRA). If $x > 0$, 1.8.iii yields $\quad L = \overline{x} \ \Rightarrow\ \mathrm{PRA} \vdash L \neq \overline{x}$

$$\Rightarrow\ L \neq \overline{x}$$

by the soundness of PRA; but this also yields a contradiction and we must conclude $L = 0$.

ii. Since $L = \overline{0}$ is true and PRA is sound, PRA + $L = \overline{0}$ is consistent. For $x > 0$, apply 1.8.i:

$$\mathrm{PRA} \vdash L = \overline{0} \wedge \overline{0} R \overline{x} \rightarrow Con_{PRA+L=\overline{x}}.$$

Since $L = \overline{0} \wedge \overline{0} R \overline{x}$ is true and since this implication is true, $Con_{PRA+L=\overline{x}}$ is true, i.e. PRA + $L = \overline{x}$ is consistent. QED

We now have a sufficiency of basic properties of F and L and can proceed to the next important step in the proof of Solovay's First Completeness Theorem-- the actual simulation of the Kripke model $\underline{K} = (\{1,\ldots,n\}, R, 1, \Vdash)$ in which $1 \nVdash A$. This is done by letting the sentences $L = \overline{x}$, for $x > 0$, assume the rôles of the nodes $x \in K = \{1,\ldots,n\}$.

Let, for any atom p,

$$p^* \ = \ \bigvee\{L = \overline{x}:\ 1 \leq x \leq n \ \&\ x \Vdash p\},$$

where the empty disjunction is $\overline{0} = \overline{1}$. (For the sake of proving Theorem 1.3, we need only interest ourselves in the sentence A, hence in the set,

$$S(A) = \{B\colon\ B \text{ is a subformula of } A\},$$

of subformulae of A; we need not concern ourselves with p^* for $p \notin S(A)$. However, it is worth our while to handle all such p and observe that the simulation of the Kripke model is complete. This will not be the case in the next section.)

The fundamental lemma is the following.

1.10. LEMMA. Let $1 \le x \le n$. For any B and for $*$ defined above,

i. $x \Vdash B \;\Rightarrow\; \mathsf{PRA} \vdash L = \overline{x} \to B^*$

ii. $x \nVdash B \;\Rightarrow\; \mathsf{PRA} \vdash L = \overline{x} \to \sim B^*$.

Before proving this, let us observe how the Completeness Theorem (1.3) follows from this lemma. The derivation is quite quick:

$$1 \nVdash A \;\Rightarrow\; \mathsf{PRA} \vdash L = \overline{1} \to \sim A^*.$$

But $\mathsf{PRA} + L = \overline{1}$ is consistent (by Lemma 1.9.ii), whence $\mathsf{PRA} + \sim A^*$ is consistent, i.e. $\mathsf{PRA} \nvdash A^*$-- which was (not) to be proven.

Proof of Lemma 1.10: This is a simple induction on the length of B.

For $B = p$ atomic, it follows practically by definition:

$$x \Vdash p \;\Rightarrow\; \mathsf{PRA} \vdash L = \overline{x} \to p^*,$$

since $L = \overline{x}$ is a disjunct of p^*. Moreover, if $x \nVdash p$, $L = \overline{x}$ contradicts every disjunct of p^*, whence

$$x \nVdash p \;\Rightarrow\; \mathsf{PRA} \vdash L = \overline{x} \to \sim p^*.$$

The cases $B = \sim C$, $C \wedge D$, $C \vee D$, and $C \to D$ are trivial.

The crucial case is $B = \Box C$. Note:

$$\begin{aligned} x \Vdash \Box C \;&\Rightarrow\; \forall y (xRy \;\Rightarrow\; y \Vdash C) \\ &\Rightarrow\; \forall y (xRy \;\Rightarrow\; \mathsf{PRA} \vdash L = \overline{y} \to C^*), \text{ by induction hypothesis} \\ &\Rightarrow\; \bigwedge_{xRy} (\mathsf{PRA} \vdash L = \overline{y} \to C^*) \\ &\Rightarrow\; \mathsf{PRA} \vdash \bigvee_{xRy} L = \overline{y} \to C^* \\ &\Rightarrow\; \mathsf{PRA} \vdash Pr(\ulcorner \bigvee_{xRy} L = \overline{y} \urcorner) \to Pr(\ulcorner C^* \urcorner), \qquad (1) \end{aligned}$$

by $D1$ and $D2$. Since $x > 0$, Lemmas 1.8.ii and 1.8.iii yield

$$\text{PRA}\vdash L = \overline{x} \to \bigwedge_{\sim x R z} Pr(\ulcorner L \neq \overline{z}\urcorner), \qquad (2)$$

whence

$$\text{PRA}\vdash L = \overline{x} \to Pr(\ulcorner \bigvee_{x R y} L = \overline{y}\urcorner). \qquad (3)$$

((3) follows from (2) by 1.5.1: $\text{PRA}\vdash Pr(\ulcorner \bigvee_{z<n} L = \overline{z}\urcorner)$.) But (1) and (3) yield

$$x \Vdash \Box C \;\Rightarrow\; \text{PRA}\vdash L = \overline{x} \to Pr(\ulcorner C^* \urcorner)$$

$$\Rightarrow\; \text{PRA}\vdash L = \overline{x} \to (\Box C)^*.$$

Moreover,

$$x \nVdash \Box C \;\Rightarrow\; \exists y\big(x R y \;\&\; y \nVdash C\big)$$

$$\Rightarrow\; \exists y\big(x R y \;\&\; \text{PRA}\vdash L = \overline{y} \to \sim C^*\big), \text{ by induction hypothesis}$$

$$\Rightarrow\; \exists y\big(x R y \;\&\; \text{PRA}\vdash C^* \to L \neq \overline{y}\big)$$

$$\Rightarrow\; \exists y\big(x R y \;\&\; \text{PRA}\vdash Pr(\ulcorner C^*\urcorner) \to Pr(\ulcorner L \neq \overline{y}\urcorner)\big). \qquad (4)$$

But, by 1.8.i, if $x R y$,

$$\text{PRA}\vdash L = \overline{x} \to \sim Pr(\ulcorner L \neq \overline{y}\urcorner).$$

With (4) this yields

$$\text{PRA}\vdash L = \overline{x} \to \sim Pr(\ulcorner C^*\urcorner),$$

i.e.

$$\text{PRA}\vdash L = \overline{x} \to \sim(\Box C)^*.$$

QED

With the completion of the proof of Lemma 1.10 we have, as remarked just prior to the proof, the completion of the proof of Solovay's First Completeness Theorem: PRL is the logic of provability of PRA in the sense that it yields all modal schemata derivable in PRA.

EXERCISES

1. Use Solovay's First Completeness Theorem to give a new proof of the closure of PRL under the Diagonalisation Rule: If

   $$\text{PRL}\vdash \boxed{s}\,\big(p \leftrightarrow A(p)\big) \to B,$$

   where p is boxed in $A(p)$ and has no occurrence in B, then $\text{PRL}\vdash B$.

2. Explain the negative assertions of Lemma 1.9: Why, for example, do we not have $\text{PRL}\vdash L = \overline{0}$?

3. Consider the Kripke frame: 2 above 1 (1 — 2)

   a. Show arithmetically: For F, L defined on the basis of this frame,

i. $\mathrm{PRA} \vdash Pr(\ulcorner L = \overline{2} \urcorner) \leftrightarrow Pr(\ulcorner \overline{0} = \overline{1} \urcorner)$

ii. $\mathrm{PRA} \vdash Pr(\ulcorner L \geq \overline{1} \urcorner) \leftrightarrow Pr(\ulcorner Pr(\ulcorner \overline{0} = \overline{1} \urcorner) \urcorner)$.

(Hints: Show: i. $\mathrm{PRA} \vdash L = \overline{2} \rightarrow Pr(\ulcorner L = \overline{2} \urcorner)$;

ii. $\mathrm{PRA} \vdash L = \overline{1} \rightarrow Pr(\ulcorner L = \overline{1} \vee L = \overline{2} \urcorner)$.)

b. Prove assertions i and ii by appealing to Lemma 1.10. (Hint: Let p_2 (p_1) be forced only at the node 2 (respectively, 1).)

c. Show: $\mathrm{PRA} \vdash L = \overline{0} \leftrightarrow \sim Pr(\ulcorner Pr(\ulcorner \overline{0} = \overline{1} \urcorner) \urcorner)$.

4. (A Version of Rosser's Theorem). Consider the unorthodox Kripke frame with no minimum node:

$$1 \bullet \quad 2 \bullet.$$

Convince yourself that the construction goes through and Lemma 1.10 holds for any forcing relation we put on this frame.

a. Show: $\mathrm{PRA} \vdash L = \overline{0} \leftrightarrow Con$.

b. Show: $\mathrm{PRA} + L = \overline{1}$ and $\mathrm{PRA} + L \neq \overline{1}$ are both consistent.

c. Show: $\mathrm{PRA} \vdash Con \rightarrow \sim Pr(\ulcorner L = \overline{1} \urcorner) \wedge \sim Pr(\ulcorner L \neq \overline{1} \urcorner)$.

c. Show: $L = \overline{1}$ is equivalent to a Σ_1-sentence.

(Remark: Exercises such as this show Solovay's First Completeness Theorem to have worthwhile applications. The Second Completeness Theorem, however, will yield such applications more directly; here we must use modal logical and arithmetical considerations.)

2. SOLOVAY'S SECOND COMPLETENESS THEOREM

Solovay's First Completeness Theorem is very pleasing; but it doesn't compare to the Second Completeness Theorem:

2.1. SOLOVAY'S SECOND COMPLETENESS THEOREM. For all modal sentences A, the following are equivalent:

i. $\mathrm{PRL}^{\omega} \vdash A$

ii. $\mathrm{PRL} \vdash \bigwedge_{\Box B \in S(A)} (\Box B \rightarrow B) \rightarrow A$

iii. A is true in all A-sound Kripke models

iv. $\forall * \, (A^*$ is true).

(Here, $S(A)$ is the set of subformulae of A.)

We don't really have to prove all of this: The equivalence of i and ii is simply Theorem 2.4.2, which we proved by appeal to Theorem 2.4.4-- the equivalence of ii and iii. The proof of Solovay's Second Completeness Theorem is, however, partially independent of the earlier proofs. Observe that implication ii ⇒ i is trivial, implication i ⇒ iv is clear, and equivalence ii ⇔ iii follows directly from the basic Completeness Theorem of PRL with respect to its Kripke models. Thus, we need only prove iv ⇒ iii to both establish Solovay's Second Completeness Theorem and derive anew Theorem 2.4.2 on the equivalence of i with ii. Needless to say, we prove the implication iv ⇒ iii by contraposition.

As in proving the First Completeness Theorem, let a Kripke model $\underline{K} = (\{1,\ldots,n\},R,1,\Vdash)$ be given. Assume the model is A-sound, i.e.

$$1 \Vdash \Box B \to B$$

for all $\Box B \in S(A)$. For the sake of the Theorem we will also assume $1 \nVdash A$; in later applications, we will assume $1 \Vdash A$. For now, let us make no assumption.

As before, we will set $0\,R\,x$ for all $x \in K$. This time, however, we go one step further and add 0 to K, i.e. we define a new model $\underline{K}'$:

$$K' = \{0,1,\ldots,n\}$$

R' extends R by assuming $0\,R'\,x$ for $x \in K$

$$\alpha_0 = 0$$

$\Vdash'$ extends $\Vdash$ by:

$$0 \Vdash' p \quad \text{iff} \quad 1 \Vdash p \text{ for } p \in S(A).$$

By the usual abuse of notation, we let R denote R' and $\Vdash$ denote $\Vdash'$ -- for, R' and $\Vdash'$ merely extend R and $\Vdash$ without changing their behaviour on their respective domains.

$\underline{K}'$ was called the *derived model* in Chapter 2, section 4, where we used the A-soundness of $\underline{K}$ to prove a slight generalisation of the following:

<u>2.2. LEMMA</u>. For all $B \in S(A)$,

$$0 \Vdash B \quad \text{iff} \quad 1 \Vdash B.$$

I refer the reader back to Lemma 2.4.7 for the proof.

Our next step, as before, is to define a function F trying very hard not to climb through R. F is defined exactly as before (since it is determined by the *frame* (K,R), not the model $\underline{K}$):

$$F0 = 0$$

$$F(x+1) = \begin{cases} y, & Prov(\overline{x} + \overline{1}, \ulcorner L \neq \overline{y} \urcorner) \ \& \ xRy \\ Fx, & \text{otherwise.} \end{cases}$$

Because of this identity of definition and because Lemmas 1.4, 1.5, 1.8 and 1.9, as well as Corollary 1.7, depended only on the frame, which has not changed, their validity remains.

The arithmetic simulation of $\Vdash$ differs slightly in that i. we must take the node 0 into account and ii. we can really only handle subformulae of A. Thus, we define

$$p^* = \bigvee\{L = \overline{x}: \ 0 \leq x \leq n \ \& \ x \Vdash p\}$$

for $p \in S(A)$ and let p^* be arbitrary for $p \notin S(A)$. Because of this difference, the analogue to Lemma 1.10 differs in content and must be stated and proved:

<u>2.3. LEMMA</u>. Let $0 \leq x \leq n$. For any $B \in S(A)$ and $*$ as defined above,

i. $x \Vdash B \ \Rightarrow \ \mathrm{PRA} \vdash L = \overline{x} \rightarrow B^*$

ii. $x \nVdash B \ \Rightarrow \ \mathrm{PRA} \vdash L = \overline{x} \rightarrow \sim B^*$.

Proof: For $x > 0$, i.e. x a node of the original model, the proof is identical to that of Lemma 1.10 and I omit it. (I note that one can even reduce this case of the present Lemma to Lemma 1.10 by noting that the new p^* differs from the old p^* only in the possible presence of the disjunct $L = \overline{0}$, which disjunct is refutable on the assumption $L = \overline{x}$. Hence, in $\mathrm{PRA} + L = \overline{x}$, the two p^*'s, whence the two B^*'s, are provably equivalent.)

The case $x = 0$ is the interesting one. It is proven by induction on the complexity of B in a manner analogous to the induction yielding Lemma 1.10. Only the subcase $B = \Box C$ is treated differently and will be presented here.

Let $B = \Box C$ and note:

$$0 \Vdash \Box C \ \Rightarrow \ \forall x(1 \leq x \leq n \ \Rightarrow \ x \Vdash C)$$

$$\Rightarrow \ \forall x(1 \leq x \leq n \ \Rightarrow \ \mathrm{PRA} \vdash L = \overline{x} \rightarrow C^*), \qquad (1)$$

by the case $x > 0$ of the Lemma. But we also have

$$0 \Vdash \Box C \;\Rightarrow\; 1 \Vdash C$$
$$\Rightarrow\; 0 \Vdash C, \text{ by Lemma 2.2}$$
$$\Rightarrow\; \mathsf{PRA} \vdash L = \overline{0} \to C^*, \qquad (2)$$

by induction hypothesis. Combining (1) and (2) we get

$$0 \Vdash \Box C \;\Rightarrow\; \bigwedge_{x \le n} (\mathsf{PRA} \vdash L = \overline{x} \to C^*)$$
$$\Rightarrow\; \mathsf{PRA} \vdash (\bigvee_{x \le n} L = \overline{x}) \to C^*$$
$$\Rightarrow\; \mathsf{PRA} \vdash Pr(\ulcorner \bigvee L = \overline{x} \urcorner) \to Pr(\ulcorner C^* \urcorner). \qquad (3)$$

But, by Corollary 1.7, $\mathsf{PRA} \vdash \bigvee L = \overline{x}$, whence $\mathsf{PRA} \vdash Pr(\ulcorner \bigvee L = \overline{x} \urcorner)$, whence (3) yields

$$0 \Vdash \Box C \;\Rightarrow\; \mathsf{PRA} \vdash Pr(\ulcorner C^* \urcorner)$$
$$\Rightarrow\; \mathsf{PRA} \vdash L = \overline{0} \to Pr(\ulcorner C^* \urcorner).$$

The other assertion is a little easier to prove:

$$0 \not\Vdash \Box C \;\Rightarrow\; \exists x(1 \le x \le n \;\&\; x \not\Vdash C)$$
$$\Rightarrow\; \exists x(1 \le x \le n \;\&\; \mathsf{PRA} \vdash L = \overline{x} \to \sim C^*)$$
$$\Rightarrow\; \exists x(1 \le x \le n \;\&\; \mathsf{PRA} \vdash C^* \to L \ne \overline{x})$$
$$\Rightarrow\; \mathsf{PRA} \vdash L = \overline{0} \to \sim Pr(\ulcorner C^* \urcorner),$$

by Lemma 1.8: $\mathsf{PRA} \vdash L = \overline{0} \to \sim Pr(\ulcorner L \ne \overline{x} \urcorner)$ for $x > 0$. QED

With this Lemma and Lemma 1.9, by which $L = \overline{0}$ is true, we can quickly deduce Solovay's Second Completeness Theorem.

Proof of Theorem 2.1: Assume A is false in the A-sound model $\underline{K}$, i.e. $1 \not\Vdash A$. By Lemma 2.2, we conclude $0 \not\Vdash A$. Lemma 2.3 then yields

$$\mathsf{PRA} \vdash L = \overline{0} \to \sim A^*.$$

Since $L = \overline{0}$ is true and PRA proves only true theorems, $\sim A^*$ is true, i.e. A^* is false. QED

As I said earlier, for the sake of Theorem 2.1 we assume A false in the A-sound model; for other applications of the construction we assume A true in an A-sound model and conclude A^* is true for some interpretation $*$.

2.4. EXAMPLE. There is an arithmetic sentence ϕ such that

i. $\mathsf{PRA} \nvdash \phi$

ii. $\mathsf{PRA} \nvdash \sim\phi$

iii. $\mathsf{PRA} \vdash Con \to \sim Pr(\ulcorner \phi \urcorner)$

iv. $\mathsf{PRA} \vdash Con \to \sim Pr(\ulcorner \sim\phi \urcorner)$.

(The Gödel sentence $\phi \leftrightarrow \sim Pr(\ulcorner\phi\urcorner)$ satisfied only i-iii. In fact, no self-referential sentence arising from the modal context satisfies i-iv (as will be proved in Chapter 6, below); something new is required.) To construct ϕ, we merely need to construct $*$ so that $\phi = p^*$ and

$$(\sim\Box p)^*, \quad (\sim\Box\sim p)^*, \quad (\Box(\sim\Box f \to \sim\Box p))^*, \quad (\Box(\sim\Box f \to \sim\Box\sim p))^*$$

are all true. Let A be the conjunction

$$\sim\Box p \wedge \sim\Box\sim p \wedge \Box(\sim\Box f \to \sim\Box p) \wedge \Box(\sim\Box f \to \sim\Box\sim p).$$

To apply Theorem 2.1 to conclude there to be an interpretation $*$ making A^*, whence i-iv, true, it suffices to find a $\sim A$-sound countermodel to $\sim A$, i.e. an A-sound model in which A is true. A-soundness requires not only that A is true, but that each of

$$\Box p \to p, \qquad \Box\sim p \to \sim p, \qquad \Box f \to f,$$

$$\Box(\sim\Box f \to \sim\Box p) \to (\sim\Box f \to \sim\Box p), \qquad \Box(\sim\Box f \to \sim\Box\sim p) \to (\sim\Box f \to \sim\Box\sim p)$$

is true at the origin. The following model works:

$\underline{K}$: 2 p 3 (nodes 2 and 3 above node 1; p at 2)
1 .

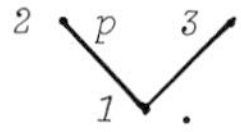

To see that A is true, note that

$1 \Vdash \sim\Box p$, since $1\,R\,3 \nVdash p$

$1 \Vdash \sim\Box\sim p$, since $1\,R\,2 \nVdash \sim p$

$1 \Vdash \Box(\sim\Box f \to C)$ for $C = \sim\Box p, \sim\Box\sim p$, since both $2,3 \Vdash \Box f$. To see that $\underline{K}$ is A-sound, observe that

$1 \Vdash \Box(\sim\Box f \to C) \to (\sim\Box f \to C)$ for $C = \sim\Box p, \sim\Box\sim p$ since $1 \Vdash C$,

$1 \Vdash \Box p \to p, \Box\sim p \to \sim p$, since $1 \Vdash \sim\Box p, \sim\Box\sim p$

$1 \Vdash \Box f \to f$, since $1 \Vdash \sim\Box f$.

As presented, Example 2.4 is an application of the Theorem as much as of the construction. We can, however, do a little better if we go back to the actual construction. The interpretation p^* has the form

$$p^* = \bigvee_{x \Vdash p} L = \overline{x}$$

a disjunction of the Σ_2-sentences,

$$L = \overline{x}: \quad \exists v_0 \forall v_1 > v_0 (Fv_1 = \overline{x}).$$

Since $\underline{K}$ is finite, we can rewrite $L = \overline{x}$ as

$$L = \overline{x}:\quad \exists v_0(Fv_0 = \overline{x}) \wedge \bigwedge_{xRy} \sim\exists v_0(Fv_0 = \overline{y}) \qquad (*)$$

a propositional combination of Σ_1-sentences. Thus, Solovay's Second Completeness Theorem can be given a more refined statement as follows: If $\mathrm{PRL}^{\omega} \nvdash A$, there is an interpretation * mapping atoms to propositional combinations of Σ_1-sentences and such that A^* is false. From this statement, we get the added information that the sentence ϕ constructed in Example 2.4 is not *too* complex; it is a propositional combination of Σ_1-sentences.

Hold on! In the Kripke model of Example 2.4, p is only forced at the terminal node 2, where (*) has the especially simple form,

$$L = \overline{2}:\quad \exists v_0(Fv_0 = \overline{2}).$$

The sentence ϕ constructed is, thus, a Σ_1-sentence and we can actually conclude the following:

2.5. ROSSER'S THEOREM. There is a Σ_1-sentence ϕ such that

i. $\mathrm{PRA} \nvdash \phi$

ii. $\mathrm{PRA} \nvdash \sim\phi$

iii. $\mathrm{PRA} \vdash Con \rightarrow \sim Pr(\ulcorner\phi\urcorner)$

iv. $\mathrm{PRA} \vdash Con \rightarrow \sim Pr(\ulcorner\sim\phi\urcorner)$.

We shall see in Chapter 6 that Rosser's Theorem has a much more elementary proof than this. That is not the point, however; the point is that Solovay's Second Completeness Theorem (or, rather: the construction behind it) is a powerful tool in obtaining refined incompleteness results: It is generally very easy to give a Kripke model illustrating the type of incompleteness phenomenon desired. With Solovay's construction, we can then conclude immediately the existence of sentences instantiating the given phenomenon. Exercises 2 and 3, below, offer a few further examples. To save the reader some small amount of labour in these exercises, let me rephrase the Second Completeness Theorem in a more applicable form.

First we need a definition:

2.6. DEFINITION. Let $\underline{K} = (K, <, \alpha_0, \Vdash)$ be a Kripke model. An atom p is *persistent relative to* $\underline{K}$ if, for all $\alpha, \beta \in K$, if $\alpha \Vdash p$ and $\alpha < \beta$ then $\beta \Vdash p$. In other words,

p is persistent relative to $\underline{K}$ if $\{a: \ a \Vdash p\}$ is upwards closed.

2.7. SOLOVAY'S SECOND COMPLETENESS THEOREM REFINED. Let A be a modal sentence and $\underline{K}$ a finite A-sound model of A. Then: There is an arithmetic interpretation * for which A^* is true. The interpretation p^* of an atom p in A is a propositional combination of Σ_1-sentences, unless the atom is persistent relative to $\underline{K}$, in which case the interpretation p^* is a Σ_1-sentence.

The point here is that, if p is persistent relative to $\underline{K} = (\{1,\ldots,n\},R,1,\Vdash)$, then

$$\bigvee\{L = \overline{x}: \ x \Vdash p\} \leftrightarrow \bigvee\{\exists v_0(\overline{F}v_0 = \overline{x}): \ x \Vdash p\}. \qquad (*)$$

For,

$$\text{PRA} \vdash L = \overline{x} \to \exists v_0(\overline{F}v_0 = \overline{x}),$$

whence the left disjunction implies the right one. But, since F is R-increasing,

$$\text{PRA} \vdash \exists v_0(\overline{F}v_0 = \overline{x}) \to L = \overline{x} \vee \bigvee_{xRy} L = \overline{y},$$

as follows quickly from Lemma 1.5.ii. Thus, each disjunct on the right of (*) implies a sub-disjunction of the left disjunction and we conclude the right-to-left implication.

EXERCISES

1. Prove the parenthetical assertion of Example 2.4: For any modal sentence A,

   $$\text{PRL} \nvdash \sim\Box f \to \sim\Box A \wedge \sim\Box\sim A.$$

2. Apply the refined form of Solovay's Second Completeness Theorem to prove the following:

   i. (First Incompleteness Theorem). There is a Σ_1-sentence ϕ such that

      a. $\text{PRA} \nvdash \phi, \sim\phi$

      b. $\text{PRA} \nvdash Con \to \sim Pr(\ulcorner\phi\urcorner)$

      c. $\text{PRA} \vdash Con \to \sim Pr(\ulcorner\sim\phi\urcorner)$

   ii. (Mostowski's Theorem). There are Σ_1-sentences ϕ,ψ such that ϕ,ψ are independent over PRA, i.e.

      a. $\text{PRA} + \phi \nvdash \psi, \sim\psi$

      b. $\text{PRA} + \sim\phi \nvdash \psi, \sim\psi$.

   iii. There are Π_1-sentences ϕ,ψ such that

      a. $\text{PRA} \vdash \phi, \psi$

      b. $\text{PRA} \vdash \phi \vee \psi$.

3. Show that there are Σ_1-sentences ϕ,ψ such that all of the following are true:

   i. $\sim\phi, \sim\psi$

   ii. $Con(\mathsf{PRA} + \sim\phi + \psi + Con(\mathsf{PRA} + \phi + \psi))$

   iii. $Con(\mathsf{PRA} + \phi + \sim\psi + Con(\mathsf{PRA} + \phi + \psi))$,

   where $Con(\mathsf{T})$ is the assertion that T is consistent.

(The interest in Exercise 3 is purely personal: In my dissertation back in 1973 I gave a hierarchical refinement of a result of de Jongh's similar in spirit to Solovay's First Completeness Theorem-- but relating the intuitionistic propositional calculus to intuitionistic arithmetic. The result I proved is that, for any propositional formula $A(p_1,\ldots,p_n)$, if A is not an intuitionistic tautology, there are Σ_1-sentences $\phi_1,\ldots,\phi_n$ such that $A(\phi_1,\ldots,\phi_n)$ is not provable in intuitionistic arithmetic. The problem of uniformity, i.e. of finding $\phi_1,\ldots,\phi_n$ independently of $A(p_1,\ldots,p_n)$ arose. For $n = 1$, de Jongh and I both proved the result fairly easily. The case for $n = 2$ reduces to the problem of Exercise 3 (but with PRA replaced by PA-- not a serious obstacle). Eventually, Daniel Leivant solved the uniformity problem; Exercise 3 shows (the expert) that Solovay's Second Completeness Theorem does also.)

As long as we are on the topic of complexity:

4. i. Show: There is no Σ_1-sentence ϕ such that $\phi + Con(\mathsf{PRA} + \sim\phi)$ is consistent. However, there is a Π_1-sentence ψ for which $\psi + Con(\mathsf{PRA} + \sim\psi)$ is consistent. Conclude that there is a Σ_1-sentence ϕ, but no such Π_1-sentence, for which $\sim\phi + Con(\mathsf{PRA} + \phi)$ is consistent.

   ii. Show: The sentences of Exercise 2.iii cannot be taken to be Σ_1.

(By Exercise 4, which can be proven arithmetically or modally (Hint: Assume $\boxed{s}\,(p \to \Box p)$.), the propositional combinations of Σ_1-sentences produced by the proofs of Solovay's First and Second Completeness Theorems are best possible-- at least with respect to the obvious, if crude, measure. In particular, a requirement like persistence is necessary for such improvements in refinements like Theorem 2.7.)

3. GENERALISATIONS, REFINEMENTS, AND ANALOGUES

Theorem 2.7 is a nice hierarchical refinement of Solovay's Second Completeness

Theorem, which itself is something of a footnote to the First Completeness Theorem. With the First Completeness Theorem we have achieved the second major goal of our study of PRL-- namely the doctrinaire vindication of our study of this modal system; with the Second Theorem we have at once surpassed this goal and returned almost to our original goal of using PRL to study self-reference-- I say "almost" because we have actually made applications to the *product* of a typical metamathematical application of self-reference, namely: incompleteness. It is time to look for new goals. This will be done in Parts II and III of this monograph. For now, we merely make a few additional footnotes to Solovay's Completeness Theorems.

The most pedestrian generalisation of Solovay's Completeness Theorems simply replaces PRA by reasonable *RE* extensions $\mathrm{T} \supseteq \mathrm{PRA}$.

3.1. DEFINITION. Let $\mathrm{T} \supseteq \mathrm{PRA}$ be an *RE* theory. A mapping $p \mapsto p^*$ of atoms to arithmetic sentences extends to an *arithmetic interpretation* * *based on* T as follows:

$$t^* \text{ is } \overline{0} = \overline{0}; \qquad f^* \text{ is } \overline{0} = \overline{1}; \qquad (\sim A)^* = \sim(A^*); \qquad (\Box A)^* = Pr_T(\ulcorner A^* \urcorner)$$

$$(A \circ B)^* = A^* \circ B^* \text{ for } \circ \in \{ \wedge, \vee, \rightarrow \},$$

where $Pr_T(\cdot)$ is the provability predicate for T.

3.2. THEOREM. Let $\mathrm{T} \supseteq \mathrm{PRA}$ be an *RE* theory.

i. Let T be Σ_1-sound. For any modal sentence A, the following are equivalent:

a. $\mathrm{PRL} \vdash A$

b. $\forall *$ based on T $(\mathrm{PRA} \vdash A^*)$

c. $\forall *$ based on T $(\mathrm{T} \vdash A^*)$.

ii. Let T be sound. For any modal sentence A, the following are equivalent:

a. $\mathrm{PRL} \vdash \bigwedge_{\Box B \in S} (\Box B \rightarrow B) \rightarrow A$

b. $\mathrm{PRL}^{\omega} \vdash A$

c. $\forall *$ based on T $(A^*$ is true$)$,

where S is the set of subformulae of A.

The proofs of these assertions are straightforward modifications of those of Solovay's First and Second Completeness Theorems. In particular, i.a $\Rightarrow$ i.b. $\Rightarrow$

i.c are straightforward, as are ii.a $\Rightarrow$ ii.b $\Rightarrow$ ii.c. The implication i.c $\Rightarrow$ i.a requires Σ_1-soundness in order to recognise that the limit L of the function F constructed in the proof is 0. The implication ii.c $\Rightarrow$ ii.a also requires only Σ_1-soundness, full soundness in this half of the Theorem being required for the implication ii.b $\Rightarrow$ ii.c.

Having said all of this, let me discount it! Σ_1-soundness is too strong a condition for the proofs of i.c $\Rightarrow$ i.a and ii.c $\Rightarrow$ ii.a. Using Lemma 1.10, we can reduce the safety assumption. To this end, recall from Definitions 2.3.6 the notion of the ordinal, $o(\alpha)$, of a node α in a Kripke model:

$$o(\alpha) = sup\{o(\beta) + 1: \alpha < \beta\},$$

so that $o(\alpha) = 0$ for terminal α. If $\underline{K} = (\{1,\ldots,n\},R,1,\Vdash)$ is the finite model used in the proof of Solovay's First Completeness Theorem and $o(1) = k$, then $x \Vdash \Box^{k+1}f$ for all $1 \le x \le n$. By Lemma 1.10, if L is once again the limit of the function F,

$$\mathrm{PRA} \vdash L = \overline{x} \rightarrow (\Box^{k+1}f)^*,$$

for each $x > 0$, whence

$$\mathrm{PRA} \vdash \sim(\Box^{k+1}f)^* \rightarrow L = \overline{0}.$$

Hence, we do not need full Σ_1-soundness to conclude $L = \overline{0}$, but merely the truth of $\sim(\Box^{k+1}f)^*$, i.e. $\mathrm{T} \nvdash Pr_T^k(\ulcorner\overline{0} = \overline{1}\urcorner)$ (using the obvious abbreviation).

These latest observations yield Albert Visser's refinement of Solovay's First Completeness Theorem:

<u>3.3. THEOREM</u>. Let $\mathrm{T} \supseteq \mathrm{PRA}$ be a consistent RE theory.

i. Assume, for all n, $\mathrm{T} \nvdash Pr_T^n(\ulcorner\overline{0} = \overline{1}\urcorner)$. Then, for any modal sentence A, the following are equivalent:

a. $\mathrm{PRL} \vdash A$

b. $\forall *$ based on T $(\mathrm{PRA} \vdash A^*)$

c. $\forall *$ based on T $(\mathrm{T} \vdash A^*)$

ii. Suppose $\mathrm{T} \vdash Pr_T^k(\ulcorner\overline{0} = \overline{1}\urcorner)$ for some $k > 0$ and let n be minimum such. For any modal sentence A, the following are equivalent:

a. $\mathrm{PRL} + \Box^n f \vdash A$

b. $\forall *$ based on T (T$\vdash A^*$)

iii. Under the assumption of ii, for any modal sentence A, the following are equivalent:

a. PRL + $\Box^{n+1}f \vdash A$

b. $\forall *$ based on T (PRA$\vdash A^*$).

The proof of part i has, essentially, been given. The proofs of parts ii and iii are left to the Exercises at the end of this section.

Before moving on to another topic, I should mention two related items. First, Visser has considered the more general problem of determining the schemata in $\Box$ provably satisfied in an arbitrary consistent theory U $\supseteq$ PRA under all interpretations * based on an *RE* theory T $\supseteq$ PRA, i.e. he has considered the problem of determining what a theory U says about $Pr_T(\cdot)$. He has made some progress in this direction and has discovered some interesting by-ways leading off therefrom. Second, Timothy Carlson and the author have considered the problem of determining the schemata in $\Box$ and a new modal operator that are variously valid under interpretations based on two (or more) provability predicates. This work is reported on in the next chapter.

Another refinement of Solovay's First Completeness Theorem-- a rather more popular one-- is its uniformisation. As I noted in Remark 1.6, this result seems to require an instance of induction not available in PRA. For this reason, we assume full induction, i.e. we state it for extensions of PA.

3.4. THEOREM. Let T $\supseteq$ PA be a Σ_1-sound *RE* theory. There is an interpretation * based on T such that, for any modal sentence A,

PRL$\vdash A$ iff PA$\vdash A^*$

iff T$\vdash A^*$.

There are two proofs of this, both uniformisations of the original. The version of Franco Montagna and Albert Visser proceeds by formalising the proof of the First Completeness Theorem and diagonalising anew to obtain the uniformisation; the version of S.N. Artyomov, Arnon Avron, and George Boolos proceeds by constructing a master Kripke model from all the finite ones and then applying Solovay's construction to this infinite structure. Although I do not intend to present a detailed proof of

Theorem 3.4, I should offer a few remarks thereon. For the sake of definiteness, I shall sketch the second proof.

Sketch of a proof of Theorem 3.4: Let $\underline{K}_1, \underline{K}_2, \ldots$ be an enumeration of finite Kripke models, say

$$\underline{K}_i = (\{\langle i,1\rangle, \ldots, \langle i,n_i\rangle\}, R_i, \langle i,1\rangle, \Vdash_i).$$

Assume everything in sight is primitive recursive and define F by

$$F0 = 0$$

$$F(x+1) = \begin{cases} \langle i,y\rangle, & Fx = 0 \ \& \ Prov(x+1, \ulcorner L \neq \langle \overline{i},\overline{y}\rangle \urcorner) \\ \langle i,y\rangle, & Fx \in K_i \ \& \ Prov(x+1, \ulcorner L \neq \langle \overline{i},\overline{y}\rangle \urcorner) \ \& \ Fx\, R_i \langle i,y\rangle \\ Fx, & \text{otherwise,} \end{cases}$$

where, as before, L is the limit of F.

Since there are infinitely many nodes $\langle i,x\rangle$, one cannot choose $p^* = \bigvee\{L = \langle \overline{i},\overline{x}\rangle : \ \langle i,x\rangle \Vdash_i p\}$; instead, one first defines

$$P = \{\langle i,x\rangle : \ \langle i,x\rangle \Vdash_i p\}.$$

The enumeration $\underline{K}_1, \underline{K}_2, \ldots$ can be made so that P is primitive recursive. One defines

$$p^*: \quad \exists v(v \in P \wedge L = v)$$

(intuitively: $L \in P$). After showing L exists (cf. Remark 3.5.ii, below), one notes that, for each i,

$$\mathsf{PA} \vdash L \in K_i \rightarrow .p^* \leftrightarrow \bigvee\{L = \langle \overline{i},\overline{x}\rangle : \ \langle i,x\rangle \Vdash_i p\}.$$

With this, the basic lemmas of the First Completeness Theorem go through and, if $\mathsf{PRL} \nvdash A$, there is some i such that $\langle i,1\rangle \nVdash_i A$, whence

$$\mathsf{PA} \vdash L = \langle \overline{i},\overline{1}\rangle \rightarrow \sim A^*.$$

From the consistency of $\mathsf{PA} + L = \langle \overline{i},\overline{1}\rangle$, it follows that $\mathsf{PA} \nvdash A^*$. QED

Now, some comments:

3.5. REMARKS. i. Theorem 3.4 cannot be proven under the weaker safety assumptions of Theorem 3.3.i. Assuming a greater uniformity than exhibited in the proof sketch, one has

$$\mathsf{PA} \vdash L > \overline{0} \rightarrow \exists v Pr_T^v(\ulcorner \overline{0} = \overline{1} \urcorner),$$

$$\mathsf{PA} \vdash L = \overline{0} \leftrightarrow \forall v \sim Pr_T^v(\ulcorner \overline{0} = \overline{1} \urcorner),$$

where $Pr_T^v(\cdot)$ indicates the v-fold application of $Pr_T(\cdot)$. Thus, one needs at least the uniform assertion of the conditions of Theorem 3.3.i.

ii. Once again, I emphasise that this proof needs more induction than PRA provides to prove that L exists. Because the model is infinite, the induction must be formalised and, again because the model is infinite, we cannot simplify the instance of induction from Σ_2 to boolean-Σ_1. This induction is the only really novel part of the proof and, perhaps, I should say more about it: Two possibilities occur. Either $\forall v(Fv = 0)$, i.e. $L = 0$, or F gets into some K_i. Now, i is a variable, not a fixed number, whence we must rely on formalisation. But, by assumption, we can get n_i primitive recursively and show by induction on $n_i - v$ that, if F reaches a node in K_i of ordinal v then L exists. I leave the details to the reader.

iii. The interpretation p^* is more complex than the boolean combination of Σ_1-sentences available for the non-uniform result; it is Σ_2. This can be simplified by noting that $(\sim p)^*$ can also be chosen Σ_2, whence p^* is Δ_2. This is the simplest possible: p^* cannot be chosen to be a boolean combination of Σ_1-sentences. (Cf. Exercise 3, below.)

iv. Solovay's Second Completeness Theorem cannot be uniformised: There is, for example, no single interpretation * under which both p^* and $(\sim p)^*$ are false. Nonetheless, some uniformisation is possible-- cf. Exercises 4, 5, below.

In addition to generalisations and refinements of Solovay's Completeness Theorems, there are analogues, i.e. completeness theorems with respect to other interpretations. Of these, the most mundane is a hierarchical relativisation.

3.6. DEFINITIONS. Let T be a consistent *RE* extension of Peano Arithmetic, PA, and let $n \geq 1$. Let $Tr_{\Sigma_n}(\cdot)$ be a Σ_n-truth definition for Σ_n-sentences. Define

$$Pr_{T,n}(v_0): \quad \exists v_1\big(Tr_{\Sigma_n}(v_1) \wedge Pr_T(v_1 \dot{\rightarrow} v_0)\big).$$

An assignment of arithmetic sentences p^* to atoms p extends to an *interpretation* * *based on* T,n as follows:

t^* is $\overline{0} = \overline{0}$; $\quad f^*$ is $\overline{0} = \overline{1}$; $\quad (\sim A)^* = (A^*)$

$(A \circ B)^* = A^* \circ B^*$ for $\circ \in \{\wedge, \vee, \rightarrow\}$

and $\quad (\Box A)^* = Pr_{T,n}(\ulcorner A^* \urcorner)$.

The current replacement of PRA by PA is not nearly so subtle as that in Theorem 3.4. Here we will need Σ_n-induction both for the definition of F and the proof that F is total: Assuming Σ_n-induction, we can show that the defining clauses of F are Σ_n, whence F is Σ_n and another Σ_n-induction will be needed to show F total.

3.7. LEMMA. Let $n \geq 1$ be given and let T be a consistent *RE* extension of PA. Further, let X be a set of Σ_n-sentences such that T + X is consistent. Then: For any sentences ϕ, ψ,

i. $T + X \vdash \phi \;\Rightarrow\; PA + X \vdash Pr_{T,n}(\ulcorner\phi\urcorner)$

ii. $PA \vdash Pr_{T,n}(\ulcorner\phi\urcorner) \wedge Pr_{T,n}(\ulcorner\phi \to \psi\urcorner) \to Pr_{T,n}(\ulcorner\psi\urcorner)$

iii. $PA \vdash Pr_{T,n}(\ulcorner\phi\urcorner) \to Pr_{T,n}(\ulcorner Pr_{T,n}(\ulcorner\phi\urcorner)\urcorner)$.

Proof: i. If $T + X \vdash \phi$, then, for some finite $X_0 \subseteq X$, $T + X_0 \vdash \phi$, whence $T \vdash \bigwedge X_0 \to \phi$. Thus

$$PA \vdash Pr_T(\ulcorner \bigwedge X_0 \to \phi\urcorner).$$

But also

$$PA \vdash \bigwedge X_0 \to Tr_{\Sigma_n}(\ulcorner \bigwedge X_0\urcorner),$$

whence

$$PA + X \vdash Tr_{\Sigma_n}(\ulcorner \bigwedge X_0\urcorner) \wedge Pr_T(\ulcorner \bigwedge X_0 \to \phi\urcorner)$$
$$\vdash Pr_{T,n}(\ulcorner\phi\urcorner).$$

ii. Formalise the following: Suppose $\chi_0, \chi_1 \in \Sigma_n$ satisfy

$$Tr_{\Sigma_n}(\ulcorner\chi_0\urcorner) \wedge Pr_T(\ulcorner\chi_0 \to \phi\urcorner)$$
$$Tr_{\Sigma_n}(\ulcorner\chi_1\urcorner) \wedge Pr_T(\ulcorner\chi_1 \to .\phi \to \psi\urcorner).$$

Then

$$Tr_{\Sigma_n}(\ulcorner\chi_0 \wedge \chi_1\urcorner) \wedge Pr_T(\ulcorner\chi_0 \wedge \chi_1 \to \psi\urcorner).$$

Thus

$$Pr_{T,n}(\ulcorner\phi\urcorner) \wedge Pr_{T,n}(\ulcorner\phi \to \psi\urcorner) \to Pr_{T,n}(\ulcorner\psi\urcorner).$$

iii. Observe

$$\begin{aligned} PA \vdash Pr_{T,n}(\ulcorner\phi\urcorner) &\to \exists \ulcorner\psi\urcorner \big(Tr_{\Sigma_n}(\ulcorner\psi\urcorner) \wedge Pr_T(\ulcorner\psi \to \phi\urcorner)\big) \\ &\to \exists \ulcorner\psi\urcorner \big(Tr_{\Sigma_n}(\ulcorner\psi\urcorner) \wedge Pr_T(\ulcorner Pr_T(\ulcorner\psi \to \phi\urcorner)\urcorner)\big) \\ &\to \exists \ulcorner\psi\urcorner \big(Tr_{\Sigma_n}(\ulcorner\psi\urcorner) \wedge Pr_T(\ulcorner\psi \to Pr_T(\ulcorner\psi \to \phi\urcorner)\urcorner)\big) \\ &\to \exists \ulcorner\psi\urcorner \big(Tr_{\Sigma_n}(\ulcorner\psi\urcorner) \wedge Pr_T(\ulcorner\psi \to Tr_{\Sigma_n}(\ulcorner\psi\urcorner) \wedge Pr_T(\ulcorner\psi \to \phi\urcorner)\urcorner)\big) \\ &\to \exists \ulcorner\psi\urcorner \big(Tr_{\Sigma_n}(\ulcorner\psi\urcorner) \wedge Pr_T(\ulcorner\psi \to Pr_{T,n}(\ulcorner\phi\urcorner)\urcorner)\big) \\ &\to Pr_{T,n}(\ulcorner Pr_{T,n}(\ulcorner\phi\urcorner)\urcorner). \end{aligned}$$

QED

3.8. COROLLARY. Let $n \geq 1$ and T be a consistent *RE* extension of PA. For any sentence ϕ,

$$PA \vdash Pr_{T,n}(\ulcorner Pr_{T,n}(\ulcorner\phi\urcorner) \to \phi\urcorner) \to Pr_{T,n}(\ulcorner\phi\urcorner).$$

Proof: Simply repeat the proof of the Formalised Löb's Theorem given in Chapter 1 (specifically, the proof of Theorem 1.2.5); for, that proof used only the Derivability Conditions (available for $Pr_{T,n}$ by 3.7) and Diagonalisation. QED

3.9. THEOREM. Let $n \geq 1$ and let T be Σ_n-sound. Let X be any set of true Σ_n-sentences. For any modal sentence A, the following are equivalent:

i. $\mathsf{PRL} \vdash A$

ii. $\forall *$ based on T, n $(\mathsf{PA} \vdash A^*)$

iii. $\forall *$ based on T, n $(\mathsf{T} + X \vdash A^*)$.

Proof sketch: The implications i $\Rightarrow$ ii $\Rightarrow$ iii are routine.

To show iii $\Rightarrow$ i it suffices to prove the result when X is as large as possible, i.e. X is the set of all true Σ_n-sentences. As before, one proves the implication contrapositively by starting with a finite Kripke countermodel $\underline{K} = (\{1,\ldots,k\},R,1,\Vdash)$ to A and defining a function F growing through $\{1,\ldots,k\}$.

The definition of F is just slightly different from what one would expect: After defining $0Ry$ for $1 \leq y \leq k$, one sets

$$F0 = 0$$

$$F(x+1) = \begin{cases} y, & \exists \ulcorner\psi\urcorner z < x\big(Tr_{\Pi_{n-1}}(\ulcorner\psi\dot{z}\urcorner) \wedge Prov_T(x, \ulcorner\exists v\psi v \to L \neq \overline{y}\urcorner)\big) \wedge FxRy \\ Fx, & \text{otherwise,} \end{cases}$$

where, as before, L is the limit of F. So defined, F is recursive in Π_{n-1}, whence it is Σ_n. Further, Σ_n-induction shows F to be total. (A quick remark: The bound $\ulcorner\psi\urcorner < x$ comes for free as the code of a derivation is larger than that of any formula or subformula of a formula appearing in it. Bounding the witness z by x is necessary for the complexity calculation. Its effect on F is merely one of slowing its possible progress through $\{1,\ldots,k\}$. If x proves $L \neq \overline{y}$ from some true Σ_n-sentence, F must still wait until a witness to the truth of the Σ_n-sentence has been provided before it can move to y.)

As before, one defines

$$p^* = \bigvee\{L = \overline{x}:\ x \Vdash p\}$$

and proves, for all sentences B,

i. $x \Vdash B \ \Rightarrow\ \mathsf{T} + true\ \Sigma_n \vdash L = \overline{x} \to B^*$

ii. $x \Vdash\!\!\!/\ B \Rightarrow \mathsf{T} + true\ \Sigma_n \vdash L = \bar{x} \to \sim B^*$,

that $L = 0$ and $\mathsf{T} + true\ \Sigma_n + L = \bar{x}$ is consistent for $1 \le x \le k$. The only non-routine part of this is the proof that $L = 0$, which largely amounts to showing T, together with all true Σ_n-sentences, is Σ_n-sound.

Toward showing $L = 0$, let $\phi, \psi \in \Sigma_n$ be such that ϕ is true and $\mathsf{T} \vdash \phi \to \psi$. Write $\phi = \exists v \phi' v$, where $\phi' \in \Pi_{n-1}$ and observe

$$\mathsf{T} \vdash \exists v \phi' v \to \psi$$

$$\vdash \forall v(\phi' v \to \psi),$$

a Π_{n+1}-sentence. I claim that this implication is true. To see this, let x be arbitrary and observe that $\phi' \bar{x} \to \psi$ is a Σ_n-consequence of T-- hence true by the assumption of the Σ_n-soundness of T. Thus $\forall v(\phi' v \to \psi)$, i.e. $\phi \to \psi$ is true. Since ϕ is true, so is ψ. Thus, any Σ_n-consequence of $\mathsf{T} + true\ \Sigma_n$ is true.

Now, suppose $L = y > 0$. Then $\exists v(Fv = \bar{y})$ is a true Σ_n-sentence. But, if $Fx = y$, we have $Prov_T(x, \ulcorner \psi \to L \ne \bar{y} \urcorner)$ for some true $\psi \in \Sigma_n$. By the argument of the last paragraph, since

$$L \ne \bar{y} \leftrightarrow \exists v(\bar{y} R Fv)$$

is Σ_n, it is true. Thus, we cannot have $L = \bar{y} > 0$ and $L = 0$.

As I said, once we know $L = 0$, the appropriate argument shows $L = \bar{1}$ is consistent with $\mathsf{T} + true\ \Sigma_n$, whence $\mathsf{T} + true\ \Sigma_n$ is consistent with $\sim A^*$, which follows from $L = \bar{1}$.

QED

I leave to the reader the task of formulating the appropriate analogue to Solovay's Second Completeness Theorem.

(Remark: One weakness of Gödel's original work was his introduction of the semantic notion of ω-consistency. I find this notion to be pointless, but I admit many proof theorists take it seriously. Roughly speaking, it is the sentence

$$\omega\text{-}Con_T\text{:}\quad \forall \ulcorner \phi v \urcorner \left(Pr_T(\ulcorner \exists v \phi v \urcorner) \to \sim \forall v Pr_T(\ulcorner \sim \phi \dot{v} \urcorner)\right).$$

I once showed

$$\mathrm{PA} \vdash \omega\text{-}Con_T \leftrightarrow RFN_{\Pi_3}(\mathsf{T} + RFN(\mathsf{T})),$$

for any RE $\mathsf{T} \supseteq \mathrm{PA}$, where $RFN(\mathsf{T})$ denotes the full *Uniform Reflexion Schema*,

$$\forall v\left(Pr_T(\ulcorner \phi \dot{v} \urcorner) \to \phi v\right),$$

and $RFN_{\Pi_3}(T)$ the restriction of this schema to $\phi \in \Pi_3$. Boolos defined

$$\omega\text{-}Pr_T(\ulcorner\phi\urcorner): \ \sim\omega\text{-}Con_{T+\sim\phi}$$

and proved the analogue to Solovay's Completeness Theorem for $\omega\text{-}Pr_T(\cdot)$: PRL axiomatises the schemata of $\omega\text{-}Pr_T(\cdot)$ provable in PA. Using my characterisation of $\omega\text{-}Con_T$, this result reduces to Theorem 3.9 by applying 3.9 to $T + RFN(T)$ and letting $n = 2$. (In proof theoretic matters, Π_{n+1} often matches up with Σ_n.) I leave the details of the reduction as an exercise for the interested reader.)

Let's finish this section with the statements of a few additional analogues to the First Completeness Theorem. These results are for set theory, ZFC, and are due to Solovay; indeed, he mentioned them in his original paper on the Completeness Theorems. Each of these results is more semantic than syntactic, but this should not deter us: By the completeness theorem for the predicate calculus, provability is equivalent to truth in all models, whence the result 3.2 is already a semantic one.

3.10. DEFINITIONS. A model of ZFC consists of a pair (a,E), where a is a "universe of sets" and E a "membership relation" on a. We write ZFC$\models \phi$ if ϕ is valid in all models of ZFC. (Recall: ZFC$\models \phi$ iff ZFC$\vdash \phi$). We have three variants of this to consider:

i. A model (a,E) of ZFC is an *ω-model* if the natural numbers in the model are all standard. We write ZFC$\models_\omega \phi$ iff ϕ is true in all ω-models of ZFC.

ii. A model (a,E) of ZFC is a *transitive model* if a is a transitive set ($x \in y \in a \Rightarrow x \in a$) and E is the restriction to a of the actual membership relation $\in$. We write ZFC$\models_t \phi$ iff ϕ is true in all transitive models of ZFC.

iii. A transitive model $(a, \in)$ of ZFC is an *inaccessible-standard model* if a is of the form V_κ for an inaccessible cardinal κ. We write ZFC$\models_{in} \phi$ iff ϕ is true in all inaccessible-standard models of ZFC.

Each of these notions of validity is expressible in the language of set theory and thus gives rise to a class of interpretations of the modal language and a corresponding analogue to Solovay's Completeness Theorem.

3.11. DEFINITIONS. We define several types of interpretations * according to the interpretation of $\Box$:

i. $*$ is an *ω-interpretation* if one always has

$$(\Box A)^* = \text{"ZFC}\models_{\omega} A^*\text{"}$$

ii. $*$ is a *t-interpretation* if one always has

$$(\Box A)^* = \text{"ZFC}\models_{t} A^*\text{"}$$

iii. $*$ is an *in-interpretation* if one always has

$$(\Box A)^* = \text{"ZFC}\models_{in} A^*\text{"}.$$

3.12. DEFINITIONS. i. $\Diamond A = \sim\Box\sim A$

ii. The following is a sort of comparability schema:

$$Com:\quad \Box(A \rightarrow \Diamond B) \vee \Box(B \rightarrow \Diamond A) \vee \Box(\Diamond A \leftrightarrow \Diamond B)$$

iii. The following linearity schema is from Chapter 2, Exercise 2.6:

$$Lin:\quad \Box(\Box A \rightarrow B) \vee \Box(\boxed{s}\, B \rightarrow A).$$

With all these definitions and notations, we can now state Solovay's results:

3.13. THEOREM. For any modal sentence A,

i. $\mathsf{PRL} \vdash A$ iff $\forall$ ω-interpretations $*$ $(\mathsf{ZFC}\models_{\omega} A^*)$

ii. $\mathsf{PRL} + Com \vdash A$ iff $\forall$ t-interpretations $*$ $(\mathsf{ZFC}\models_{t} A^*)$

iii. $\mathsf{PRL} + Lin \vdash A$ iff $\forall$ in-interpretations $*$ $(\mathsf{ZFC}\models_{in} A^*)$.

EXERCISES

1. Prove parts ii and iii of Theorem 3.3.

2. This Exercise verifies that Theorem 3.3.i offers a genuine improvement over Theorem 3.2. (Cf. also Exercise 5.)

i. Let T be a consistent RE extension of PRA. Show, for any Π_1-sentence π, $\mathsf{T}\vdash \pi \Rightarrow \mathsf{PRA} + Con_T \vdash \pi$.

ii. Let $RFN_{\Sigma_1}(\mathsf{T})$ be the sentence,

$$RFN_{\Sigma_1}(\mathsf{T}):\quad \forall \ulcorner\phi\urcorner \in \Sigma_1\big(Pr_T(\ulcorner\phi\urcorner) \rightarrow Tr_{\Sigma_1}(\ulcorner\phi\urcorner)\big).$$

Show: For all n,

$$\mathsf{PRA} + RFN_{\Sigma_1}(\mathsf{T}) \vdash \sim Pr_T^n(\ulcorner \overline{0} = \overline{1}\urcorner).$$

iii. Let $\phi = Pr_{PRA}(\ulcorner \sim RFN_{\Sigma_1}(\mathsf{PRA})\urcorner)$. Show:

a. $\mathsf{PRA} + \phi$ is not Σ_1-sound

b. $\mathsf{PRA} + \phi \nvdash Pr^{n}_{PRA}(\ulcorner \overline{0} = \overline{1} \urcorner)$ for any n.

iv. Show by induction on n:

$$\mathsf{PRA} + Pr_{PRA}(\ulcorner \psi \urcorner) \vdash Pr^{n}_{PRA+Pr(\ulcorner \psi \urcorner)}(\ulcorner \chi \urcorner)$$

$$\Rightarrow \quad \mathsf{PRA} + Pr_{PRA}(\ulcorner \psi \urcorner) \vdash Pr^{n}_{PRA}(\ulcorner \chi \urcorner)$$

for any ψ, χ. Conclude: $\mathsf{PRA} + \phi \nvdash Pr^{n}_{PRA+\phi}(\ulcorner \overline{0} = \overline{1} \urcorner)$.

(Hints: i. use demonstrable Σ_1-soundness; iii.b. use contraposition, minimal n, part ii, and Gödel's Second Incompleteness Theorem.)

3. (Gaifman, Efron). This Exercise verifies Remark 3.5.iii that the uniform substitutions of Theorem 3.4 cannot be chosen to be boolean combinations of Σ_1-sentences.

i. a. Define $\Box_q A = \Box(q \to A)$. Let n, atoms $p_1, \ldots, p_{n+1}$, and p be given. Show by means of Kripke models that

$$\mathsf{PRL} \nvdash \bigwedge_{1 \le i \le n} \Box_{p_{i+1}}(\sim\Box_{p_i} p \wedge \sim\Box_{p_i} \sim p) \to . \Box_{p_{i+1}} p \vee \Box_{p_{i+1}} \sim p.$$

b. Do the same with p_i replaced by $\sim\Box^{i} f$.

c. Conclude there to exist a sentence ϕ and an infinite sequence $\mathsf{T}_1, \mathsf{T}_2, \ldots$ of RE extensions of PRA such that, for each n,

$$\mathsf{PRA} \nvdash \bigwedge_{1 \le i \le n} Pr_{T_{i+1}}(\ulcorner \sim Pr_{T_i}(\ulcorner \phi \urcorner) \wedge \sim Pr_{T_i}(\ulcorner \sim\phi \urcorner) \urcorner) \to . Pr_{T_{i+1}}(\ulcorner \phi \urcorner) \vee Pr_{T_{i+1}}(\ulcorner \sim\phi \urcorner).$$

ii. In the following, π, σ denote arbitrary Π_1- and Σ_1-sentences, respectively. Define a sentence ϕ to be *n-determined* if, for any sequence $\mathsf{T}_1, \mathsf{T}_2, \ldots$ of RE extensions of PRA

$$\mathsf{PRA} \vdash \bigwedge_{1 \le i \le n} Pr_{T_{i+1}}(\ulcorner \sim Pr_{T_i}(\ulcorner \phi \urcorner) \wedge \sim Pr_{T_i}(\ulcorner \sim\phi \urcorner) \urcorner) \to . Pr_{T_{i+1}}(\ulcorner \phi \urcorner) \vee Pr_{T_{i+1}}(\ulcorner \sim\phi \urcorner).$$

a. Show: σ is 1-determined. In fact, for any T,

$$\mathsf{PRA} \vdash \sim Pr_T(\ulcorner \sigma \urcorner) \to \sim\sigma.$$

b. Show: π is 1-determined. In fact, for any T,

$$\mathsf{PRA} \vdash \sim Pr_T(\ulcorner \sim\pi \urcorner) \to \pi.$$

c. Show: $\pi \vee \sigma$ is 2-determined.

d. Show that any boolean combination of Σ_1-sentences can be put in the form $\sigma_1 \wedge \pi_1 \vee \sigma_2 \wedge \pi_2 \vee \ldots \vee \sigma_k \wedge \pi_k$.

e. Show: If ϕ is n-determined, then $\phi \vee \sigma$ is $(n + 1)$-determined.

f. Show: If ϕ is n-determined, then $\phi \vee \sigma \wedge \pi$ is $(2n+3)$-determined.

Conclude: Any boolean combination of Σ_1-sentences is m-determined for some m.

(Hints: ii.e. Given $T_1,\ldots,T_{n+2}$, show $\bigwedge\!\!\bigwedge_{1\le i\le n+1} Pr_{T_{i+1}}(\ulcorner\sim\sigma\urcorner)$, whence

$\bigwedge\!\!\bigwedge_{1\le i\le n+1} Pr_{T_{i+1}}(\ulcorner\phi \vee \sigma_{.} \leftrightarrow \phi\urcorner)$. ii.f. My proof of this is horribly inelegant:

Let $T_1,\ldots,T_n,T_{n+1},\ldots,T_{2n+3}$ be given. From the hypothesis,

$$\bigwedge\!\!\bigwedge Pr_{T_{i+1}}(\ulcorner\sim Pr_{T_i}(\ulcorner\phi \vee \sigma\wedge \pi\urcorner)\urcorner),$$

one concludes

$$\bigwedge\!\!\bigwedge Pr_{T_{i+1}}(\ulcorner\sim Pr_{T_i}(\ulcorner\phi\urcorner)\urcorner).$$

From the other hypothesis, one concludes

$$\bigwedge\!\!\bigwedge Pr_{T_{i+1}}(\ulcorner\sim Pr_{T_i}(\ulcorner\sim\phi \wedge \sim\pi\urcorner)\urcorner),$$

whence

$$\sim\pi \to \bigwedge\!\!\bigwedge Pr_{T_{i+1}}(\ulcorner\sim Pr_{T_i}(\ulcorner\sim\phi\urcorner)\urcorner).$$

Thus, from T_{n+1} on,

$$\sim\pi \to Pr_{T_{i+1}}(\ulcorner\phi\urcorner) \vee Pr_{T_{i+1}}(\ulcorner\sim\phi\urcorner).$$

Thus, from T_{n+2} on,

$$Pr_{T_{n+i+1}}(\ulcorner\pi\urcorner)$$

and, for $T_{n+2},\ldots,T_{2n+3}$, $\phi \vee \pi\wedge\sigma$ can be replaced by $\phi \vee \sigma$ and one can appeal to ii.e.)

As noted in Remark 3.5.iv, Solovay's Second Completeness Theorem does not uniformise. Some infinite models can, however, be simulated and applications can be made. The next two Exercises explore this possibility.

4. Let a sequence $A_1,A_2,\ldots$ of sentences be given and define

$$S = \{B\colon\ B \text{ is a subformula of some } A_i\}.$$

Let $\underline{K}_0,\underline{K}_1,\ldots$ be a primitive recursive enumeration of finite Kripke models,

$$\underline{K}_i = (\{\langle i,1\rangle,\ldots,\langle i,n_i\rangle\}, R_i, \langle i,1\rangle, \Vdash_i).$$

Assume everything in sight is primitive recursive, i.e. $n_i = fi$ is a primitive recursive function and the relations

$$R(i,x,y)\colon\ x,y \in K_i \ \&\ x R_i y$$

$$For(x,i,\ulcorner p\urcorner)\colon\ x \in K_i \ \&\ x \Vdash_i p$$

are primitive recursive. (This latter need only be primitive recursive for each fixed p.)

i. Show: For each $B \in S$, the relation

$For(x, i, \ulcorner B \urcorner)$: $x \in K_i \ \& \ x \Vdash_i B$ is primitive recursive.

Let $\underline{K}^1$ be formed by affixing below all the $\underline{K}_i$'s a new node 1:

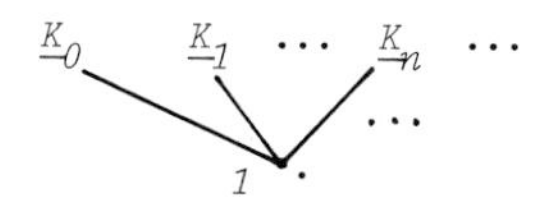

Let 1 force or fail to force atoms in some primitive recursive fashion so that

a. $1 \Vdash A_n$, for all n

b. $1 \Vdash \Box B \to B$, for all $\Box B \in S$.

Define $\underline{K} = (K, R, 0, \Vdash)$ by affixing 0 below 1:

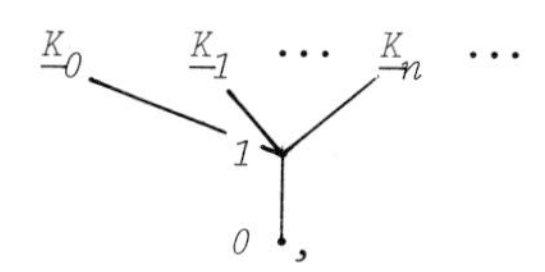

and having $0 \Vdash p$ iff $1 \Vdash p$. Finally, define F in terms of its limit L by the Recursion Theorem by:

$$F0 = 0$$

$$F(x+1) = \begin{cases} y, & FxRy \ \& \ Prov(x, \ulcorner L \neq \overline{y} \urcorner) \\ Fx, & \text{otherwise.} \end{cases}$$

ii. Prove in PA:

a. $\forall v_0 v_1 \big(Fv_0 = v_1 \to \forall v_2 > v_0 (Fv_2 = v_1 \vee v_1 R Fv_2)\big)$

b. $\exists v_1 \exists v_2 \forall v_0 > v_2 (Fv_0 = v_1)$, i.e. L exists

c. $\forall v_0 v_1 \big(L = v_0 \wedge v_0 R v_1 \to Con(\text{PA} + L = \dot{v}_1)\big)$

d. $\forall v_0 v_1 \big(L = v_0 \wedge v_0 \neq v_1 \wedge \sim v_0 R v_1 \to Pr(\ulcorner L \neq \dot{v}_1 \urcorner)\big)$

e. $\forall v_0 \big(L = v_0 \wedge v_0 > \overline{0} \to Pr(\ulcorner L \neq \dot{v}_0 \urcorner)\big)$.

iii. Show: $L = \overline{0}$ is true.

Define, for each variable p, a predicate P by:

Pv: $v \Vdash p$.

(Observe that P is primitive recursive.) Define an interpretation $*$ by

p^*: $\exists v \big(L = v \wedge Pv\big)$.

iv. For each $B \in S$, prove in PA:

a. $\forall v \in \bigcup K_i (v \Vdash B \wedge L = v \to B^*)$

b. $\forall v \in \bigcup K_i (v \nVdash B \wedge L = v \to \sim B^*)$.

This means the analogue to Lemma 1.10 is proven uniformly for the nodes of $\underline{K}_0, \underline{K}_1, \ldots$. To handle 1 and 0, we need an extra pair of assumptions: For all $\Box B \in S$, there is a number n_B such that

$$\mathrm{PA} \vdash \exists v \in \bigcup K_i (v \Vdash B) \to \exists v \in \bigcup_{i<n_B} K_i (v \Vdash B)$$

$$\mathrm{PA} \vdash \forall v \in \bigcup_{i<n_B} K_i (v \Vdash B) \to \forall v \in \bigcup K_i (v \Vdash B).$$

v. Show: For each $B \in S$, the assertions

$$1 \Vdash B, \quad 0 \Vdash B, \quad v \Vdash B \text{ (for } v \in \{0,1\} \cup \bigcup K_i)$$

are primitive recursive.

vi. For each $B \in S$, prove in PA,

a. $\overline{1} \Vdash B \wedge L = \overline{1} \to B^*$

b. $\overline{1} \nVdash B \wedge L = \overline{1} \to \sim B^*$.

vii. For each $B \in S$, show: $0 \Vdash B$ iff $1 \Vdash B$.

viii. For each $B \in S$, prove in PA:

a. $\overline{0} \Vdash B \wedge L = \overline{0} \to B^*$

b. $\overline{0} \nVdash B \wedge L = \overline{0} \to \sim B^*$.

ix. Conclude: $A_0^*, A_1^*, \ldots$ are all true.

x. Show: If p is provably upward persistent, then p^* is Σ_1.

(Hints: iv. & vi. By i and v, one has $\forall v \in \bigcup K_i (v \Vdash B \to Pr(\ulcorner \dot{v} \Vdash B \urcorner))$ and $\overline{1} \Vdash B \to Pr(\ulcorner \overline{1} \Vdash B \urcorner)$ for $B \in S$.)

5. Apply Exercise 4 to construct:

i. an infinite sequence of Σ_1-sentences $\phi_0, \phi_1, \ldots$ such that

$$\mathrm{PA} \vdash Con \to Con(\mathrm{PA} + \bigwedge_{i \in X} \phi_i + \bigwedge_{j \in Y} \sim \phi_j)$$

for any disjoint finite sets $X, Y \subseteq \omega$.

ii. a sentence $\phi \in \Sigma_1$ such that, for $\mathrm{T} = \mathrm{PA} + \phi$ and all n,

$\mathrm{T} \nvdash Pr_T^n(\ulcorner \overline{0} = \overline{1} \urcorner)$ and $\mathrm{PA} \nvdash \phi, \sim\phi$.

iii. a sentence ϕ such that, for all n, if $\mathrm{T}_n = \mathrm{PA} + \sim Pr_{PA}^n(\ulcorner \overline{0} = \overline{1} \urcorner)$,

$\mathrm{T}_{n+1} \vdash \sim Pr_{T_n}(\ulcorner \phi \urcorner) \wedge \sim Pr_{T_n}(\ulcorner \sim\phi \urcorner)$

iv. an infinite sequence of Σ_1-sentences $\phi_0,\phi_1,\ldots$ that are *very strongly independent*, where this last concept is defined as follows: First, for finite disjoint $X,Y \subseteq \omega$, define

$$Con^1_{X,Y}(\mathrm{T}): \quad Con(\mathrm{T} + \bigwedge_{i \in X} \phi_i + \bigwedge_{j \in Y} \sim\phi_j)$$

$$Con^{n+1}_{X,Y}(\mathrm{T}): \quad Con\Big(\mathrm{T} + \bigwedge_{i \in X} \phi_i + \bigwedge_{j \in Y} \sim\phi_j +$$

$$+ \bigwedge_{j_0 \in Y} Con^n_{X+,Y-}(\mathrm{T} + \bigwedge_{i \in X+} \phi_i + \bigwedge_{j \in Y-} \sim\phi_j)\Big),$$

where $X+ = X \cup \{j_0\}$, $Y- = Y - \{j_0\}$. $\phi_0,\phi_1,\ldots$ is very strongly independent if, for all finite X,Y and all appropriate n, $Con^n_{X,Y}(\mathrm{PA})$ is true.

(Remark: The condition cited between parts iv and v of Exercise 4, being a nuisance to verify, it is worth noting that it is only needed in handling *truth*. For mere consistency, i.e. underivability, like parts ii and iv of Exercise 5, one can dispense with the nodes *0*,*1*, hence this condition, and even simplify the proof of this special case of Exercise 4.)

Part II
Multi-Modal Logic and Self-Reference

Chapter 4
Bi-Modal Logics and Their Arithmetic Interpretations

If we have a metamathematical predicate other than provability, yet strong enough (if "strong" is the right word) to satisfy the axioms of PRL, then some of our preceding modal analysis carries over from $Pr(\cdot)$ to the predicate in question. It could happen that the analogue to Solovay's First Completeness Theorem holds, i.e. that PRL is the logic of the given predicate, or it could happen that additional axioms are required and one must find these and prove completeness. This last is important if we wish to obtain applications like those we made of Solovay's Second Completeness Theorem in Chapter 3, section 2, above. Even without this, however, we have some applications-- particularly, the explicit definability and uniqueness of the fixed points.

What if we have a metamathematical predicate of definite interest, but which is not sufficiently strong to yield analogues to all the Derivability Conditions? What can we do with such a predicate? The explicit definability and uniqueness theorems hold for the fixed points based on the weak predicate

$$\rho(\ulcorner\phi\urcorner) \;=\; Pr_{T_1}(\ulcorner\phi\urcorner) \vee Pr_{T_2}(\ulcorner\phi\urcorner),$$

for any consistent *RE* theories T_1, T_2 provably extending PRA. But, how do we prove these results where ρ obviously will *not* satisfy

$$\text{PRA} \vdash \rho(\ulcorner\phi\urcorner) \wedge \rho(\ulcorner\phi \to \psi\urcorner) \to \rho(\ulcorner\psi\urcorner),$$

(since the implication need not even be true), which axiom was certainly used in proving, say, the Substitution Lemmas which were central to the derivations of the results? The answer is simple: When $\rho(\cdot)$ is too weak to stand alone, we prop it up with $Pr(\cdot)$. Modally, this means we consider a modal logic with two operators-- the usual box and a new one to represent ρ.

Before outlining the contents of this chapter, let me give a pre-modal illustration of the interplay between $Pr(\cdot)$ and a new predicate $\rho(\cdot)$. To this end, let $\rho(v)$ be any *substitutable* Σ_1-formula, i.e. suppose, for all sentences ϕ,ψ,

$$\mathsf{PRA} \vdash Pr(\ulcorner\phi \leftrightarrow \psi\urcorner) \rightarrow .\rho(\ulcorner\phi\urcorner) \leftrightarrow \rho(\ulcorner\psi\urcorner).$$

(I am tempted to use the word "extensional" together with some modifier. A substitutable predicate is certainly *extensional* in that

$$\mathsf{PRA} \vdash \phi \leftrightarrow \psi \;\Rightarrow\; \mathsf{PRA} \vdash \rho(\ulcorner\phi\urcorner) \leftrightarrow \rho(\ulcorner\psi\urcorner)$$

for all sentences ϕ,ψ. Moreover, such a predicate is also *provably extensional* in that it also schematically satisfies

$$\mathsf{PRA} \vdash Pr(\ulcorner\phi \leftrightarrow \psi\urcorner) \rightarrow Pr(\ulcorner\rho(\ulcorner\phi\urcorner) \leftrightarrow \rho(\ulcorner\psi\urcorner)\urcorner).$$

In studying PRL, we noted two non-arithmetically equivalent, but modally equivalent, formalisations of closure rules-- one of the form $\Box A \rightarrow \Box B$, like provable extensionality, and one of the form $\boxed{s} A \rightarrow B$, which is not quite like substitutability, which bears the stronger form $\Box A \rightarrow B$. The temptation is to use the term "provable extensionality", which isn't quite right. Hence, I use "substitutable" even though it also isn't quite right: It is not ρ that is substitutable for something, but inside ρ that equivalents are substitutable. My wretched choice of terminology at least has the advantage of referring, if obliquely, to a key property of the predicate.)

<u>THEOREM</u>. Let ρ be a substitutable Σ_1-formula. Then the "Formalised Löb's Theorem" holds for ρ:

$$\mathsf{PRA} \vdash \rho(\ulcorner\rho(\ulcorner\phi\urcorner) \rightarrow \phi\urcorner) \leftrightarrow \rho(\ulcorner\phi\urcorner).$$

<u>REMARK</u>. The unformalised "Löb's Theorem",

$$\mathsf{PRA} \vdash \rho(\ulcorner\phi\urcorner) \rightarrow \phi \;\Rightarrow\; \mathsf{PRA} \vdash \phi, \qquad (*)$$

need not hold. To see this, let ρ be a refutable constant:

$$\rho(v):\quad \overline{0} = \overline{1} \wedge v = v.$$

Trivially, $\mathsf{PRA} \vdash \rho(\ulcorner\phi\urcorner) \rightarrow \phi$ for any sentence ϕ, but $\mathsf{PRA} \nvdash \phi$ unless ϕ is a theorem. To conclude that (*) also holds, we must know $\mathsf{PRA} \vdash \rho(\ulcorner\psi\urcorner)$ for some theorem ψ. (Exercise.)

Proof of the Theorem: The proof of this is at first reminiscent of Kreisel's preferred proof of Löb's Theorem. Let ϕ be given and let δ be the analogue to Kreisel's fixed point:

$$\text{PRA} \vdash \delta \leftrightarrow \rho(\ulcorner \delta \to \phi \urcorner).$$

Since ρ is Σ_1,

$$\text{PRA} \vdash \delta \to Pr(\ulcorner \delta \urcorner). \qquad (1)$$

Toying with a tautology, the Derivability Conditions for Pr, and the substitutability of ρ, we get

$$\begin{aligned} \text{PRA} &\vdash \delta \to (\delta \to \phi. \leftrightarrow \phi) \\ &\vdash Pr(\ulcorner \delta \urcorner) \to Pr(\ulcorner \delta \to \phi. \leftrightarrow \phi \urcorner) \\ &\vdash Pr(\ulcorner \delta \urcorner) \to .\rho(\ulcorner \delta \to \phi \urcorner) \leftrightarrow \rho(\ulcorner \phi \urcorner) \\ &\vdash Pr(\ulcorner \delta \urcorner) \to .\delta \leftrightarrow \rho(\ulcorner \phi \urcorner), \qquad (2) \end{aligned}$$

by the choice of δ.

Combining (1) and (2), we get

$$\begin{aligned} \text{PRA} &\vdash \delta \to .\delta \leftrightarrow \rho(\ulcorner \phi \urcorner) \\ &\vdash \delta \to \rho(\ulcorner \phi \urcorner). \qquad (3) \end{aligned}$$

We now need the converse to (3) so that we can substitute $\rho(\ulcorner \phi \urcorner)$ for δ in the definition of δ and obtain the "Formalised Löb's Theorem" for ρ. Observe,

$$\text{PRA} \vdash \rho(\ulcorner \phi \urcorner) \to Pr(\ulcorner \rho(\ulcorner \phi \urcorner) \urcorner)$$

since $\rho(\ulcorner \phi \urcorner)$ is Σ_1. With an instance of *D2* this yields

$$\begin{aligned} \text{PRA} &\vdash Pr(\ulcorner \rho(\ulcorner \phi \urcorner) \to \delta \urcorner) \wedge \rho(\ulcorner \phi \urcorner) \to Pr(\ulcorner \delta \urcorner) \\ &\vdash Pr(\ulcorner \rho(\ulcorner \phi \urcorner) \to \delta \urcorner) \wedge \rho(\ulcorner \phi \urcorner) \to .\delta \leftrightarrow \rho(\ulcorner \phi \urcorner), \text{ by (2)} \\ &\vdash Pr(\ulcorner \rho(\ulcorner \phi \urcorner) \to \delta \urcorner) \to .\rho(\ulcorner \phi \urcorner) \to \delta, \end{aligned}$$

whence Löb's Theorem (for $Pr(\cdot)$) yields

$$\text{PRA} \vdash \rho(\ulcorner \phi \urcorner) \to \delta. \qquad (4)$$

From (3) and (4), we get

$$\begin{aligned} \text{PRA} &\vdash \delta \leftrightarrow \rho(\ulcorner \phi \urcorner) \qquad (5) \\ &\vdash \delta \to \phi. \leftrightarrow .\rho(\ulcorner \phi \urcorner) \to \phi \\ &\vdash \rho(\ulcorner \delta \to \phi \urcorner) \leftrightarrow \rho(\ulcorner \rho(\ulcorner \phi \urcorner) \to \phi \urcorner), \text{ by substitutability} \\ &\vdash \rho(\ulcorner \phi \urcorner) \leftrightarrow \rho(\ulcorner \rho(\ulcorner \phi \urcorner) \to \phi \urcorner), \end{aligned}$$

by (5) and the definition of δ. QED

The latter half of this derivation can be considerably simplified if we assume ρ satisfies a monotonicity condition, which we call *provable monotonicity* (although this term is not really any more appropriate for the given concept than "provable extensionality" was for "substitutability"): For all ϕ,ψ,

$$\text{PRA}\vdash Pr(\ulcorner\phi \to \psi\urcorner) \wedge \rho(\ulcorner\phi\urcorner) \to \rho(\ulcorner\psi\urcorner).$$

Given the assumption of provable monotonicity, we can derive $\rho(\ulcorner\phi\urcorner) \to \delta$ as follows: First, by monotonicity,

$$\text{PRA}\vdash \rho(\ulcorner\phi\urcorner) \wedge Pr(\ulcorner\phi \to .\delta \to \phi\urcorner) \to \rho(\ulcorner\delta \to \phi\urcorner)$$

$$\vdash \rho(\ulcorner\phi\urcorner) \to \rho(\ulcorner\delta \to \phi\urcorner), \text{ since } \phi \to .\delta \to \phi \text{ is a tautology}$$

$$\vdash \rho(\ulcorner\phi\urcorner) \to \delta,$$

by choice of δ. qed Note that this proof does not appeal to Löb's Theorem and, in fact, when $\rho = Pr$, this is the proof of the Formalised Löb's Theorem.

In the most interesting applications, ρ will be provably monotone and not merely substitutable. However, it is nice to know the minimal conditions needed for a basic result. Besides, there are applications in which the operator ρ is only substitutable The following example is from the literature and is a generalisation of the key lemma (1.3.2) used in proving the Explicit Definability of fixed points:

COROLLARY. (Sambin). Let $\alpha(\phi)$ denote any sentence built up from ϕ and other arithmetic sentences by means of $\sim, \wedge, \vee, \to$, and $Pr(\cdot)$. Then:

$$\text{PRA}\vdash Pr(\ulcorner\alpha(\phi)\urcorner) \to \phi \quad\Rightarrow\quad \text{PRA}\vdash Pr(\ulcorner\alpha(\phi)\urcorner) \leftrightarrow Pr(\ulcorner\alpha(\overline{0} = \overline{0})\urcorner).$$

Proof: Let $\rho(\ulcorner\phi\urcorner) = Pr(\ulcorner\alpha(\phi)\urcorner)$ and observe that ρ is substitutable. By the Theorem,

$$\text{PRA}\vdash \rho(\ulcorner\rho(\ulcorner\phi\urcorner) \to \phi\urcorner) \leftrightarrow \rho(\ulcorner\phi\urcorner). \qquad (*)$$

Set this aside for a moment and examine the hypothesis:

$$\text{PRA}\vdash Pr(\ulcorner\alpha(\phi)\urcorner) \to \phi \quad\Rightarrow\quad \text{PRA}\vdash \rho(\ulcorner\phi\urcorner) \to \phi$$

$$\Rightarrow\quad \text{PRA}\vdash (\rho(\ulcorner\phi\urcorner) \to \phi) \leftrightarrow \overline{0} = \overline{0}$$

$$\Rightarrow\quad \text{PRA}\vdash Pr(\ulcorner(\rho(\ulcorner\phi\urcorner) \to \phi) \leftrightarrow \overline{0} = \overline{0}\urcorner)$$

$$\Rightarrow\quad \text{PRA}\vdash \rho(\ulcorner\rho(\ulcorner\phi\urcorner) \to \phi\urcorner) \leftrightarrow \rho(\ulcorner\overline{0} = \overline{0}\urcorner),$$

by substitutability. But now, an application of (*) yields the result:

$$\text{PRA}\vdash \rho(\ulcorner\phi\urcorner) \leftrightarrow \rho(\ulcorner\overline{0} = \overline{0}\urcorner). \qquad \text{QED}$$

So, what are we going to do in this chapter? In rough terms, we are going to lift our discussion of $Pr(\cdot)$ to the bi-modal context of $Pr(\cdot)$ and $\rho(\cdot)$. This is not quite accurate as there is not a complete parallel. In section 1, we encounter a system SR with which we can modally analyse self-reference involving Pr and ρ. Unlike the situation with Pr alone, we have no fixed interpretation of ρ in mind; indeed, there are many interpretations of ρ to choose from and we will encounter some of these along with applications in this section. Section 1 is syntactic in nature; in sections 2 and 3 we cover semantics. The phrase "model theory" is probably more descriptive than "semantics", but this latter word has its meaning as well. The system SR of section 1 resulted from the analysis of a proof, not of any particular concept. It doesn't seem to have a nice model theory. We obtain theories with nice model theories by adding some new axioms. The two chief theories are called MOS (after Mostowski) and PRL_1 (PRL augmented by 1 additional, stronger, proof predicate). Section 2 contains completeness theorems with respect to Kripke models. The goal, of course, is to use the model theories to obtain completeness results with respect to arithmetic interpretations. This is done in section 3 by transforming the Kripke models into models of a new kind that more intimately relate MOS to PRL_1 and PRL_1 to PRL and allow us thereby to reduce the problem to the case already settled by Solovay's results. In section 4, we encounter Timothy Carlson's bi-modal analysis of the provabilities in a theory and a very strong extension.

It should be emphasised that we only offer complete bi-modal analyses of three out of many possible pairs of predicates. These are, at the time of writing, the only examples that have been worked out. Most problems in this area are, thus, open.

Another point that should be emphasised, more properly later, but also here, is this: When we talk about an extension T of PRA, we will need to know T to be provably in PRA an extension of PRA, i.e. that

$$PRA \vdash \forall \ulcorner\phi\urcorner \left(Pr(\ulcorner\phi\urcorner) \rightarrow Pr_T(\ulcorner\phi\urcorner)\right). \qquad (*)$$

Now, PRA can be reformulated as a finitely axiomatised theory (as discussed in Chapter 0, section 6) and, under such a reformulation, (*) follows for any extension T of PRA by the Derivability Conditions for T. In the sequel, whenever he sees the word "extension", the reader can either assume this reformulation of PRA or (*) as the meaning of the word. If we wish to replace PRA by a demonstrably non-finite theory like PA, only the latter option remains viable.

1. BI-MODAL SELF-REFERENCE

We might as well begin with a definition.

1.1. DEFINITION. The system SR is the system of bi-modal logic with language, axiom schemata, and rules of inference as follows:

LANGUAGE:

PROPOSITIONAL VARIABLES: $p, q, r, \ldots$

TRUTH VALUES: t, f

PROPOSITIONAL CONNECTIVES: $\sim, \wedge, \vee, \rightarrow$

MODAL OPERATORS: $\Box, \nabla$

AXIOMS: *A1.* All (boolean) tautologies

A2. $\Box A \wedge \Box(A \rightarrow B) \rightarrow \Box B$

A3. $\Box A \rightarrow \Box\Box A;\quad \nabla A \rightarrow \Box \nabla A$

A4. $\Box(\Box A \rightarrow A) \rightarrow \Box A$

A5. $\Box(A \leftrightarrow B) \rightarrow .\nabla A \leftrightarrow \nabla B$

RULES: *R1.* $A,\ A \rightarrow B\ /\ B$

R2. $A\ /\ \Box A.$

Most of this is, of course, familiar to us as constituting PRL. What is new is: i. the addition of a single new modal operator ∇ intended to be interpreted by a predicate ρ, ii. an extra clause in *A3* mirroring the demonstrable completeness of any Σ_1-formula ρ chosen to interpret ∇, and iii. the axiom schema *A5* simulating the intended substitutability of ρ. What is missing is also of interest: There is no explicit mechanism to handle self-reference involving ∇, not even the Formalised Löb's Theorem for ∇,

$$\nabla(\nabla A \rightarrow A) \leftrightarrow \nabla A,$$

which we know from the introduction to be derivable for the intended substitutable Σ_1-interpretations ρ of ∇ via self-reference. It turns out that this schema is derivable in SR-- not, of course, by the proof given in the introduction, but by Sambin's original proof of his Corollary thereto discussed in the introduction: The Formalised Löb's Theorem for ∇ follows modally from that for $\Box$ in SR. We prove this by repeating and generalising our earlier fixed point analysis.

The first step in our earlier analysis was to establish a couple of Substitution Lemmas. The first (*FSL*), which easily implies the second (*SSL*), generalises as follows:

<u>1.2. SUBSTITUTION LEMMA</u>. Let $A(p)$ be given.

$$\mathrm{SR} \vdash \boxed{s}(B \leftrightarrow C) \to .A(B) \leftrightarrow A(C).$$

Proof: The proof is by induction on the construction of A. The only case not treated exactly as in the proof of the *FSL* is the new one: $A = \nabla D(p)$. But this proof is easy:

$$\begin{aligned} \mathrm{SR} \vdash \boxed{s}(B \leftrightarrow C) &\to .D(B) \leftrightarrow D(C), \\ &\to \Box\big(D(B) \leftrightarrow D(C)\big) \\ &\to .\nabla D(B) \leftrightarrow \nabla D(C), \text{ by } A5. \end{aligned}$$

QED

It is also an easy matter to derive a generalisation of Sambin's Theorem by applying an analogue to the proof of the key fixed point lemma (Lemma 1.3.2):

<u>1.3. LEMMA</u>. For all E,F,

$$\mathrm{SR} \vdash \nabla E \to .E \leftrightarrow F \quad \Rightarrow \quad \mathrm{SR} \vdash \nabla E \leftrightarrow \nabla F.$$

Proof: First, observe

$$\begin{aligned} \mathrm{SR} \vdash \nabla E \to .E \leftrightarrow F \quad &\Rightarrow \quad \mathrm{SR} \vdash \Box \nabla E \to \Box(E \leftrightarrow F) \\ &\Rightarrow \quad \mathrm{SR} \vdash \nabla E \to \Box(E \leftrightarrow F), \quad \text{by } A3 \\ &\Rightarrow \quad \mathrm{SR} \vdash \nabla E \to .\nabla E \leftrightarrow \nabla F, \quad \text{by } A5 \\ &\Rightarrow \quad \mathrm{SR} \vdash \nabla E \to \nabla F. \end{aligned}$$

Conversely,

$$\begin{aligned} \mathrm{SR} \vdash \nabla E \to .E \leftrightarrow F \quad &\Rightarrow \quad \mathrm{SR} \vdash \Box \nabla E \to \Box(E \leftrightarrow F) \\ &\Rightarrow \quad \mathrm{SR} \vdash \Box \nabla E \to .\nabla E \leftrightarrow \nabla F, \quad \text{by } A5 \\ &\Rightarrow \quad \mathrm{SR} \vdash \nabla F \to .\Box \nabla E \to \nabla E \qquad (*) \\ &\Rightarrow \quad \mathrm{SR} \vdash \Box \nabla F \to \Box(\Box \nabla E \to \nabla E) \\ &\Rightarrow \quad \mathrm{SR} \vdash \Box \nabla F \to \Box \nabla E, \quad \text{by } A4 \\ &\Rightarrow \quad \mathrm{SR} \vdash \nabla F \to \Box \nabla E, \quad \text{by } A3 \\ &\Rightarrow \quad \mathrm{SR} \vdash \nabla F \to \nabla E, \quad \text{by } (*). \end{aligned}$$

QED

<u>1.4. COROLLARY</u>. $\mathrm{SR} \vdash \nabla(\nabla E \to E) \leftrightarrow \nabla E.$

Proof: Let F be $\nabla E \to E$ in 1.3:

$$SR \vdash \nabla E \to .E \leftrightarrow (\nabla E \to E)$$

$$\vdash \nabla E \leftrightarrow \nabla(\nabla E \to E).$$

QED

As I have already remarked, both just prior to the statement of Lemma 1.3 and in the introduction, this result is a generalisation of the basic lemma on the explicit definability of fixed points. A corollary to such a remark is that this latter lemma follows (or, *should* follow) from Lemma 1.3. Indeed it does:

1.5. LEMMA ON EXPLICIT DEFINABILITY. Let $C(p)$ be given. Then:

i. $SR \vdash \Box C(t) \leftrightarrow \Box C(\Box C(t))$

ii. $SR \vdash \nabla C(t) \leftrightarrow \nabla C(\nabla C(t))$.

Proof: i. Copy the old proof or mimic the following case.

ii. By the Sambin result:

$$SR \vdash \nabla C(t) \to .t \leftrightarrow \nabla C(t) \quad \Rightarrow \quad SR \vdash \nabla C(t) \to .C(t) \leftrightarrow C(\nabla C(t)), \text{ by 1.2}$$

$$\Rightarrow \quad SR \vdash \nabla C(t) \leftrightarrow \nabla C(\nabla C(t)), \text{ by 1.3.}$$

QED

From this Lemma, one can conclude, as before, the full result:

1.6. EXPLICIT DEFINABILITY OF FIXED POINTS. Let $A(p)$ be a formula in which every occurrence of p lies within the scope of either a $\Box$ or a ∇. Then: There is a sentence D possessing only the propositional variables of $A(p)$ other than p (and, incidentally, possessing only $\Box$ or only ∇ if only $\Box$ or only ∇ occur) such that:

$$SR \vdash D \leftrightarrow A(D).$$

The reduction of Theorem 1.6 to Lemma 1.5 is entirely analogous to that of the original fixed point calculation (1.3.5) to the corresponding lemma (1.3.2) and I will simply leave the derivation to the reader as an exercise. In its place, I cite merely the following example.

1.7. EXAMPLE. Let $A(p) = \sim\nabla p$. Then, for $D = \sim\nabla f$, $SR \vdash D \leftrightarrow \sim\nabla D$.

Proof: Let $C(p) = \sim p$ and apply the Lemma on Explicit Definability to conclude

$$SR \vdash \nabla\sim t \leftrightarrow \nabla\sim\nabla\sim t$$

Since $\sim t \leftrightarrow f$, *A5* or Lemma 1.2 yields

$$SR \vdash \nabla f \leftrightarrow \nabla\sim\nabla f,$$

whence

$$SR \vdash \sim\nabla f \leftrightarrow \sim\nabla(\sim\nabla f). \qquad \text{QED}$$

As important as the existence of explicitly defined fixed points is their uniqueness, a result that still holds.

1.8. UNIQUENESS OF FIXED POINTS. Let $A(p)$ be a formula in which every occurrence of p lies within the scope of either a $\Box$ or a ∇. Then:

$$SR \vdash \boxed{s}(p \leftrightarrow A(p)) \wedge \boxed{s}(q \leftrightarrow A(q)) \rightarrow .p \leftrightarrow q.$$

For PRL we had both syntactic and semantic proofs of the uniqueness theorem. For SR, we have no semantics and must adapt Bernardi's syntactic proof.

Proof of 1.8: Write

$$A(p) \;=\; B(\Box C_1(p),\ldots,\Box C_k(p),\nabla D_1(p),\ldots,\nabla D_m(p)),$$

where $B(q_1,\ldots,q_k,r_1,\ldots,r_m)$ is propositional in $q_1,\ldots,q_k,r_1,\ldots,r_m$ and contains no occurrence of p. For all i,j,

$$SR \vdash \boxed{s}(p \leftrightarrow q) \rightarrow (C_i(p) \leftrightarrow C_i(q)) \wedge (D_j(p) \leftrightarrow D_j(q)), \quad \text{by 1.2}$$

$$\vdash \Box(p \leftrightarrow q) \rightarrow \Box(C_i(p) \leftrightarrow C_i(q)) \wedge \Box(D_j(p) \leftrightarrow D_j(q))$$

$$\vdash \Box(p \leftrightarrow q) \rightarrow (\Box C_i(p) \leftrightarrow \Box C_i(q)) \wedge (\nabla D_j(p) \leftrightarrow \nabla D_j(q)).$$

Thus,

$$SR \vdash \Box(p \leftrightarrow q) \rightarrow .A(p) \leftrightarrow A(q),$$

whence

$$SR \vdash \boxed{s}(p \leftrightarrow A(p)) \wedge \boxed{s}(q \leftrightarrow A(q)) \rightarrow .\Box(p \leftrightarrow q) \rightarrow (p \leftrightarrow q) \qquad (*)$$

$$\rightarrow \Box(\Box(p \leftrightarrow q) \rightarrow (p \leftrightarrow q))$$

$$\rightarrow \Box(p \leftrightarrow q), \quad \text{by } A4$$

$$\rightarrow .p \leftrightarrow q, \quad \text{by } (*) \qquad \text{QED}$$

With this last, we have completed the basic modal analysis of self-reference in SR. But it is not the last of our discussion of self-reference and SR. We can, e.g., apply this to the discussion of self-reference in PRL:

1.9. EXAMPLE. (Exercise 7 of Chapter 1, section 3). Define a sentence C of the language of PRL to be *almost boxed* if $PRL \vdash C \rightarrow \Box C$. Suppose $C(p)$ is almost boxed and p is boxed in $C(p)$. Then:

$$PRL \vdash C(t) \leftrightarrow C(C(t)).$$

(Thus, we can base our fixed point calculation on almost boxed, rather than boxed, components and might, thus, reduce the effort involved.) To see this, interpret SR into PRL as follows:

$$p^C = p, \qquad t^C = t, \qquad f^C = f, \qquad (\sim A)^C = \sim(A^C)$$

$$(A \circ B)^C = A^C \circ B^C \text{ for } \circ \in \{\wedge, \vee, \rightarrow\}$$

$$(\Box A)^C = \quad (A^C), \quad (\nabla A)^C = C(A^C).$$

It is easy to see that theorems of SR map to theorems of PRL: Instances of $\Box$-axioms and rules translate to instances of the same axioms and rules; the instance, $\nabla A \rightarrow \Box \nabla A$, of *A3* translates to the derivable $C(A^C) \rightarrow \Box C(A^C)$; and the instance, $\Box(A \leftrightarrow B) \rightarrow .\nabla A \leftrightarrow \nabla B$, translates, almost, to an instance of the Second Substitution Lemma. One must write $C(p) = D(\Box C_1(p), \ldots, \Box C_k(p))$, with p absent from $D(q_1, \ldots, q_k)$ and D propositional in $q_1, \ldots, q_n$. Then,

$$\mathsf{PRL} \vdash \Box(A \leftrightarrow B) \rightarrow \bigwedge(\Box C_i(A) \leftrightarrow \Box C_i(B)), \quad \text{by } SSL$$

$$\vdash \Box(A \leftrightarrow B) \rightarrow (D(\Box C_1(A), \ldots, \Box C_k(A)) \leftrightarrow D(\Box C_1(B), \ldots, \Box C_k(B))),$$

by the substitution lemma for the propositional calculus. Then one gets

$$\mathsf{PRL} \vdash \Box(A \leftrightarrow B) \rightarrow .C(A) \leftrightarrow C(B),$$

which guarantees the validity of *A5* under the given interpretation. The translation

of $\quad \mathsf{SR} \vdash \nabla t \leftrightarrow \nabla(\nabla t)$

is just $\quad \mathsf{PRL} \vdash C(t) \leftrightarrow C(C(t))$.

One instance of this example worth mentioning is this: Let $C(p) = \Box(r_1 \rightarrow p) \vee \Box(r_2 \rightarrow p)$. (Cf. Exercise 4 of Chapter 2, section 2.) The fixed point is just $C(t) = \Box(r_1 \rightarrow t) \vee \Box(r_2 \rightarrow t)$, which is equivalent to t. The fixed point to $\sim C(p)$ is $\sim C(f)$, since

$$\mathsf{SR} \vdash \sim\nabla f \leftrightarrow \sim\nabla(\sim\nabla f).$$

But $\sim C(f) = \sim(\Box(r_1 \rightarrow f) \vee \Box(r_2 \rightarrow f))$ is equivalent to $\sim(\Box\sim r_1 \wedge \Box\sim r_2)$, whence to $\sim\Box\sim r_1 \wedge \sim\Box\sim r_2$.

This is a fairly pedestrian sort of interpretation of ∇. In arithmetic, we can find many interesting interpretations of SR. I mention here several:

1.10. SOME ARITHMETIC INTERPRETATIONS OF SR. In each of the following, we interpret $\Box$ as $Pr(\cdot)$ and ∇ as some substitutable Σ_1-formula ρ:

i. (Provability). $\rho(v) = Pr_T(v)$, where T is some fixed consistent *RE* extension of PRA, e.g. T = PA, ZF, etc. *N.B.* To guarantee the validity of *A5*, it must be assumed here, and throughout the remainder of the Chapter, that PRA proves T to be an extension thereof, i.e. we must restrict ourselves to extensions T such that

$$\text{PRA} \vdash \forall \ulcorner\phi\urcorner \left(Pr(\ulcorner\ulcorner\phi\urcorner\urcorner) \to Pr_T(\ulcorner\ulcorner\phi\urcorner\urcorner)\right). \qquad (*)$$

As noted in the introduction, PRA can be assumed replaced by a finitely axiomatised version for which (*) is automatic; this will not happen if we make, say, PA our base theory.

ii. (Mostowski operator). $\rho(v) = \bigvee\!\!\!\bigvee_i Pr_{T_i}(v)$, where $T_0, T_1, \ldots$ is an *RE* sequence of consistent extensions of PRA and $\bigvee\!\!\!\bigvee_i Pr_{T_i}(v)$ is a reasonable Σ_1-formula defining the union of the sets of theorems of the various T_i's and (for the sake of the interpretation of *A5*) where we assume PRA uniformly proves each T_i to extend PRA.

iii. (Provable relative consistency). $\rho(\ulcorner\phi\urcorner) = Pr(\ulcorner Con_T \to Con_{T+\phi}\urcorner)$, where T is a consistent *RE* extension of PRA (and, of course, provably so).

iv. (Relative interpretability). $\rho(\ulcorner\phi\urcorner) = Relint_{GB}(\ulcorner\phi\urcorner)$, where $Relint_{GB}(v)$ is a Σ_1-formula defining relative interpretability in GB, i.e. $Relint_{GB}(\ulcorner\phi\urcorner)$ asserts GB + ϕ is interpretable in GB, and where GB is the usual finitely axiomatised set theory. (It is important in this example that GB be finitely axiomatised; relative interpretability in PA or, say, ZF is not an *RE*-predicate.)

v. (Provable intersection). $\rho(v) = Pr(\ulcorner \bigwedge\!\!\!\bigwedge_i Pr_{T_i}(v)\urcorner)$, where $T_0, T_1, \ldots$ is an *RE* sequence of consistent extensions of PRA and, as in ii, we assume PRA uniformly proves each T_i to extend PRA.

Different interpretations ρ of ∇ will have different properties, i.e. different modal schemata will be valid. Of the interpretations given, the new schemata have only been characterised for the interpretation $\rho = Pr_T$ for "generic" T and an occasional specific T and the interpretation $\rho = \bigvee\!\!\!\bigvee Pr_{T_i}$ for "generic" $T_0, T_1, \ldots$. These results will be proved in sections 3 and 4, below. For now, let me note merely that the interpretations of Example 1.10 all validate a schema not valid under the almost boxed interpretation of Example 1.9, namely,

$$\Box A \to \nabla A.$$

We finish this section by taking a brief look at this schema:

<u>1.11. LEMMA</u>. Let $T \supseteq SR$ be closed under $R2$. Then, the following are equivalent:

i. $T \vdash \Box A \to \nabla A$

ii. $T \vdash \nabla t$

iii. if $T \vdash \nabla A \to A$, then $T \vdash A$.

Proof: i $\Rightarrow$ ii. This is trivial: $T \vdash \Box t \to \nabla t$ and $T \vdash \Box t$ yield $T \vdash \nabla t$.

ii $\Rightarrow$ iii. Observe,

$$\begin{aligned}
T \vdash \nabla A \to A \quad &\Rightarrow \quad T \vdash (\nabla A \to A) \leftrightarrow t \\
&\Rightarrow \quad T \vdash \Box\big((\nabla A \to A) \leftrightarrow t\big) \\
&\Rightarrow \quad T \vdash \nabla(\nabla A \to A) \leftrightarrow \nabla t, \quad \text{by } A5 \\
&\Rightarrow \quad T \vdash \nabla(\nabla A \to A), \quad \text{by ii} \\
&\Rightarrow \quad T \vdash \nabla A, \quad \text{by Corollary 1.4} \\
&\Rightarrow \quad T \vdash A, \text{ by assumption } T \vdash \nabla A \to A.
\end{aligned}$$

(A quicker proof appeals to Lemma 1.3:

$$\begin{aligned}
T \vdash \nabla A \to A \quad &\Rightarrow \quad T \vdash \nabla A \to .A \leftrightarrow t \\
&\Rightarrow \quad T \vdash \nabla A \leftrightarrow \nabla t, \quad \text{by 1.3} \\
&\Rightarrow \quad T \vdash \nabla A \\
&\Rightarrow \quad T \vdash A.
\end{aligned}$$

I prefer the former proof because the Formalised Löb's Theorem for ∇, 1.4, is more intuitively true than 1.3.)

iii $\Rightarrow$ ii. Let $A = \nabla t$:

$$\begin{aligned}
T &\vdash \Box \nabla t \to \Box(\nabla t \leftrightarrow t) \\
&\vdash \Box \nabla t \to .\nabla\nabla t \leftrightarrow \nabla t, \quad \text{by } A5 \\
&\vdash \nabla\nabla t \to .\Box \nabla t \to \nabla t \qquad (*) \\
&\vdash \nabla\nabla t \to \Box(\Box \nabla t \to \nabla t), \quad \text{by } R1, R2, A2, A3 \\
&\vdash \nabla\nabla t \to \Box \nabla t, \quad \text{by } A4 \\
&\vdash \nabla\nabla t \to \nabla t, \quad \text{by } (*) \\
&\vdash \nabla t, \quad \text{by iii.}
\end{aligned}$$

(Note the similarity of this proof to that of 1.3 or, back in Chapter 1, 1.3.2.)

ii $\Rightarrow$ i. Observe,

$$T \vdash \Box A \to \Box(A \leftrightarrow t)$$

$\vdash \Box A \to . \nabla A \leftrightarrow \nabla t$

$\vdash \Box A \to \nabla A$, by ii.

QED

EXERCISES

1. Do one of the following:

   i. Carry out the details of the proof of 1.6.

   ii. Define a notion of "almost boxed" formulae of SR and prove for SR an analogue to the result of Example 1.9.

   iii. Consider the system SR$(\nabla_1,\ldots,\nabla_n)$ with n new modal operators and, in place of axioms $A3$ and $A5$ of SR, the new schemata

   $A3_n$. $\Box A \to \Box\Box A$; $\nabla_i A \to \Box \nabla_i A$, $i = 1,\ldots,n$

   $A5_n$. $\Box(A \leftrightarrow B) \to . \nabla_i A \leftrightarrow \nabla_i B$, $i = 1,\ldots,n$.

   State and prove the Uniqueness and Explicit Definability of Fixed Points in SR$(\nabla_1,\ldots,\nabla_n)$.

2. Choose one or more of the interpretations of Example 1.10 and prove it or them to validate the schemata of SR.

3. Let T $\supseteq$ SR be closed under $R2$.

   i. Prove the following to be equivalent to any of the conditions of Lemma 1.11: T is closed under the rule $R2^{\nabla}$: $A \,/\, \nabla A$.

   ii. Assume T satisfies any of the conditions of Lemma 1.11. Show: T$\vdash \Box(\nabla A \to A) \to \Box A$.

4. (Solovay, Smoryński, Friedman). Let $T_0, T_1, \ldots$ be an RE sequence of consistent extensions of PRA so that the relation

   $Pr_{v_0}(v_1)$: "the formula with code v_1 is provable in T_{v_0}"

   is RE. Assume PRA handles these theories uniformly, i.e., for all sentences ϕ,ψ,

   PRA$\vdash Pr(\ulcorner\phi\urcorner) \to \forall v_0 Pr_{v_0}(\ulcorner\phi\urcorner)$

   PRA$\vdash \forall v_0\big(Pr_{v_0}(\ulcorner\phi\urcorner) \wedge Pr_{v_0}(\ulcorner\phi \to \psi\urcorner) \to Pr_{v_0}(\ulcorner\psi\urcorner)\big)$

   PRA$\vdash \forall v_0\big(Pr_{v_0}(\ulcorner\phi\urcorner) \to Pr(\ulcorner Pr_{\dot{v}_0}(\ulcorner\phi\urcorner)\urcorner)\big)$.

   (The second and third of these come for free; the first depends on the exact presentations of the T_i's.) In 1974, Haim Gaifman asked the question: Does there exist an RE sequence of consistent extensions $T_0, T_1, \ldots$ of PRA such that, for all n,

$$\mathsf{T}_n \vdash Con(\mathsf{T}_{n+1})\ ?$$

The answer is yes, but not if one assumes uniformity of the provability of consistency:

$$\mathsf{PRA} \vdash \forall v_0 Pr_{v_0}(\ulcorner Con(\mathsf{T}_{\dot{v}_0+\bar{1}})\urcorner). \qquad (*)$$

Let $\rho(\ulcorner\phi\urcorner) \;=\; Pr(\ulcorner \forall v_0 Pr_{v_0}(\ulcorner\phi\urcorner)\urcorner)$.

i. Assume $\mathsf{T}_0, \mathsf{T}_1, \ldots$ consistent. Using SR, show

$$\mathsf{PRA} \vdash \rho(\ulcorner\sim\rho(\ulcorner\bar{0} = \bar{1}\urcorner)\urcorner) \rightarrow \rho(\ulcorner\bar{0} = \bar{1}\urcorner).$$

ii. Assuming (*), show arithmetically

$$\mathsf{PRA} \vdash \rho(\ulcorner\sim\rho(\ulcorner\bar{0} = \bar{1}\urcorner)\urcorner).$$

iii. Assuming (*), conclude that *all* theories $\mathsf{T}_0, \mathsf{T}_1, \ldots$ are inconsistent.

iv. By diagonalisation, define ϕv so that

$$\mathsf{PRA} \vdash \phi v \leftrightarrow \big(v = \bar{0} \wedge Con(\mathsf{PRA})\big) \vee \big(v > \bar{0} \wedge Con(\mathsf{PRA} + \phi(v \dot{-} \bar{1}))\big).$$

Let ψ_n be the sentence

$$\forall v_0 \big(Prov_{\mathsf{PRA}+\forall v\phi v}(v_0, \ulcorner\bar{0} = \bar{1}\urcorner) \wedge \forall v_1 < v_0 \sim Prov_{\mathsf{PRA}+\forall v\phi v}(v_1, \ulcorner\bar{0} = \bar{1}\urcorner) \rightarrow$$
$$\rightarrow \phi(v_0 \dot{-} \bar{n})\big).$$

Show: a. $\mathsf{PRA} + \psi_n \vdash Con(\mathsf{PRA} + \psi_{n+1})$

b. For each n, $\mathsf{PRA} + \psi_n$ is consistent.

Hence, there is a consistent sequence $\mathsf{T}_0, \mathsf{T}_1, \ldots$ of extensions of PRA such that $\mathsf{T}_n \vdash Con(\mathsf{T}_{n+1})$.

5. Extend SR by the axiom schema asserting the *decidability* of ∇,

$$\sim\nabla A \rightarrow \Box\sim\nabla A.$$

Call the resulting theory SR^D. Prove: For any A, B,

$$\mathsf{SR}^D \vdash \nabla A \leftrightarrow \nabla B.$$

(Hint: Assume, e.g., $\nabla A \wedge \sim\nabla B$. Define $D \leftrightarrow .B \wedge \nabla D \vee A \wedge \sim\nabla D$, and see what $\nabla D, \sim\nabla D$ imply.)

(REMARK: Exercise 5 is a modal analogue to a theorem of Recursion Theory known as Rice's Theorem: There are no non-trivial recursive extensional sets of codes of Σ_1-formulae.)

2. KRIPKE MODELS

The present and immediately following sections set several goals for themselves. In the end, the main goals will turn out to have been the proofs of analogues to Solovay's First Completeness Theorem for two modal logics, MOS and PRL_1, relative to their respective arithmetic interpretations. Another goal is to exhibit some variations in the notion of a Kripke model-- hence the titles of these sections. Finally, we would like to apply the model theory to the study of MOS, PRL_1, and PRL.

Let us first introduce the system MOS.

2.1. DEFINITION. The system MOS (for *Mos*towski) is the system of bi-modal logic with axioms and rules as follow:

AXIOMS. *A1-A4* as in SR

A6. $\Box A \rightarrow \nabla A$

A7. $\Box(A \rightarrow B) \rightarrow .\nabla A \rightarrow \nabla B.$

RULES. *R1, R2* as usual.

Note that MOS differs from SR in two respects. First, it has the additional axiom schema which we discussed in the last section and was was equivalent to ∇t and to the closure under an analogue to Löb's Theorem: $\vdash \nabla A \rightarrow A$ implies $\vdash A$. Second, *A5* has been replaced by the stronger schema *A7*: Substitutability has been replaced by provable monotonicity.

2.2. DEFINITION. The system PRL_1 (*Pr*ovability *L*ogic with *1* extra provability predicate) is the system of bi-modal logic with axioms and rules as follow:

AXIOMS. *A1-A4* as in SR

A6. $\Box A \rightarrow \nabla A$ as in MOS

A8. $\nabla(A \rightarrow B) \rightarrow .\nabla A \rightarrow \nabla B.$

RULES. *R1, R2* as usual.

The axiomatisation of PRL_1 is a bit easier to understand than that for MOS: ∇ is supposed to simulate $Pr_T(\cdot)$ for some extension $T \supseteq PRA$. That T extends PRA is manifested in *A6*. The rest of the axioms of PRL_1 are just those of $PRL(\Box)$ and $PRL(\nabla)$-- minus those proven redundant in the last section.

Alternatively, one could explain that PRL_1 extends MOS by the mere strengthening of provable monotonicity to provable closure (of T) under modus ponens.

We might as well be introduced to one more aystem.

2.3. DEFINITION. The system PRL_{ZF} is the system of bi-modal logic with axioms and rules as follow:

AXIOMS. *A1-A4, A6, A8* as in PRL_1

A9. $\nabla(\Box A \rightarrow A)$

RULES. *R1, R2* as usual.

The nomenclature is only slightly suggestive. One could as well use PRL_S ("*S*" for "strong") or PRL^+ or some such. The presence of "*ZF*" is simply intended to indicate that ∇ refers to Pr_T for a *much stronger* theory T (e.g. ZF) than PRA. As we saw in discussing Solovay's Second Completeness Theorem, those schemata of $\Box$ true in arithmetic were axiomatised over (the *R2*-free version of) PRL by the schema $\Box A \rightarrow A$ of reflexion. Now, the theory ZF of Zermelo-Fraenkel set theory (and even Peano's arithmetic, PA) is strong enough to prove reflexion for PRA. Indeed, this fact can be proven in PRA:

$$\mathsf{PRA} \vdash Pr_{ZF}(\ulcorner Pr_{PRA}(\ulcorner \phi \urcorner) \rightarrow \phi \urcorner).$$

Hence, *A9* is valid in PRA under interpretations * for which

$$(\Box A)^* = Pr_{PRA}(\ulcorner A^* \urcorner)$$

$$(\nabla A)^* = Pr_{ZF}(\ulcorner A^* \urcorner).$$

Carlson's Completeness Theorem, which we will study in section 4, below, asserts that *A9* is all we need to add to PRL_1 to axiomatise the schemata about PRA- and ZF-provability that are valid in PRA.

Most syntactic matters regarding these theories were already settled in the previous section. Just to be official, let me cite a few syntactic results anyway.

2.4. LEMMA. i. $\mathsf{SR} \subseteq \mathsf{MOS} \subseteq \mathsf{PRL}_1 \subseteq \mathsf{PRL}_{ZF}$, i.e. for any modal sentence A,

a. $\mathsf{SR} \vdash A \Rightarrow \mathsf{MOS} \vdash A$

b. $\mathsf{MOS} \vdash A \Rightarrow \mathsf{PRL}_1 \vdash A$

c. $\mathsf{PRL}_1 \vdash A \Rightarrow \mathsf{PRL}_{ZF} \vdash A$

ii. For T = MOS, PRL_1, or PRL_{ZF}

a. $\mathsf{T} \vdash \nabla t$

b. For any A, $T \vdash \nabla A \to A \Rightarrow T \vdash A$

c. For any A, $T \vdash A \Rightarrow T \vdash \nabla A$.

I leave the proofs as simple exercises for the reader.

Because $SR \subseteq T$ for each of $T = MOS, PRL_1$, and PRL_{ZF}, the Substitution Lemma and Fixed Point analysis hold for these theories T. A particular part of this latter analysis, obviously used in proving 2.4.ii.b, is the derivability of the analogue to *A4* for ∇:

$$T \vdash \nabla(\nabla A \to A) \to \nabla A.$$

Arithmetically, the strengthening of the assumption of substitutability to provable monotonicity simplified the derivation,via self-reference, of this analogue to the Formalised Löb's Theorem; modally, I don't see how to do this, but we can simplify this derivation for $T \supseteq PRL_1$:

<u>2.5. LEMMA</u>. $PRL_1 \vdash \nabla(\nabla A \to A) \to \nabla A$.

More direct proof: Observe

$$PRL_1 \vdash \nabla(\nabla A \to A) \to .\nabla^2 A \to \nabla A. \tag{1}$$

by *A8*. Applying *R2,A2,A3*, we get

$$PRL_1 \vdash \Box\nabla(\nabla A \to A) \to \Box(\nabla^2 A \to \nabla A)$$

$$\vdash \nabla(\nabla A \to A) \to \Box(\nabla^2 A \to \nabla A), \tag{2}$$

by *A3*. But, for any B,

$$PRL_1 \vdash \Box(\nabla B \to B) \to .\Box\nabla B \to \Box B, \text{ by } A2$$

$$\vdash \Box(\nabla B \to B) \to .\Box\nabla B \to \nabla B, \tag{3}$$

by *A6*. Again *R2,A2,A3* yield

$$PRL_1 \vdash \Box(\nabla B \to B) \to \Box(\Box\nabla B \to \nabla B)$$

$$\vdash \Box(\nabla B \to B) \to \Box\nabla B, \text{ by } A4$$

$$\vdash \Box(\nabla B \to B) \to \nabla B, \tag{4}$$

by (3). Now, letting $B = \nabla A$ in (2), (4) yields

$$PRL_1 \vdash \nabla(\nabla A \to A) \to \nabla^2 A$$

$$\vdash \nabla(\nabla A \to A) \to \nabla A, \text{ by (1)}$$

QED

So much for syntax. Semantics concerns us here. There are two natural ways of treating a modal operator in Kripke model theory. The first, familiar from the model theory of PRL, associates to the modal operator an accessibility relation R on the set of possible worlds α and asserts the ∇-necessity of a sentence at α if the sentence is true at all α-accessible worlds:

$$\alpha \Vdash \nabla A \quad \text{iff} \quad \forall \beta(\alpha R \beta \Rightarrow \beta \Vdash A).$$

The second approach assigns to each world α a family F_α of sets of possible worlds and declares ∇A true at α if the set of worlds at which A is true contains or is an element of F_α:

$$\alpha \Vdash \nabla A \quad \text{iff} \quad \exists X(\{\beta : \beta \Vdash A\} \supseteq X \in F_\alpha).$$

With two modal operators, $\square$ and ∇, this gives us four possible approaches to a Kripke model theory. We will consider two: For both MOS and PRL_1, we interpret $\square$ via its familiar accessibility relation; for PRL_1, we base our treatment of ∇ on an accessibility relation; and, for MOS, we base our treatment of ∇ on the families F_α.

Because of the distinct interpretations of ∇ in the two model theories, it will be convenient to have two distinct names for the modal operator.

2.6. NOTATIONAL CONVENTION. For the rest of this chapter, "∇" will be reserved for the new modal operator of MOS and "Δ" for that of PRL_1 and PRL_{ZF}.

Another consequence of the distinct model theoretic treatments of ∇ and Δ is that the model theories do not build on top of each other in the sense that, back in Chapter 2, we derived a model theory for PRL from that for the weaker BML by specialising to models in which the extra axioms were valid. Hence, we will not first obtain a model theory for MOS and then specialise it to one for the stronger PRL_1. In fact, we will first discuss the model theory for PRL_1 and then that for MOS.

One final comment before proceeding: After having proven the completeness of MOS and PRL_1 with respect to their Kripke models, we will proceed in section 3 to transform these models into Carlson models-- further variants of Kripke models due to Timothy Carlson. With the Carlson models, we will be able to compare MOS and PRL_1 more readily and to obtain arithmetic completeness results as corollaries to Solovay's First Completeness Theorem. The Kripke models are, thus, a mere preparation for the Carlson models; they are, in fact, a detour that need not be cited explicitly. Nevertheless, I chose to expound on them so that the reader would be exposed to as many variants of the model theory as possible: We are presenting the only bi-modal arithmetic completeness theorems known at the time of writing and this sample may be too small to use to judge the efficiency of any particular variant of the Kripke model.

Now, we may begin our discussion of model theory. Even once we decide on which Kripke models we want, we must decide the route to take to get to them. In Chapter 2, we first presented a strong completeness theorem for BML and then specialised down to finite models for PRL. Since it is the class of finite models that, by virtue of the finiteness of their frames, are readily recognised to satisfy *A4*, we shall this time prove completeness with respect to finite models directly. This means we will define "model" to mean *finite* model.

2.7. DEFINITION. A *Kripke model for* PRL_1 is given by $\underline{K} = (K,<,R,\alpha_0,\Vdash)$, where

i. $(K,<,\alpha_0)$ is a finite partially ordered set with minimum element α_0

ii.a. R is a partial subordering of $<$ on a subset of K, i.e. $\alpha R \beta \Rightarrow \alpha < \beta$

ii.b. $\alpha < \beta R \gamma \Rightarrow \alpha R \beta$

iii. $\Vdash$ satisfies, in addition to the usual boolean laws,

a. $\alpha \Vdash \Box A$ iff $\forall \beta(\alpha < \beta \Rightarrow \beta \Vdash A)$

b. $\alpha \Vdash \Delta A$ iff $\forall \beta(\alpha R \beta \Rightarrow \beta \Vdash A)$.

2.8. THEOREM. For every (bi-)modal sentence A,

$$\mathsf{PRL}_1 \vdash A \text{ iff } \alpha_0 \Vdash A \text{ for all finite Kripke models } \underline{K} \text{ for } \mathsf{PRL}_1.$$

Proof: $\Rightarrow$. Routine. Let me merely note how 2.7.ii.a yields *A6* and 2.7.ii.b yields the second clause of *A3*:

A6: Observe

$$\alpha \Vdash \Box A \Rightarrow \forall \beta(\alpha < \beta \Rightarrow \beta \Vdash A)$$
$$\Rightarrow \forall \beta(\alpha R \beta \Rightarrow \beta \Vdash A)$$
$$\Rightarrow \alpha \Vdash \Delta A$$

whence $\alpha \Vdash \Box A \rightarrow \Delta A$.

A3: Again observe

$$\alpha \Vdash \Delta A \Rightarrow \forall \beta(\alpha R \beta \Rightarrow \beta \Vdash A)$$

$$\Rightarrow \quad \forall\gamma\forall\beta(\alpha < \gamma R \beta \Rightarrow \beta \Vdash A),$$

since, by 2.7.ii.b, $\alpha < \gamma R \beta \Rightarrow \alpha R \beta$. Thus

$$\begin{aligned} \alpha \Vdash \Delta A \quad &\Rightarrow \quad \forall\gamma(\alpha < \gamma \Rightarrow \forall\beta(\gamma R \beta \Rightarrow \beta \Vdash A)) \\ &\Rightarrow \quad \forall\gamma(\alpha < \gamma \Rightarrow \gamma \Vdash \Delta A) \\ &\Rightarrow \quad \alpha \Vdash \Box \Delta A, \end{aligned}$$

whence $\alpha \Vdash \Delta A \to \Box\Delta A$. (Note the similarity of condition 2.7.ii.b to transitivity, $\alpha < \beta < \gamma \Rightarrow \alpha < \gamma$, which accounts for the validity of the first clause, $\Box A \to \Box\Box A$, of $A3$.)

$\Leftarrow$ We prove the converse by contraposition. Suppose $\mathsf{PRL}_1 \nvdash A$.

First, we define some notation that will be used in proofs throughout this section:

$$\begin{aligned} S &= \{B:\ B \text{ is a subformula of } A\} \cup \{\Box B:\ \Delta B \text{ is a subformula of } A\} \\ S^+ &= S \cup \{B:\ \sim B \in S\}. \end{aligned}$$

A subset $\alpha \subseteq S^+$ is *S-complete* if i. $\mathsf{PRL}_1 + \bigwedge\alpha$ is consistent, and ii. for all $B \in S$ $B \in \alpha$ or $\sim B \in \alpha$.

Since $\mathsf{PRL}_1 \nvdash A$, $\mathsf{PRL}_1 + \sim A$ is consistent. Let α_0 be any S-completion of $\{\sim A\}$.

Define, for S-complete α,β: $\alpha < \beta$ iff

i. $\forall \Box B \in S(\Box B \in \alpha \Rightarrow B, \Box B \in \beta)$

ii. $\forall \Delta B \in S(\Delta B \in \alpha \Rightarrow \Delta B \in \beta)$

and iii. $\exists \Box B \in S(\Box B \in \beta \ \&\ B \notin \alpha)$ or $\exists \Delta B \in S(\Delta B \in \beta \ \&\ \Delta B \notin \alpha)$.

Let $K = \{\alpha_0\} \cup \{\beta:\ \alpha_0 < \beta \text{ is } S\text{-complete}\}$.

Claim 1. $(K,<,\alpha_0)$ is a finite partially ordered set with minimum element α_0.

Before proving the Claim, let me comment briefly on the definitions so far given. Except for replacing "PRL_1" by "MOS" and "Δ" by "∇" above, these definitions carry over to the construction in the completeness theorem for MOS and are similar to those for a proof for PRL. The novel points (so far), with respect to such a proof for PRL, are the extra set $\{\Box B:\ \Delta B$ is a subformula of $A\}$ in the definition of S, condition ii in the definition of $<$, and the disjunction in condition iii of the definition of $<$. These are related to the new schemata $A3$ and $A6$. (But: The addition to S isn't needed until we discuss MOS.)

Proof of the Claim: Finiteness is obvious as K is a collection of S-complete sets, i.e. subsets of the fixed finite set S. Minimality of α_0 is clear by the definition of <-- once we've established that < is indeed a partial ordering. Toward this latter end, we have:

Asymmetry. Suppose $\alpha < \beta$. By clauses i and ii of the definition of <,

$$\Box B \in \alpha \;\Rightarrow\; \Box B \in \beta, \qquad \Delta C \in \alpha \;\Rightarrow\; \Delta C \in \beta, \qquad (*)$$

for $\Box B$, $\Delta C \in S$. If $\beta < \alpha$ also held, then, by clause iii of the definition, there would either be $D = \Box B \in S$ or $D = \Delta C \in S$ such that $D \in \beta$ and $D \notin \alpha$, contrary to (*).

Transitivity. Suppose $\alpha < \beta$ and $\beta < \gamma$. Observe:

i. For $\Box B \in S$,

$$\Box B \in \alpha \;\Rightarrow\; \Box B \in \beta, \text{ since } \alpha < \beta$$

$$\Rightarrow\; B, \Box B \in \gamma, \text{ since } \beta < \gamma$$

ii. For $\Delta C \in S$,

$$\Delta C \in \alpha \;\Rightarrow\; \Delta C \in \beta \;\Rightarrow\; \Delta C \in \gamma.$$

iii. Let $D = \Box B \in S$ or $D = \Delta C \in S$ be such that $D \in \gamma$ and $D \notin \beta$. Then $D \notin \alpha$, for: $D \in \alpha \Rightarrow D \in \beta$. Thus we see that clauses i-iii of the definition of $\alpha < \gamma$ are satisfied and we conclude $\alpha < \beta \;\&\; \beta < \gamma \Rightarrow \alpha < \gamma$.

This proves the Claim. qed

Returning to the proof of 2.8, we next define R:

$$\alpha R \beta \quad \text{iff} \quad \alpha < \beta \;\&\; \forall \Delta B \in S(\Delta B \in \alpha \Rightarrow B \in \beta).$$

Claim 2. $R \subseteq <$ is a partial ordering and satisfies the extra transitivity property,

$$\alpha < \beta R \gamma \;\Rightarrow\; \alpha R \gamma.$$

Proof: That $R \subseteq <$ follows by definition. Because of this, R is automatically asymmetric.

To see the transitivity of R, note that it suffices to prove the stronger transitivity property also desired: $\alpha < \beta R \gamma \Rightarrow \alpha R \gamma$. For, $\alpha R \beta R \gamma \Rightarrow \alpha < \beta R \gamma$. Thus, assume $\alpha < \beta$ and $\beta R \gamma$. Since $\beta R \gamma \Rightarrow \beta < \gamma$, we have $\alpha < \beta$ and $\beta < \gamma$, whence $\alpha < \gamma$ (for, we already know that < is transitive). But also, for $\Delta B \in S$

$$\Delta B \in \alpha \;\Rightarrow\; \Delta B \in \beta, \text{ since } \alpha < \beta$$

$\Rightarrow \quad B \in \gamma$, since $\beta R \gamma$.

Thus, both conditions making $\alpha R \gamma$ have been verified. qed

Proof of Theorem 2.8 (continued): So far we have defined $K, <, R$, and α_0 and have verified all the frame properties. To obtain a model, we need only define $\Vdash$: For any $\alpha \in K$ and any atom p, define

$$\alpha \Vdash p \quad \text{iff} \quad p \in \alpha.$$

This completes the definition of the model $\underline{K} = (K, <, R, \alpha_0, \Vdash)$. Having verified all the crucial frame properties of Definition 2.7 which are required, by part i of the Theorem, to hold to guarantee $\underline{K}$ to be a model of PRL_1, we need only verify that $\alpha_0 \nVdash A$ to complete the proof of the Theorem. Since $PRL_1 + \bigwedge \alpha_0 \nvdash A$, this reduces, as usual, to a lemma of the following sort.

Claim 3. For all $B \in S$ and all $\alpha \in K$,

$$\alpha \Vdash B \quad \text{iff} \quad B \in \alpha.$$

Proof: The proof is by induction on the complexity of B. The atomic case follows by definition and the propositional cases follow from the S-completeness of each α. The proof for the boxed case, $B = \Box C$, is similar to those given in Chapter 2 for completeness proofs of PRL with respect to various model theories and I am tempted to omit it. However, it is a tiny bit more involved; so we shall examine it here, along with the case $B = \Delta C$.

$B = \Box C$. First, observe

$$\begin{aligned} \Box C \in \alpha \quad &\Rightarrow \quad \forall \beta(\alpha < \beta \Rightarrow C \in \beta), \text{ by definition of } < \\ &\Rightarrow \quad \forall \beta(\alpha < \beta \Rightarrow \beta \Vdash C), \text{ by induction hypothesis} \\ &\Rightarrow \quad \alpha \Vdash \Box C. \end{aligned}$$

Inversely,

$$\begin{aligned} \Box C \notin \alpha \quad &\Rightarrow \quad PRL_1 + \bigwedge \alpha \nvdash \Box C \\ &\Rightarrow \quad PRL_1 + \bigwedge \alpha \nvdash \Box(\Box C \to C), \text{ by } A4 \end{aligned}$$

Let $X = \{D: \ \Box D \in \alpha\} \cup \{\Box D: \ \Box D \in \alpha\} \cup \{\Delta D: \ \Delta D \in \alpha\}$. Observe

$$\begin{aligned} PRL_1 + X \vdash \Box C \to C \quad &\Rightarrow \quad PRL_1 \vdash \bigwedge X \to (\Box C \to C) \\ &\Rightarrow \quad PRL_1 \vdash \Box \bigwedge X \to \Box(\Box C \to C) \\ &\Rightarrow \quad PRL_1 \vdash \bigwedge \Box X \to \Box C, \text{ by } A4 \end{aligned}$$

$$\Rightarrow \quad \mathrm{PRL}_1 \vdash \bigwedge\alpha \rightarrow \Box C, \qquad (*)$$

since

$$\bigwedge\Box X \leftrightarrow \bigwedge\{\Box D:\ \Box D \in \alpha\} \wedge \bigwedge\{\Box^2 D:\ \Box D \in \alpha\} \wedge \bigwedge\{\Box\Delta D:\ \Delta D \in \alpha\}$$

and each conjunct on the right follows from an element of α. But (*) contradicts the assumption that $\Box C \notin \alpha$ and we conclude $\mathrm{PRL}_1 + X + \Box C + \sim C$ to be consistent. Let β be any S-completion of $X + \Box C + \sim C$. Evidently, $\alpha < \beta$:

i. $\Box D \in \alpha \Rightarrow D, \Box D \in \beta$, by choice of X

ii. $\Delta D \in \alpha \Rightarrow \Delta D \in \beta$, by choice of X

iii. $\Box C \in \beta$ and $\Box C \notin \alpha$.

Since $\alpha_0 < \alpha < \beta$, transitivity puts β into K. We are almost done: By induction hypothesis, since $C \notin \beta$, $\beta \not\Vdash C$. Thus: $\alpha < \beta \ \&\ \beta \not\Vdash C \Rightarrow \alpha \not\Vdash \Box C$, completing the proof in the case $B = \Box C$.

$B = \Delta C$. The proof is somewhat similar. First, observe

$$\Delta C \in \alpha \quad \Rightarrow \quad \forall\beta(\alpha R \beta \Rightarrow C \in \beta), \text{ by definition of } R$$

$$\Rightarrow \quad \forall\beta(\alpha R \beta \Rightarrow \beta \Vdash C), \text{ by induction hypothesis}$$

$$\Rightarrow \quad \alpha \Vdash \Delta C.$$

The converse is again proven contrapositively:

$$\Delta C \notin \alpha \quad \Rightarrow \quad \mathrm{PRL}_1 + \bigwedge\alpha \nvdash \Delta(\Delta C \rightarrow C). \qquad (*)$$

Let $X = \{D:\ \Box D \in \alpha\} \cup \{\Box D:\ \Box D \in \alpha\} \cup \{\Delta D:\ \Delta D \in \alpha\} \cup \{D:\ \Delta D \in \alpha\}$ and observe,

$$\mathrm{PRL}_1 + X \vdash \Delta C \rightarrow C \quad \Rightarrow \quad \mathrm{PRL}_1 \vdash \bigwedge X \rightarrow .\Delta C \rightarrow C$$

$$\Rightarrow \quad \mathrm{PRL}_1 \vdash \Delta\bigwedge X \rightarrow \Delta(\Delta C \rightarrow C)$$

$$\Rightarrow \quad \mathrm{PRL}_1 \vdash \bigwedge\alpha \rightarrow \Delta(\Delta C \rightarrow C), \qquad (**)$$

since each sentence ΔE for $E \in X$ is derivable from $\bigwedge\alpha$. But (**) is contrary to (*), whence $\mathrm{PRL}_1 + X + \Delta C + \sim C$ is consistent and $X + \Delta C + \sim C$ has an S-completion β. By choice of X, $\alpha R \beta \in K$:

i-iii. $\alpha < \beta$ follows as before

iv. $\Delta D \in \alpha \Rightarrow D \in X \Rightarrow D \in \beta$.

Thus,

$$\Delta C \notin \alpha \quad \Rightarrow \quad \exists\beta(\alpha R \beta \ \&\ C \notin \beta)$$

$$\Rightarrow \quad \exists\beta(\alpha R \beta \ \&\ \beta \not\Vdash C)$$

$$\Rightarrow \quad \alpha \not\Vdash \Delta C.$$

This completes the proof ot the Claim and therewith the proof of the Theorem. QED

If one reviews the above, rather longish, proof, one will find that there is not much new in it-- just an excess of detail. Our completeness theorem for MOS will at least have some novelty. Before going on to this, however, let us pause and consider a generalisation of PRL_1 and the above proof that will be of some use in discussing MOS.

2.9. DEFINITION. The system PRL_n is the multi-modal system in a language with $n+1$ modal operators $\Box, \Delta_1, \ldots, \Delta_n$ and axioms and rules as follow:

AXIOMS. *A1-A2,A4* as usual

$A3^n$. $\Box A \rightarrow \Box\Box A$; $\Delta_i A \rightarrow \Box\Delta_i A$, $i = 1,\ldots,n$

$A6^n$. $\Box A \rightarrow \Delta_i A$, $i = 1,\ldots,n$

$A8^n$. $\Delta_i A \wedge \Delta_i(A \rightarrow B) \rightarrow \Delta_i B$, $i = 1,\ldots,n$.

RULES. *R1, R2* as usual.

Intuitively: $\Delta_1,\ldots,\Delta_n$ stand for $Pr_{T_1},\ldots,Pr_{T_n}$ for n totally unrelated consistent *RE* extensions $\mathrm{T}_1,\ldots,\mathrm{T}_n$ of PRA. It is possible (cf. the exercises) to consider $\Delta_1,\ldots,\Delta_n$ representing theories related by some configuration of inclusions and non-inclusions.

2.10. DEFINITION. A *Kripke model for* PRL_n is given by $\underline{K} = (K,<,R_1,\ldots,R_n,\alpha_0,\Vdash)$, where

i. $(K,<,\alpha_0)$ is a finite partially ordered set with minimum element α_0

ii. a. each R_i is a partial subordering of $<$, i.e.

$\alpha R_i \beta \Rightarrow \alpha < \beta$, for each i

b. for each i, $\alpha < \beta R_i \gamma \Rightarrow \alpha R_i \gamma$

iii. $\Vdash$ satisfies, in addition to the usual boolean laws,

a. $\alpha \Vdash \Box A$ iff $\forall\beta(\alpha < \beta \Rightarrow \beta \Vdash A)$

b. for each i, $\alpha \Vdash \Delta_i A$ iff $\forall\beta(\alpha R_i \beta \Rightarrow \beta \Vdash A)$.

2.11. THEOREM. For any modal sentence A,

$\mathrm{PRL}_n \vdash A$ iff $\alpha_0 \Vdash A$ for all finite Kripke models $\underline{K}$ for PRL_n.

I leave the proof as an exercise (cf. Exercise 1) for the reader.

Let us now consider MOS. For it, we use an alternate Kripke modelling.

2.12. DEFINITION. A *Kripke model for* MOS is given by $\underline{K} = (K,<,\mathrm{F},\alpha_0,\Vdash)$, where

i. $(K,<,\alpha_0)$ is a finite partially ordered set with minimum element α_0

ii. F is a map $K \to \mathsf{PP}(K)$ (where P is the power set operation) satisfying

a. F_α is a collection of subsets of $K_\alpha = \{\beta \in K\colon \ \alpha < \beta\}$

b. $\alpha < \beta \ \& \ X \in \mathsf{F}_\alpha \Rightarrow X \cap K_\beta \in \mathsf{F}_\beta$

c. $K_\alpha \in \mathsf{F}_\alpha$

iii. $\Vdash$ satisfies, in addition to the usual boolean laws,

a. $\alpha \Vdash \Box A$ iff $\forall \beta(\alpha < \beta \Rightarrow \beta \Vdash A)$

b. $\alpha \Vdash \nabla A$ iff $\exists X \in \mathsf{F}_\alpha \forall \beta(\beta \in X \Rightarrow \beta \Vdash A)$.

(A variant: Replace ii.c by

ii.c. $X \in \mathsf{F}_\alpha \ \& \ X \subseteq Y \subseteq K_\alpha \Rightarrow Y \in \mathsf{F}_\alpha$; $\mathsf{F}_\alpha \neq \emptyset$

and iii.b by

iii.b. $\alpha \Vdash \nabla A$ iff $\{\beta \in K_\alpha\colon \ \beta \Vdash A\} \in \mathsf{F}_\alpha$.)

2.13. THEOREM. For any modal sentence A,

$$\mathsf{MOS} \vdash A \text{ iff } \alpha_0 \Vdash A \text{ for all finite Kripke models } \underline{K} \text{ for } \mathsf{MOS}.$$

Proof: $\Rightarrow$ Let us merely check the new axioms *A3*, *A6*, and *A7*.

A3. Suppose $\alpha \Vdash \nabla B$. Let $X \in \mathsf{F}_\alpha$ be such that for all $\gamma \in X$, $\gamma \Vdash B$. Let $\beta > \alpha$ and $X_\beta = X \cap K_\beta \in \mathsf{F}_\beta$. Obviously, for all $\gamma \in X_\beta$, $\gamma \Vdash B$. Thus $\beta \Vdash \nabla B$. Since $\beta > \alpha$ was arbitrary, $\alpha \Vdash \Box \nabla B$.

A6. Observe

$$\begin{aligned} \alpha \Vdash \Box B &\Rightarrow \forall \beta > \alpha(\beta \Vdash B) \\ &\Rightarrow \forall \beta \in K_\alpha \in \mathsf{F}_\alpha(\beta \Vdash B) \\ &\Rightarrow \alpha \Vdash \nabla B. \end{aligned}$$

A7. Suppose $\alpha \Vdash \nabla B$ and $\alpha \Vdash \Box(B \to C)$. By the former, there is some $X \in \mathsf{F}_\alpha$ such that $\beta \Vdash B$ for all $\beta \in X$. By the latter supposition, for all $\beta > \alpha$-- in particular for all $\beta \in X$-- $\beta \Vdash B \Rightarrow \beta \Vdash C$, whence $\forall \beta \in X(\beta \Vdash C)$. Thus $\alpha \Vdash \nabla C$.

$\Leftarrow$ As usual, we prove the contrapositive: We assume $\mathsf{MOS} \nvdash A$ and construct a model $\underline{K} = (K,<,\mathsf{F},\alpha_0,\Vdash)$ of MOS in which $\alpha_0 \nVdash A$. The construction of $(K,<,\alpha_0)$ is exactly as in the proof of Theorem 2.8 and the fact that $(K,<,\alpha_0)$ is a finite partially ordered set with minimum element α_0 is established as before.

To define F_α, we first consider each $\nabla B \in S$. If $\nabla B \in \alpha$, define

$$X(\alpha,\nabla B) = \{\beta > \alpha:\ B \in \beta\};$$

if $\nabla B \notin \alpha$, leave $X(\alpha,\nabla B)$ undefined. We then set

$$F_\alpha = \{X(\alpha,\nabla B):\ X(\alpha,\nabla B) \text{ has been defined}\} \cup \{K_\alpha\}.$$

Of conditions 2.12.ii governing the behaviour of F, the only nontrivial one is ii.b. Suppose $\alpha < \beta$ and $X \in F_\alpha$. If $X = K_\alpha$, then $X \cap K_\beta = K_\beta \in F_\beta$. Thus, assume $X = X(\alpha,\nabla B)$ for some $\nabla B \in S$. Now

$$X \cap K_\beta = \{\gamma > \alpha:\ B \in \gamma\} \cap \{\gamma:\ \gamma > \beta\}$$
$$= \{\gamma > \beta:\ B \in \gamma\} = X(\beta,\nabla B),$$

if this last is defined, i.e. if $\nabla B \in \beta$. But:

$$\nabla B \in \alpha \ \&\ \alpha < \beta \implies \nabla B \in \beta,$$

by the definition of $<$ given back in the proof of Theorem 2.8. Hence, $X(\beta,\nabla B)$ is defined and $X \cap K_\beta = X(\beta,\nabla B) \in F_\beta$.

Finally, define $\Vdash$ by

$$\alpha \Vdash p \quad \text{iff} \quad p \in \alpha$$

for all atoms $p \in S$ and all S-completions $\alpha \in K$.

Claim. For all $B \in S$,

$$\alpha \Vdash B \quad \text{iff} \quad B \in \alpha.$$

Proof: As usual, this is an induction on the complexity of B. The only new case is $B = \nabla C$.

First, observe that, if $\nabla C \in \alpha$, then $X(\alpha,\nabla C) \in F_\alpha$ is defined. Now, for any $\beta \in X(\alpha,\nabla C)$, $C \in \beta$, whence the induction hypothesis yields $\beta \Vdash C$. Thus, $\forall \beta \in X(\alpha,\nabla C) \in F_\alpha(\beta \Vdash C)$, i.e. $\alpha \Vdash \nabla C$.

The converse is, of course, more involved. Observe,

$$\alpha \Vdash \nabla C \implies \exists X \in F_\alpha \forall \beta \in X(\beta \Vdash C). \qquad (*)$$

If $X = K_\alpha$, then $\alpha \Vdash \Box C$, whence $\Box C \in \alpha$, whence $\nabla C \in \alpha$ by *A3*. (*N.B.* This is where we use the extra set $\{\Box C:\ \nabla C$ is a subformula of $A\}$ in the definition of S.) Otherwise $X = X(\alpha,\nabla D)$ for some D such that $\nabla D \in \alpha$. I make the

Subclaim. $\mathrm{MOS} + \bigwedge\alpha \vdash \Box(D \to .\nabla C \to C)$.

For, otherwise

$$\{E \wedge \Box E:\ \Box E \in \alpha\} + \{\nabla E:\ \nabla E \in \alpha\} + D + \nabla C + \sim C$$

is consistent with MOS. Let β be any S-completion of this. If $\nabla C \in \alpha$, we can forget this whole argument, so assume $\nabla C \notin \alpha$. Thus $\alpha < \beta$. By the definition of $X(\alpha,\nabla D)$, $\beta \in X(\alpha,\nabla D)$, whence $\beta \Vdash C$ by (*). But $C \notin \beta$ and the induction hypothesis yields $\beta \nVdash C$, a contradiction. Thus, we get the subclaim.

Now, we have

$$\begin{aligned} \mathsf{MOS} + \bigwedge\alpha &\vdash \Box(D \to .\nabla C \to C), \quad \text{by the subclaim} \\ &\vdash \nabla D \to \nabla(\nabla C \to C), \quad \text{by } A7 \\ &\vdash \nabla D \to \nabla C, \quad \text{by } 1.4 \\ &\vdash \nabla C, \quad \text{since } \nabla D \in \alpha. \end{aligned}$$

Since α is S-complete, this means $\nabla C \in \alpha$ and we have established

$$\alpha \Vdash \nabla C \Rightarrow \nabla C \in \alpha, \quad \text{for } \nabla C \in S.$$

This completes the proof of the Theorem. QED

And what about PRL_{ZF}? This theory has no finite model theory of its own. As with PRL^{ω} and PRL, however, PRL_{ZF} is reducible to PRL_1: For any modal sentence A,

$$\mathsf{PRL}_{ZF} \vdash A \quad \text{iff} \quad \mathsf{PRL}_1 \vdash \bigwedge_{B \in S} \Delta(\Box B \to B) \to A,$$

where S is the set of subformulae of A. Unlike the situation with PRL^{ω}, however, there seems to be no simple model theoretic proof of this fact.

EXERCISES

1. i. Give a detailed proof of Theorem 2.11 on the completeness of PRL_n with respect to its models.

   ii. Let $\underline{P} = (\{1,\ldots,n\},<_P)$ be a partial ordering. Define $\mathsf{PRL}(\underline{P})$ to be the extension of PRL_n by the axiom schemata

   $$\Delta_i A \to \Delta_j A \quad \text{for } i <_P j.$$

   Define a model $\underline{K} = (K,<,R_1,\ldots,R_n,\alpha_0,\Vdash)$ for PRL_n to be a model for $\mathsf{PRL}(\underline{P})$ if one has

   $$R_i \subseteq R_j \quad \text{whenever } i <_P j$$

   Prove the completeness of $\mathsf{PRL}(\underline{P})$ with respect to these models.

   iii. Prove the soundness of $\mathsf{PRL}(\underline{P})$ with respect to arithmetic interpretations in which Δ_i is interpreted by Pr_{T_i}, where $T_1,\ldots,T_n$ are consistent RE extensions of PRA such that

$$\mathsf{PRA} \vdash Pr_{T_i}(\ulcorner\phi\urcorner) \to Pr_{T_j}(\ulcorner\phi\urcorner), \text{ whenever } i <_P j.$$

Show that, if each T_i is a finite extension of PRA, this last condition can be relaxed to (i.e. is implied by)

$$\mathsf{T}_i \subseteq \mathsf{T}_j, \text{ whenever } i <_P j.$$

2. Use Kripke models to derive the following closure rules:

    i. $\mathsf{PRL}_1 \vdash \Box A \Rightarrow \mathsf{PRL}_1 \vdash A$

    ii. $\mathsf{PRL}_1 \vdash \Delta A \Rightarrow \mathsf{PRL}_1 \vdash A$

    iii. $\mathsf{PRL}_1 \vdash \Box A \to \Box B \Rightarrow \mathsf{PRL}_1 \vdash \boxed{s} A \to B$

    iv. $\mathsf{PRL}_1 \vdash \Delta A \to \Delta B \Rightarrow \mathsf{PRL}_1 \vdash A \wedge \Delta A \to B$

    v. $\mathsf{PRL}_1 \vdash \Delta A \to \Box B \Rightarrow \mathsf{PRL}_1 \vdash \Delta A \to B$

    vi. $\mathsf{PRL}_{ZF} \vdash \Box A \Rightarrow \mathsf{PRL}_{ZF} \vdash A$

    vii. Does $\mathsf{PRL}_{ZF} \vdash \Delta A \Rightarrow \mathsf{PRL}_{ZF} \vdash A$?

(Hints: i-v: Given countermodels $\underline{K}$ to the right-hand assertions, construct countermodels to the left-hand ones by tacking new nodes α_{-1} on below the original minima α_0: $\alpha_{-1} < \alpha_0$. Sometimes one will have to assume $\alpha_{-1} R \alpha_0$ and sometimes that $\alpha_{-1} \not R \alpha$ for any $\alpha \in K$. vi. Reduce this to an earlier result. Do not use the characterisation of provability in PRL_{ZF} cited in the last paragraph of text preceding the exercises.)

3. Prove the arithmetic completeness of PRL_1:

    i. Let T be a consistent *RE* extension of PRA, provably so over PRA, i.e., for all sentences ϕ,

$$\mathsf{PRA} \vdash Pr(\ulcorner\phi\urcorner) \to Pr_T(\ulcorner\phi\urcorner).$$

Extend our earlier notion of an arithmetic interpretation by mapping: $(\Delta B)^* = Pr_T(\ulcorner B^* \urcorner)$. Show: For all A, $\mathsf{PRL}_1 \vdash A \Rightarrow \mathsf{PRA} \vdash A^*$

    ii. Assume $\underline{K} = (\{1,\ldots,n\}, R_1, R_2, 1, \Vdash)$ is a finite model of PRL_1, where we write "R_1" and "R_2" for "<" and "R", respectively, to avoid confusion with the usual ordering of the integers. Define two functions F, G by

$$F0 = 0; \qquad G0 = 0$$

$$F(x+1) = \begin{cases} y, & Prov(x+1, \ulcorner L \neq \overline{y} \urcorner) \ \& \ Fx\, R_1\, y \\ Fx, & \text{otherwise,} \end{cases}$$

$$G(x+1) = \begin{cases} y, & Prov(x+1, \ulcorner F = G \to M \neq \overline{y}\urcorner) \ \& \ Gx R_2 y \\ Gx, & \text{otherwise}, \end{cases}$$

where $L = \lim_{x\to\infty} Fx$, $M = \lim_{x\to\infty} Gx$. Formulate and prove the basic properties of F,G, L, and M analogous to those of F,L used in proving Solovay's First Completeness Theorem.

iii. Let $T = PRA + F = G$. Define $*$ by

$$p^* = \bigvee\{L = \overline{i}: \ i \Vdash p\}$$

$$(\square B)^* = Pr(\ulcorner B^*\urcorner), \qquad (\Delta B)^* = Pr_T(\ulcorner B^*\urcorner), \qquad \text{etc.}$$

iv. Prove: $PRL_1 \vdash A$ iff $PRA \vdash A^*$ for all $*$ based on finitely axiomatised extensions $T \supseteq PRA$ (i.e. extensions $T = PRA + \phi$).

(REMARKS: In the next section, we will prove the arithmetic completeness of PRL_1 by a much simpler device-- reduction to Solovay's First Completeness Theorem. Note that the present proof shows PRL_1 is complete with respect to PRA and finitely axiomatised extensions $T \supseteq PRA$. What is the model theoretic interpretation of this? To find the answer, read on.)

3. CARLSON MODELS

The Kripke models of the previous section are okay-- even moderately useful (as illustrated in the exercises at the end of that section). But they do not reveal fully the relations among PRL, PRL_1, PRL_n, and MOS. This is accomplished by a new model theory due to Timothy Carlson. To explain this, recall the intended interpretations:

$$(\Delta A)^* = Pr_T(\ulcorner A^*\urcorner)$$

$$(\Delta_i A)^* = Pr_{T_i}(\ulcorner A^*\urcorner)$$

$$(\nabla A)^* = \bigvee Pr_{T_i}(\ulcorner A^*\urcorner).$$

If T and each T_i is finitely axiomatised over PRA-- and don't forget that a conjunction yields a single axiom-- each $Pr_{T_i}(\ulcorner\psi\urcorner)$, say, assumes the form

$$Pr(\ulcorner \phi_i \to \psi\urcorner),$$

i.e. $\quad (\square(p_i \to q))^*$,

for some fixed p_i. And how is p_i interpreted in a model $\underline{K} = (K,<,\alpha_0,\Vdash)$ of PRL? Answer: Let $D_i = \{\alpha \in K: \ \alpha \Vdash p_i\}$. Thus:

3.1. DEFINITION. A *Carlson model for* PRL$_1$ is given by $\underline{K} = (K,<,D,\alpha_0,\Vdash)$, where

i. $(K,<,\alpha_0)$ is a finite partially ordered set with minimum element α_0

ii. D is a subset of K

iii. $\Vdash$ satisfies, in addition to the usual boolean laws,

a. $\alpha \Vdash \Box A$ iff $\forall\beta(\alpha < \beta \Rightarrow \beta \Vdash A)$

b. $\alpha \Vdash \Delta A$ iff $\forall\beta(\alpha < \beta \;\&\; \beta \in D \Rightarrow \beta \Vdash A)$.

3.2. DEFINITION. A *Carlson model for* PRL$_n$ is given by $\underline{K} = (K,<,D_1,\ldots,D_n,\alpha_0,\Vdash)$, where

i. $(K,<,\alpha_0)$ is a finite partially ordered set with minimum element α_0

ii. each D_i is a subset of K

iii. $\Vdash$ satisfies, in addition to the usual boolean laws,

a. $\alpha \Vdash \Box A$ iff $\forall\beta(\alpha < \beta \Rightarrow \beta \Vdash A)$

b. $\alpha \Vdash \Delta_i A$ iff $\forall\beta(\alpha < \beta \;\&\; \beta \in D_i \Rightarrow \beta \Vdash A)$.

3.3. DEFINITION. A *Carlson model for* MOS is given by $\underline{K} = (K,<,D_1,\ldots,D_n,\alpha_0,\Vdash)$, where

i. $(K,<,\alpha_0)$ is a finite partially ordered set with minimum element α_0

ii. each D_i is a subset of K

iii. $\Vdash$ satisfies, in addition to the usual boolean laws,

a. $\alpha \Vdash \Box A$ iff $\forall\beta(\alpha < \beta \Rightarrow \beta \Vdash A)$

b. $\alpha \Vdash \nabla A$ iff $\exists i \forall\beta(\alpha < \beta \;\&\; \beta \in D_i \Rightarrow \beta \Vdash A)$.

Our immediate goal is, obviously, to prove the completeness of PRL$_1$, PRL$_n$, and MOS with respect to their respective Carlson models. We shall accomplish this simply by reducing these results to those of the previous section by transforming Kripke models into Carlson models.

3.4. THEOREM. Let A be a sentence of the language of PRL$_1$. Then: PRL$_1 \vdash A$ iff $\alpha_0 \Vdash A$ for all Carlson models $\underline{K} = (K,<,D,\alpha_0,\Vdash)$ for PRL$_1$ in which $(K,<,\alpha_0)$ is a finite tree.

Proof: $\Rightarrow$ The soundness of the interpretation is a routine matter best left to the reader as an exercise. Note that the proof (as well as those of soundness for the ensuing completeness theorems) does not require the frame $(K,<,\alpha_0)$ to be a tree.

$\Leftarrow$ Suppose PRL$_1 \nvdash A$. Then there is a finite Kripke model $\underline{K} = (K,<,R,\alpha_0,\Vdash)$

in which $\alpha_0 \not\Vdash A$. We transform $\underline{K}$ into a Carlson model as follows: First, let K_T ("T" for "tree") be the set of all finite sequences $(\alpha_0,\alpha_1,\ldots,\alpha_{k-1})$ such that $\alpha_0 < \alpha_1 < \ldots < \alpha_{k-1}$ and α_0 is the minimum element of K. Order these sequences by proper extendability, so

$$(\alpha_0,\ldots,\alpha_{k-1}) <_T (\alpha_0,\ldots,\alpha_{k-1},\alpha_k,\ldots,\alpha_{m-1}).$$

$(K_T,<_T,(\alpha_0))$ is obviously a finite tree with origin (α_0).

Define D by

$$(\alpha_0,\ldots,\alpha_{k-1},\alpha_k) \in D \quad \text{iff} \quad \alpha_{k-1} R \alpha_k.$$

(One can arbitrarily add (α_0) to D or refrain from doing so.) Finally, define $\Vdash_T$ by:

$$(\alpha_0,\ldots,\alpha_{k-1}) \Vdash_T p \quad \text{iff} \quad \alpha_{k-1} \Vdash p$$

for all atoms p.

Claim. For all sentences B,

$$(\alpha_0,\ldots,\alpha_{k-1}) \Vdash_T B \quad \text{iff} \quad \alpha_{k-1} \Vdash B.$$

As usual, the only novel case in the inductive proof of the Claim is that involving the new operator. Thus, let $B = \Delta C$.

First, observe

$$(\alpha_0,\ldots,\alpha_{k-1}) \Vdash_T \Delta C \;\Rightarrow\; \forall\beta\big(\alpha_{k-1} R \beta \Rightarrow (\alpha_0,\ldots,\alpha_{k-1},\beta) \Vdash_T C\big)$$

since $\alpha_{k-1} R \beta \Rightarrow (\alpha_0,\ldots,\alpha_{k-1},\beta) \in D$. But the induction hypothesis yields $\beta \Vdash C$ for all β such that $\alpha_{k-1} R \beta$, whence $\alpha_{k-1} \Vdash \Delta C$.

The converse is, of course, trickier. First note that

$$\alpha_{k-1} \Vdash \Delta C \;\Rightarrow\; \forall\beta(\alpha_{k-1} R \beta \Rightarrow \beta \Vdash C).$$

Now let $(\alpha_0,\ldots,\alpha_{k-1},\ldots,\alpha_{m-1},\alpha_m) \in D$ extend $(\alpha_0,\ldots,\alpha_{k-1})$. (I leave the case of an immediate extension $(\alpha_0,\ldots,\alpha_{k-1},\alpha_k)$ to the reader.) By the mixed transitivity of $<$ and R, it follows that $\alpha_{k-1} R \alpha_m$. Thus $\alpha_m \Vdash C$ and the induction hypothesis yields $(\alpha_0,\ldots,\alpha_{k-1},\ldots,\alpha_{m-1},\alpha_m) \Vdash_T C$. Since this was an arbitrary extension of $(\alpha_0,\ldots,\alpha_{k-1})$ in D, it follows that $(\alpha_0,\ldots,\alpha_{k-1}) \Vdash_T \Delta C$.

An n-fold iteration of this proof yields:

<u>3.5. THEOREM</u>. Let A be a sentence of the language of PRL_n. Then: $PRL_n \vdash A$ iff $\alpha_0 \Vdash A$ for all Carlson models $\underline{K} = (K,<,D_1,\ldots,D_n,\alpha_0,\Vdash)$ for PRL_n in which $(K,<,\alpha_0)$

is a finite tree.

I leave the proof as an exercise to the reader.

For MOS, let me separate soundness from completeness. First, soundness:

3.6. LEMMA. Let A be a sentence of the language of MOS. Then: $\mathrm{MOS}\vdash A \Rightarrow \alpha_0 \Vdash A$ for all n and all Carlson n-models $\underline{K} = (K,<,D_1,\ldots,D_n,\alpha_0,\Vdash)$ for MOS.

The proof is routine and I omit it.

The completeness result is numerically refined:

3.7. THEOREM. Let the sentence A of the language of MOS have exactly $n \geq 1$ subformulae of type ∇B. Then: $\mathrm{MOS}\vdash A$ iff $\alpha_0 \Vdash A$ in all Carlson n-models $\underline{K} = (K,<,D_1,\ldots,D_n,\alpha_0,\Vdash)$ in which $(K,<,\alpha_0)$ is a finite tree.

From this refined information, it is easy to prove the completeness theorem in its more familiar form:

3.8. THEOREM. Let A be a sentence of the language of MOS. Then: $\mathrm{MOS}\vdash A$ iff $\alpha_0 \Vdash A$ for all n and all Carlson n-models $\underline{K} = (K,<,D_1,\ldots,D_n,\alpha_0,\Vdash)$ for MOS in which $(K,<,\alpha_0)$ is a finite tree.

I leave this reduction to the reader (but cf. Exercise 1, below).

Proof of Theorem 3.7: $\Rightarrow$ By Lemma 3.6.

$\Leftarrow$ By contraposition. Assume $\mathrm{MOS}\nvdash A$. Let $\underline{K} = (K,<,F,\alpha_0,\Vdash)$ be a finite Kripke countermodel to A: $\alpha_0 \nVdash A$. The transformation of $\underline{K}$ into a Carlson n-model $\underline{K}_C$ is performed in two steps. First we transform $\underline{K}$ into a Kripke tree model $\underline{K}_T$; then we transform $\underline{K}_T$ into the desired model $\underline{K}_C$.

The construction of $\underline{K}_T$ is routine: K_T consists of all nontrivial finite increasing sequences,

$$\alpha_0 < \alpha_1 < \ldots < \alpha_{k-1},$$

beginning at the origin α_0. $<_T$ is the usual ordering by extension. So defined, $(K_T,<_T,(\alpha_0))$ is clearly a finite tree with origin (α_0).

To define $F^T_{(\alpha_0,\ldots,\alpha_{k-1})}$, start with $X(\alpha_{k-1},\nabla B)$ (for $\nabla B \in S$) and replace it by what is intended to be $X((\alpha_0,\ldots,\alpha_{k-1}),\nabla B)$:

$$\{(\alpha_0,\ldots,\alpha_{k-1},\ldots,\alpha_{m-1}): \ \alpha_{k-1} < \ldots < \alpha_{m-1} \in X(\alpha_{k-1},\nabla B)\}.$$

Similarly, replace $K_{\alpha_{k-1}}$ by $K_{T,(\alpha_0,\ldots,\alpha_{k-1})}$. $F^T_{(\alpha_0,\ldots,\alpha_{k-1})}$ consists only of these replacements.

Finally, define $\Vdash_T$ by

$$(\alpha_0,\ldots,\alpha_{k-1}) \Vdash_T p \quad \text{iff} \quad \alpha_{k-1} \Vdash p,$$

for all atoms p.

Claim 1. For all sentences $B \in S$,

$$(\alpha_0,\ldots,\alpha_{k-1}) \Vdash_T B \quad \text{iff} \quad \alpha_{k-1} \Vdash B.$$

The proof is routine and I omit it. (I do note, however, the restriction to $B \in S$: We have ignored $X(\alpha_{k-1},\nabla C)$ for $\nabla C \notin S$. This is not necessary for this step of the proof, but is necessary for the transition from $\underline{K}_T$ to $\underline{K}_C$. (S, incidentally, is the slight extension of the set of subformulae of A introduced in the proof of Theorem 2.8.))

Now: For the sake of notation, we can delete all subscripts "T" and assume our initial model $\underline{K}$ to be a finite tree *with the special property*: For any α and any $\beta > \alpha$, there is an immediate successor α' to α which behaves exactly like β: For any $B \in S$, $\alpha' \Vdash B$ iff $\beta \Vdash B$. The reason $\underline{K}_T$ has this property is simple: α,β are finite sequences the behaviours of which are determined by their last entries. Tacking the last element of $\beta > \alpha$ onto the end of α yields α'.)

The transformation is very nearly complete already: Each F_α consists of K_α and at most n other sets of the form $X(\alpha,\nabla B) = \{\beta > \alpha\colon\ \beta \Vdash B\}$ for $\nabla B \in S$. The final step is the construction of sets $D_{\nabla B_1},\ldots,D_{\nabla B_n}$, for $\nabla B_1,\ldots,\nabla B_n \in S$, as conglomerations of the appropriate $X(\alpha,\nabla B)$'s. We do this as follows: For each $\nabla B \in S$, we start at the bottom of the tree and upwardly proceed to determine which elements $\beta > \alpha$ are in $D_{\nabla B}$. For a while we specify only the immediate successors; suddenly we hit an α at which we determine all successors (immediate or otherwise) which belong to $D_{\nabla B}$.

Enumerate all ∇-formulae of S: $\nabla B_1,\ldots,\nabla B_n$.

Stage ∇B_1. This stage of the construction proceeds inductively up the tree. Suppose we are at node α (in the beginning, $\alpha = \alpha_0$). If $\alpha \Vdash \nabla B_1$, set $K_\alpha \cap D_{\nabla B_1} = X(\alpha,\nabla B_1)$, declare the job completed for all $\beta \geq \alpha$ and move on to any untreated incomparable nodes.

If $\alpha \Vdash\!\!\!/\ \nabla B_1$, there is no $X(\alpha, \nabla B_1)$ but $K_\alpha \in F_\alpha$. However, K_α is too large to put into $D_{\nabla B_1}$ since we may later force ∇B_1. So we only put all immediate successors to α into $D_{\nabla B_1}$. The decision at α has been made and one can now pass to the successors of α and any remaining incomparable nodes.

Successively treat $\nabla B_2, \ldots, \nabla B_n$ similarly.

Define $\underline{K}_C = (K, <, D_{\nabla B_1}, \ldots, D_{\nabla B_n}, \alpha_0, \Vdash_C)$, where

$$\alpha \Vdash_C p \text{ iff } \alpha \Vdash p$$

for all atomic p.

Claim 2. For any α and any sentence $B \in S$,

$$\alpha \Vdash_C B \quad \text{iff} \quad \alpha \Vdash B.$$

Proof: As always, the proof is by induction on the complexity of B. The only nontrivial case is $B = \nabla C$, i.e. $B = \nabla B_i$ for some i.

First observe,

$$\begin{aligned} \alpha \Vdash \nabla B_i \quad &\Rightarrow \quad D_{\nabla B_i} \cap K_\alpha = X(\alpha, \nabla B_i), \text{ by construction} \\ &\Rightarrow \quad \forall \beta > \alpha(\beta \in D_{\nabla B_i} \Rightarrow \beta \Vdash B_i) \\ &\Rightarrow \quad \forall \beta > \alpha(\beta \in D_{\nabla B_i} \Rightarrow \beta \Vdash_C B_i), \text{ by induction hypothesis} \\ &\Rightarrow \quad \alpha \Vdash_C \nabla B_i. \end{aligned}$$

The converse is handled contrapositively. Suppose $\alpha \Vdash\!\!\!/\ \nabla B_i$. Consider any ∇B_j. We want $\alpha < \beta \in D_{\nabla B_j}$ such that $\beta \Vdash\!\!\!/_C\ B_i$.

Case 1. $\alpha \Vdash \nabla B_j$: Then $D_{\nabla B_j} \cap K_\alpha = X(\alpha, \nabla B_j)$. By the definition of $\Vdash$, there is some $\beta \in X(\alpha, \nabla B_j)$ for which $\beta \Vdash\!\!\!/\ B_i$, whence $\beta \Vdash\!\!\!/_C\ B_i$ (by induction hypothesis). Thus, $\exists \beta > \alpha\big(\beta \in D_{\nabla B_j} \ \&\ \beta \Vdash\!\!\!/_C\ B_i\big)$.

Case 2. $\alpha \Vdash\!\!\!/\ \nabla B_j$: Since $\alpha \Vdash\!\!\!/\ \nabla B_i$, there is some $\beta \in K_\alpha$ such that $\beta \Vdash\!\!\!/\ B_i$. By our assumed special property of $\underline{K}$, we can assume β is an immediate successor of α. But then $\beta \in D_{\nabla B_j}$ is such that $\beta \Vdash\!\!\!/_C\ B_i$ (this latter by the induction hypothesis).

Thus, we have shown that, if $\alpha \Vdash\!\!\!/\ \nabla B_i$, then

$$\forall j\, \exists \beta > \alpha\big(\beta \in D_{\nabla B_j} \ \&\ \beta \Vdash\!\!\!/_C\ B_i\big)$$

whence $\alpha \Vdash\!\!\!/_C\ \nabla B_i$.

This completes the proof of the Claim and therewith the proof of the Theorem. QED

A good model theory has many uses. With the model theories for PRL_1, PRL_n, and MOS, we could extend the semantic analysis of self-reference from PRL to these theories. Or, we could derive closure under various rules of inference. However, the real beauty of the Carlson semantics is that they allow several instantaneous interpretability results.

3.9. DEFINITION. Let $n \geq 1$. We define the *n-interpretation*, A^n, of a sentence of the language of MOS inductively on the construction of A as follows:

$$f^n = f; \qquad t^n = t; \qquad p^n = p$$

$$(B \circ C)^n = B^n \circ C^n \text{ for } \circ \in \{\wedge, \vee, \rightarrow\}$$

$$(\sim B)^n = \sim B^n; \qquad (\Box B)^n = \Box B^n$$

$$(\nabla B)^n = \Delta_1 B^n \vee \ldots \vee \Delta_n B^n.$$

3.10. THEOREM. Let $n \geq 1$ be given.

i. Let A be any sentence of the language of MOS. Then:

$$\mathsf{MOS} \vdash A \Rightarrow \mathsf{PRL}_n \vdash A^n.$$

ii. Let A be a sentence of the language of MOS with at most n occurrences of ∇. Then:

$$\mathsf{MOS} \vdash A \text{ iff } \mathsf{PRL}_n \vdash A^n.$$

The (syntactic) proof of part i is a routine induction on the length of a derivation in MOS; the (semantic) proof of part ii is a routine induction on the length of A. The key step in the latter induction is, of course, that for formulae ∇B. Here there is little problem: MOS and PRL_n have the same structure underlying their n-models-- namely, $D_1,\ldots,D_n$. If Δ_i is determined by D_i, then

$$\alpha \Vdash \nabla A \text{ iff } \alpha \Vdash \Delta_1 A^n \vee \ldots \vee \Delta_n A^n,$$

assuming the obvious induction hypothesis on A. I leave the details to the reader.

Somewhat more interesting are the interpretations of PRL_1 and PRL_n (whence, by composition, of MOS) into PRL: D (or, for PRL_n: $D_1,\ldots,D_n$) is (are) just the set(s) of nodes at which some new propositional variable q_0 (variables $q_1,\ldots,q_n$, respectively) can be forced.

3.11. DEFINITION. Let the languages of PRL_1 and PRL_n (as well as that of MOS) have as atoms the variables $p_0, p_1, \ldots$. Add to PRL the variables $q_0, q_1, \ldots$. Then, A^Δ ambiguously denotes the interpretations of PRL_1 and PRL_n into PRL under the following schemata:

$$t^\Delta = t; \qquad f^\Delta = f; \qquad p^\Delta = p$$

$$(A \circ B)^\Delta = A^\Delta \circ B^\Delta \text{ for } \circ \in \{ \wedge , \vee , \rightarrow \}$$

$$(\sim A)^\Delta = \sim A^\Delta; \qquad (\Box A)^\Delta = \Box A^\Delta$$

$$(\Delta A)^\Delta = \Box(q_0 \rightarrow A^\Delta), \text{ where applicable}$$

$$(\Delta_i A) = \Box(q_i \rightarrow A^\Delta), \text{ where applicable.}$$

3.12. THEOREM. Let A be a sentence of the appropriate language. Then:

i. $\mathsf{PRL}_1 \vdash A$ iff $\mathsf{PRL} \vdash A^\Delta$

ii. $\mathsf{PRL}_n \vdash A$ iff $\mathsf{PRL} \vdash A^\Delta$

iii. $\mathsf{MOS} \vdash A$ iff $\mathsf{PRL} \vdash (A^n)^\Delta$, where A has at most n occurrences of ∇.

Again, I omit the proof as the result is model theoretically obvious.

Because these translations simulate the arithmetic relations among the interpretations of $\Box, \Delta, \Delta_i$, and ∇, we can read arithmetic completeness results for PRL_1, PRL_n, and MOS directly off that for PRL. For: The interpretations of the atoms $q_0, q_1, \ldots$ are merely sentences axiomatising extensions of PRA. If * denotes an arithmetic interpretation of PRL,

$$(\Delta_i A)^* = \Box(q_i \rightarrow A^\Delta)^*$$

$$= Pr(\ulcorner q_i^* \rightarrow A^{\Delta *}\urcorner)$$

$$= Pr_{T_i}(\ulcorner A^{\Delta *}\urcorner),$$

where $\mathsf{T}_i = \mathsf{PRA} + q_i^*$. Similarly, under the n-interpretation,

$$(\nabla A)^* = \bigvee_{i=1}^{n} Pr_{T_i}(\ulcorner A^{n\Delta *}\urcorner).$$

More formally:

3.13. DEFINITION. An arithmetic interpretation * of the modal langauge including $\Box, \Delta, \Delta_1, \ldots, \Delta_n, \nabla$ is given firstly by choosing a consistent *RE* theory T (provably) extending PRA, consistent *RE* theories $\mathsf{T}_1, \ldots, \mathsf{T}_n$ (provably) extending PRA, and a (possibly infinite) *RE* sequence of (appropriately behaved--cf. 1.10.ii) consistent theories $\mathsf{T}_1', \mathsf{T}_2', \ldots$ extending PRA, and secondly by assigning arithmetical sentences A^* to modal ones A by:

$$t^* \text{ is } \overline{0} = \overline{0}; \qquad f^* \text{ is } \overline{0} = \overline{1}; \qquad p^* \text{ is arbitrary}$$

$$(A \circ B)^* = A^* \circ B^* \text{ for } \circ \in \{ \wedge , \vee , \rightarrow \}$$

$$(\sim A)^* = \sim A^*; \qquad (\Box A)^* = Pr(\ulcorner A^* \urcorner) = Pr_{PRA}(\ulcorner A^* \urcorner)$$

$$(\Delta A)^* = Pr_T(\ulcorner A^* \urcorner); \qquad (\Delta_i A)^* = Pr_{T_i}(\ulcorner A^* \urcorner)$$

$$(\nabla A)^* = \bigvee_i Pr_{T_i'}(\ulcorner A^* \urcorner).$$

(This definition is a cheat: We really want three definitions, one for each of PRL_1, PRL_n, and MOS. Because we use different modal operators for the differing theories, however, the above specialises to distinct definitions when we restrict the various languages.)

3.14. SOUNDNESS LEMMA. For A in the appropriate language,

i. $PRL_1 \vdash A \Rightarrow \forall * (PRA \vdash A^*)$

ii. $PRL_n \vdash A \Rightarrow \forall * (PRA \vdash A^*)$

iii. $MOS \vdash A \Rightarrow \forall * (PRA \vdash A^*)$.

The proofs are routine and I omit them.

The converses to these statements also hold:

3.15. ARITHMETIC COMPLETENESS THEOREMS. For A in the appropriate language,

i. $PRL_1 \nvdash A \Rightarrow \exists T \exists *$ based on T $(PRA \nvdash A^*)$

ii. $PRL_n \nvdash A \Rightarrow \exists T_1, \ldots, T_n \exists *$ based on $T_1, \ldots, T_n$ $(PRA \nvdash A^*)$

iii. $MOS \nvdash A \Rightarrow \exists T_1, \ldots, T_n \exists *$ based on $T_1, \ldots, T_n$ $(PRA \nvdash A^*)$,

where n is the number of occurrences of ∇ in A.

The Theorem follows immediately from Solovay's First Completeness Theorem after showing, by yet another induction on the length of A, that

$$A^* = (A^{\Delta})^* \quad \text{or} \quad A^* = (A^{n\Delta})^*,$$

as appropriate.

Appealing to the uniformisation of Solovay's First Completeness Theorem (Chapter 3, section 4), we obtain uniform versions of 3.15.i and 3.15.ii-- provided we replace PRA by PA because of the extra induction needed in the uniformisation: There is a fixed T and a fixed $*$ such that

$$PRL_1 \vdash A \quad \text{iff} \quad PA \vdash A^*;$$

and there are fixed $T_1, \ldots, T_n$ and a fixed $*$ such that

$$PRL_1 \vdash A \quad \text{iff} \quad PA \vdash A^*.$$

For MOS, this proof only supplies a fixed sequence $T_1, T_2, \ldots$ and a fixed interpretation $*$ of atoms, but for which the interpretations $(\nabla A)^*$ vary over disjunctions of provabilities in the truncated sequences $T_1, \ldots, T_n$. One cannot restrict one's attention to a fixed finite sequence $T_1, \ldots, T_n$ (Exercise 3).

EXERCISES

1. Prove, without presupposing 3.8, that MOS is conservative over PRL, i.e. for A containing no occurrence of ∇, MOS$\vdash A$ iff PRL$\vdash A$. Give the details of the reduction of 3.8 to 3.7.

2. Give detailed proofs of some of the interpretation results of the end of the section.

3. Let $n \geq 1$. i. Show that the following schema is valid in all Carlson n-models for MOS, but not in all $(n+1)$-models:

$$A10_n.\quad \bigwedge_{0 \leq i \leq n} A_i \rightarrow \bigvee_{0 \leq i < j \leq n} \nabla(A_i \wedge A_j).$$

ii. For $n = 1$, this schema is simply

$$A10_1.\quad \nabla A \wedge \nabla B \rightarrow \nabla(A \wedge B).$$

Show syntactically: $A10_1$ axiomatises PRL_1 over MOS.

iii. Let $\mathrm{MOS}^{<n}$ be the extension of MOS by the axiom schema $A10_n$. Show: $\mathrm{MOS}^{<n}$ is complete with respect to Carlson n-models for MOS. (Hint: Modify the proof of completeness for MOS; but add to S all sentences $\nabla(B_1 \wedge \ldots \wedge B_k)$ such that $\nabla B_1, \ldots, \nabla B_k$ are (distinct) subformulae of A. The Kripke model of MOS will then satisfy crucial instances of $A10_n$. In transforming it into a Carlson n-model, proceed up the tree trying to define $D_{\nabla B_1}, \ldots, D_{\nabla B_k}$. Whenever more than n subformulae ∇B_i are forced, replace $\nabla B_1, \ldots, \nabla B_m$ by $\nabla B_1^*, \ldots, \nabla B_n^*$, each B_i^* a conjunction of B_i's. *N.B.* Here, where $\alpha \Vdash \nabla B_i$, one declares all appropriate *immediate* successors to belong to $D_{\nabla B_i}$. Later, one may have to pare down to $D_{\nabla(B_i \wedge B_j)}$.)

iv. Prove arithmetic completeness: $\mathrm{MOS}^{<n} \vdash A$ iff, for all $T_1, \ldots, T_n$ provably extending PRA and all $*$ based on $T_1, \ldots, T_n$ (i.e. $(\nabla B)^* = Pr_{T_1}(\ulcorner B^* \urcorner) \vee \ldots \vee Pr_{T_n}(\ulcorner B^* \urcorner)$), PRA$\vdash A^*$.

4. State and prove arithmetic completeness for the modal system $PRL(\underline{P})$ of Exercise 1 of section 2.

5. Prove the following analogue to Solovay's Second Completeness Theorem: The following are equivalent:

a. $PRL_1 \vdash \bigwedge_{\Box B \in S} (\Box B \rightarrow B) \wedge \bigwedge_{\Delta B \in S} (\Delta B \rightarrow B) \rightarrow A$,

where S is the set of subformulae of A,

b. A^* is true for all *sound* T and all interpretations $*$ based on T.

(Note the restriction to sound theories T. This is needed even to prove a $\Rightarrow$ b. To prove the converse by reducing it to Solovay's Second Completeness Theorem you will have to guarantee that q_0^* is true. This extra difficulty can be avoided by extending the treatment, in Exercise 3 of the previous section, of the First Completeness Theorem-- the new axiom used there, $\forall x(Fx = Gx)$, is already true. Finally, I remark that similar analogues for PRL_n and MOS also obtain.)

4. CARLSON'S ARITHMETIC COMPLETENESS THEOREM

Back in Chapter 3, we saw that Solovay's Second Completeness Theorem was the more useful of the two arithmetic Completeness Theorems; in the last section, I relegated the analogue to the Second Completeness Theorem to the Exercises. Why? Well, look at any application: Suppose, given $PRA \subseteq T$ (provably so), we want a sentence ϕ such that ϕ is independent of PRA, but T doesn't prove this, i.e. we want an interpretation $*$ under which

$$\sim\Box p \wedge \sim\Box\sim p \wedge \sim\Delta\sim\Box p \wedge \sim\Delta\sim\Box\sim p$$

comes out true. If we apply the given construction, we construct $*$-- *and* T! In short, we only conclude that there is a sound *RE* theory T with the desired property and not that every consistent *RE* $T \supseteq PRA$ has the property. For this particular result, we can use the methods of Chapter 7, below; but what can we do in the general situation?

One might suppose the problem lies in the rather cheap proof of the analogues to Solovay's Completeness Theorems we gave; maybe we should start with PRA and T and attempt to prove the analogues anew. Alas, this cannot be so easily accomplished:

PRL_1 is *not* always the theory of Pr_{PRA} and Pr_T. As already remarked in introducing PRL_{ZF} (2.3, above), if T is a lot stronger than PRA, the interpretation of * as Pr_T validates the new schema,

$$\Delta(\Box A \to A).$$

Even more simply, if, say $T = PRA + \sim Con(PRA)$, the interpretation validates

$$\Delta\Box f \quad \text{and} \quad \Delta\Delta f.$$

For a given T, it might not be easy to list all the new schemata validated, much less prove completeness.

There is one example of a theory T for which we have a good idea of what the additional bi-modal schemata should be-- namely, ZF (or any other strong sound theory like ZF, e.g. PA). ZF is powerful enough to prove the soundness of PRA. From this fact and Solovay's Second Completeness Theorem, by which the soundness of PRA accounted for all true modal schemata of $Pr(\cdot)$, I naturally conjectured that the theory PRL_{ZF} axiomatised the modal schemata provable in PRA when $\Box$ was interpreted as Pr_{PRA} and Δ as Pr_{ZF}. In this section, we will study Tim Carlson's affirmation of this conjecture.

4.1. DEFINITION. Given an assignment $p \mapsto p^*$ of arithmetic sentences to propositional atoms, we define an interpretation * extending this inductively as follows:

$$f^* \text{ is } \overline{0} = \overline{1}; \quad t^* \text{ is } \overline{0} = \overline{0}; \quad (\sim A)^* = \sim A^*$$

$$(A \circ B)^* = A^* \circ B^* \text{ for } \circ \in \{\wedge, \vee, \to\}$$

$$(\Box A)^* = Pr_{PRA}(\ulcorner A^* \urcorner); \quad (\Delta A)^* = Pr_{ZF}(\ulcorner A^* \urcorner).$$

4.2. CARLSON'S COMPLETENESS THEOREM. Let A be a modal sentence and let S be the set of subformulae of A.

i. The following are equivalent:

a. $\mathsf{PRL}_1 \vdash \bigwedge_{\Box B \in S} \Delta(\Box B \to B) \to A$

b. $\mathsf{PRL}_{ZF} \vdash A$

c. $\mathsf{PRA} \vdash A^*$ for all interpretations *

ii. The following are equivalent:

a. $\mathsf{PRL}_1 \vdash \bigwedge_{\Box B \in S} \Delta(\Box B \to B) \wedge \bigwedge_{\Box B \in S} (\Box B \to B) \to A$

b. $\mathsf{PRL}_{ZF} + Reflexion_{\Box} \vdash A$

c. $ZF \vdash A^*$ for all interpretations $*$

iii. The following are equivalent:

a. $PRL_1 \vdash \bigwedge_{\Box B \in S} \Delta(\Box B \to B) \wedge \bigwedge_{\Box B \in S} (\Box B \to B) \wedge \bigwedge_{\Delta B \in S} (\Delta B \to B) \to A$

b. $PRL_{ZF} + Reflexion_{\Box} + Reflexion_{\Delta} \vdash A$

c. A^* is true for all interpretations $*$.

Before setting out to prove this, I must explain what I mean by $Reflexion_{\Box}$ and $Reflexion_{\Delta}$.

4.3. DEFINITION-REMARK. Recall that the soundness of PRA is schematically represented by the formulae,

$$Pr(\ulcorner\phi\urcorner) \to \phi.$$

Modally, this is rendered in PRL^{ω} over PRL by the schema of *reflexion*,

$$Reflexion: \quad \Box A \to A.$$

With two theories and their respective provability predicates, we get two schemata,

$$Reflexion_{\Box}: \quad \Box A \to A$$

$$Reflexion_{\Delta}: \quad \Delta A \to A.$$

Of course, neither schema is consistent with the rule $R2$. Thus, when citing such theories as $PRL_{ZF} + Reflexion_{\Box}$ or $PRL_{ZF} + Reflexion_{\Box} + Reflexion_{\Delta}$, we assume PRL_{ZF} given in an $R2$-free formulation. (The theory $PRL_{ZF} + Reflexion_{\Box} + Reflexion_{\Delta}$ has, incidentally, a redundancy-- cf. Exercise 1, below. I have simply chosen the axioms as they occur in the proof of Theorem 4.2.)

Without further ado, let us prove Carlson's Theorem.

Proof of Theorem 4.2: In all three parts of the Theorem, the implications a $\Rightarrow$ b and b $\Rightarrow$ c are routine and I leave the details to the reader. The implications c $\Rightarrow$ a are proven contrapositively: Assume A is not derivable over PRL_1 from the given conjunction and let $\underline{K} = (\{1,\ldots,n\},R,D,1,\Vdash)$ be an appropriate Carlson counter-model. As in the proof of Solovay's First and Second Completeness Theorems, we shall construct an arithmetic interpretation by constructing a function F that doesn't ascend the partial ordering.

i. As just announced, assume $1 \Vdash \bigwedge_{\Box B \in S} \Delta(\Box B \to B)$, $1 \nVdash A$ in $\underline{K}$, where S is the set of subformulae of A. Define $0\,R\,x$ for all $1 \leq x \leq n$, but do not yet bother to add 0 to the model, i.e. do not extend $\Vdash$ to 0. Because of this, we can assume $1 \notin D$-- membership or non-membership of the minimum node 1 in D has no effect on $\Vdash$.

Define F by the Recursion Theorem so that

$$F0 = 0$$

$$F(x+1) = \begin{cases} y, & Prov(\overline{x}, \ulcorner L \neq \overline{y} \urcorner) \ \&\ Fx\,R\,y \\ y, & Prov_{ZF}(\overline{x}, \ulcorner L \neq \overline{y} \urcorner) \ \&\ Fx\,R\,y \ \&\ y \in D \\ Fx, & \text{otherwise,} \end{cases}$$

where $L = \lim_{x \to \infty} Fx$. (We assume here that i. any proof x is a proof of only one formula, and ii. any proof belongs to only one system. The former holds automatically for the coding discussed in Chapter 0; the latter holds if we index each proof by the name of the system it is intended to be a proof in. Thus, for each $x + 1$, there is a unique clause in the definition that applies.) The definition of F is analogous to that of the function constructed in the proof of Solovay's First Completeness Theorem and F has, therefore, many similar properties, which properties we now proceed to list.

First, there are some basic facts.

4.4. LEMMA. i. $\mathsf{PRA} \vdash \forall v_0 \exists ! v_1 (Fv_0 = v_1)$, i.e. PRA proves F is a total function

ii. $\mathsf{PRA} \vdash \forall v_0 (Fv_0 \leq \overline{n})$

iii. for any $x \in \omega$,

$$\mathsf{PRA} \vdash \forall v_0 \big(Fv_0 = \overline{x} \rightarrow \forall v_1 > v_0 (Fv_1 = \overline{x} \vee \overline{x}\,R\,Fv_1)\big)$$

iv. $\mathsf{PRA} \vdash \exists v_0 v_1 \forall v_2 \geq v_0 (Fv_2 = v_1)$, i.e. $\mathsf{PRA} \vdash \exists v_1 (L = v_1)$

v. $\mathsf{PRA} \vdash L \leq n$, i.e. $\mathsf{PRA} \vdash \bigvee_{x \leq n} L = \overline{x}$

vi. for any $x \in \omega$,

$$\mathsf{PRA} \vdash \exists v_0 (Fv_0 = \overline{x}) \rightarrow L = \overline{x} \vee \overline{x}\,R\,L.$$

The proof of Lemma 4.4 is routine and I omit it.

The next Lemma relates the properties of F and L with their formal counterparts.

4.5. LEMMA. For all $x,y \in \omega$,

i. a. $\mathsf{PRA} \vdash L = \overline{x} \wedge \overline{x}\,R\,\overline{y} \rightarrow Con(\mathsf{PRA} + L = \overline{y})$

b. $\mathsf{PRA} \vdash L = \overline{x} \wedge \overline{x}\,R\,\overline{y} \wedge \overline{y} \in D \rightarrow Con(\mathsf{ZF} + L = \overline{y})$

ii. a. $\mathrm{PRA} \vdash L = \overline{x} \wedge \overline{x} > \overline{0} \wedge \overline{x} \notin \mathrm{D} \to Pr(\ulcorner L \neq \overline{x} \urcorner)$

b. $\mathrm{PRA} \vdash L = \overline{x} \wedge \overline{x} > \overline{0} \to Pr_{ZF}(\ulcorner L \neq \overline{x} \urcorner)$

c. $\mathrm{PRA} \vdash L = \overline{x} \wedge \overline{x} \notin \mathrm{D} \to Pr_{ZF}(\ulcorner L \neq \overline{x} \urcorner)$

iii. a. $\mathrm{PRA} \vdash L = \overline{x} \wedge \overline{x} \neq \overline{y} \wedge \sim \overline{x} R \overline{y} \to \sim Con(\mathrm{PRA} + L = \overline{y})$

b. $\mathrm{PRA} \vdash L = \overline{x} \wedge \overline{x} \neq \overline{y} \wedge (\sim \overline{x} R \overline{y} \vee \overline{y} \notin \mathrm{D}) \to \sim Con(\mathrm{ZF} + L = \overline{y})$.

Proof: i. Immediate by definition of F, as the reader can quickly verify.

ii.a. If $L = x > 0$, then $Fv_0 = x$ for some v_0. Choose v_0 minimum with this property. Now, since $Fv_0 = x \notin \mathrm{D}$, it follows that $Prov(v_0 \dot{-} \overline{1}, \ulcorner L \neq \overline{x} \urcorner)$, whence $Pr(\ulcorner L \neq \overline{x} \urcorner)$.

ii.b. Let $L = x > 0$. If $x \notin \mathrm{D}$, part a yields $Pr(\ulcorner L \neq \overline{x} \urcorner)$, whence $Pr_{ZF}(\ulcorner L \neq \overline{x} \urcorner)$. If $x \in \mathrm{D}$, and v_0 is minimum such that $Fv_0 = L$, then $Prov(v_0 \dot{-} 1, \ulcorner L \neq \overline{x} \urcorner)$ or $Prov_{ZF}(v_0 \dot{-} 1, \ulcorner L \neq \overline{x} \urcorner)$. The former case gives $Pr(\ulcorner L \neq \overline{x} \urcorner)$, whence $Pr_{ZF}(\ulcorner L \neq \overline{x} \urcorner)$; the latter gives $Pr_{ZF}(\ulcorner L \neq \overline{x} \urcorner)$ directly.

ii.c. Observe

$$\mathrm{PRA} \vdash \overline{x} > \overline{0} \wedge \overline{x} \notin \mathrm{D} \to .L = \overline{x} \to Pr(\ulcorner L \neq \overline{x} \urcorner), \quad \text{by ii.a}$$

$$\vdash \overline{x} > \overline{0} \wedge \overline{x} \notin \mathrm{D} \to Pr_{ZF}(\ulcorner L = \overline{x} \to Pr(\ulcorner L \neq \overline{x} \urcorner) \urcorner), \quad (*)$$

by the Derivability Conditions and the simple nature of the antecedent. But (*) and the reflexion schema for PRA in ZF yield

$$\mathrm{PRA} \vdash \overline{x} > \overline{0} \wedge \overline{x} \notin \mathrm{D} \to Pr_{ZF}(\ulcorner L = \overline{x} \to L \neq \overline{x} \urcorner)$$

$$\vdash \overline{x} > \overline{0} \wedge \overline{x} \notin \mathrm{D} \to Pr_{ZF}(\ulcorner L \neq \overline{x} \urcorner).$$

iii.a. Observe

$$\mathrm{PRA} \vdash L = \overline{x} \to \exists v_0 (Fv_0 = \overline{x})$$

$$\vdash L = \overline{x} \to Pr(\ulcorner \exists v_0 (Fv_0 = \overline{x}) \urcorner)$$

$$\vdash L = \overline{x} \to Pr(\ulcorner L = \overline{x} \vee \overline{x} R L \urcorner), \quad \text{by 4.4.vi}$$

$$\vdash L = \overline{x} \wedge \overline{x} \neq \overline{y} \wedge \sim \overline{x} R \overline{y} \to Pr(\ulcorner L \neq \overline{y} \urcorner)$$

$$\vdash L = \overline{x} \wedge \overline{x} \neq \overline{y} \wedge \sim \overline{x} R \overline{y} \to \sim Con(\mathrm{PRA} + L = \overline{y}).$$

iii.b. Similar. QED

Note that we have not proven

$$\mathrm{PRA} \vdash L = \overline{x} \wedge \overline{x} > \overline{0} \wedge \overline{x} \in \mathrm{D} \to Pr(\ulcorner L \neq \overline{x} \urcorner).$$

We will not need this because the D-nodes handle ZF, which proves $Pr(\ulcorner L \neq \overline{x} \urcorner) \to L \neq \overline{x}$, while in the models the D-nodes will force $\Box C \to C$.

To complete the proof of Theorem 4.2.i, let us now define the interpretation $*$ and prove its key properties. For each atom $p \in S$, set

$$p^* = \bigvee\{L = \overline{x}:\ 1 \le x \le n \ \&\ x \Vdash p\}.$$

4.6. LEMMA. For all $B \in S$,

i. $x \Vdash B \Rightarrow \text{PRA} \vdash L = \overline{x} \rightarrow B^*$

ii. $x \nVdash B \Rightarrow \text{PRA} \vdash L = \overline{x} \rightarrow \sim B^*$.

Proof: Parts i and ii are proven simultaneously by induction on the complexity of B. The atomic case follows by definition and the boolean cases are routine.

Let $B = \Box C$.

i. $x \Vdash \Box C \Rightarrow \forall y(xRy \Rightarrow y \Vdash C)$

$\Rightarrow \forall y(xRy \Rightarrow \text{PRA} \vdash L = \overline{y} \rightarrow C^*)$, by induction hypothesis

$\Rightarrow \forall y(xRy \Rightarrow \text{PRA} \vdash Pr(\ulcorner L = \overline{y} \rightarrow C^* \urcorner))$. (1)

But, by 4.5.iii.a,

$$\text{PRA} \vdash L = \overline{x} \rightarrow Pr(\ulcorner L = \overline{x} \vee \bigvee_{xRy} L = \overline{y} \urcorner). \tag{2}$$

If $x \notin D$,

$\text{PRA} \vdash L = \overline{x} \rightarrow Pr(\ulcorner L \neq \overline{x} \urcorner)$, by 4.5.ii.a

$\vdash L = \overline{x} \rightarrow Pr(\ulcorner \bigvee_{xRy} L = \overline{y} \urcorner)$, by (2)

$\vdash L = \overline{x} \rightarrow Pr(\ulcorner C^* \urcorner)$, by (1)

$\vdash L = \overline{x} \rightarrow (\Box C)^*$.

If $x \in D$, then $x > 1$ and, since $1 \Vdash \Delta(\Box C \rightarrow C)$, it follows that $x \Vdash C$, whence

$\text{PRA} \vdash L = \overline{x} \rightarrow C^*$, by induction hypothesis.

This, (1), and (2) yield

$\text{PRA} \vdash L = \overline{x} \rightarrow Pr(\ulcorner C^* \urcorner)$

$\vdash L = \overline{x} \rightarrow (\Box C)^*$.

(Remark: It is for precisely this step that we assumed $1 \notin D$. In parts ii and iii of the Theorem, we shall put 1 into D-- but we will also have $1 \Vdash \Box C \rightarrow C$ by assumption.)

ii. $x \Vdash \sim\Box C \Rightarrow \exists y(xRy \ \&\ y \nVdash C)$

$\Rightarrow \exists y(xRy \ \&\ \text{PRA} \vdash L = \overline{y} \rightarrow \sim C^*)$, by induction hypothesis

$\Rightarrow \exists y(xRy \ \&\ \text{PRA} \vdash Pr(\ulcorner C^* \rightarrow L \neq \overline{y} \urcorner))$

$$\Rightarrow \text{PRA} \vdash L = \overline{x} \rightarrow \sim Pr(\ulcorner C^* \urcorner).$$

since $\text{PRA} \vdash L = \overline{x} \wedge \overline{x} R \overline{y} \rightarrow \sim Pr(\ulcorner L \neq \overline{y} \urcorner)$.

Let $B = \Delta C$.

i. $x \Vdash \Delta C \Rightarrow \forall y(xRy \;\&\; y \in D \Rightarrow y \Vdash C)$

$\Rightarrow \forall y(xRy \;\&\; y \in D \Rightarrow \text{PRA} \vdash L = \overline{y} \rightarrow C^*)$, by ind. hyp.

$\Rightarrow \text{PRA} \vdash Pr_{ZF}(\ulcorner \bigvee_{xRy \in D} L = \overline{y} \rightarrow C^* \urcorner)$

$\Rightarrow \text{PRA} \vdash L = \overline{x} \rightarrow Pr_{ZF}(\ulcorner C^* \urcorner)$,

since $\text{PRA} \vdash L = \overline{x} \rightarrow Pr_{ZF}(\ulcorner \bigvee_{xRy \in D} L = \overline{y} \urcorner)$, by 4.5.iii.b.

ii. $x \not\Vdash \Delta C \Rightarrow \exists y(xRy \in D \;\&\; y \not\Vdash C)$

$\Rightarrow \exists y(xRy \in D \;\&\; \text{PRA} \vdash L = \overline{y} \rightarrow \sim C^*)$, by induction hypothesis

$\Rightarrow \exists y(xRy \in D \;\&\; \text{PRA} \vdash Pr_{ZF}(\ulcorner C^* \rightarrow L \neq \overline{y} \urcorner))$

$\Rightarrow \text{PRA} \vdash L = \overline{x} \rightarrow \sim Pr_{ZF}(\ulcorner C^* \urcorner)$,

since $\text{PRA} \vdash L = \overline{x} \wedge \overline{x} R \overline{y} \wedge \overline{y} \in D \rightarrow \sim Pr_{ZF}(\ulcorner L \neq \overline{y} \urcorner)$. QED

To complete the proof of part i of Theorem 4.2, we need only observe that $\text{PRA} \vdash L = \overline{1} \rightarrow \sim A^*$ and $\text{PRA} + L = \overline{1}$ is consistent, whence $\text{PRA} \nvdash A^*$. To this end, we need the following lemma.

4.7. LEMMA. The following are true, though unprovable in PRA:

i. $L = \overline{0}$

ii. for $0 \leq x \leq n$, $\text{PRA} + L = \overline{x}$ is consistent

iii. for $0 \leq x \leq n$, $x \in D$, $\text{ZF} + L = \overline{x}$ is consistent.

I leave the proof as an exercise to the reader.

Proof of Theorem 4.2 continued: As remarked, Lemma 4.7 completes the proof of Theorem 4.2.i characterising the schemata in $\square, \Delta$ provable in PRA.

To prove part ii, assume K satisfies

$$1 \Vdash \bigwedge_{\square B \in S} \Delta(\square B \rightarrow B), \quad 1 \Vdash \bigwedge_{\square B \in S} (\square B \rightarrow B), \quad 1 \not\Vdash A,$$

where S is the set of subformulae of A. Put 1 into D. Lemmas 4.4 and 4.5 remain valid without change. The proof of Lemma 4.6 needs adjustment only in the place cited, and this adjustment was also provided at that spot. Finally, Lemma 4.7 remains valid without change. Now, put everything together:

$$1 \Vdash\!\!\!\not\;\; A \quad\Rightarrow\quad \mathrm{PRA} \vdash L = \overline{1} \rightarrow \sim A^*$$
$$\Rightarrow\quad \mathrm{ZF} \vdash L = \overline{1} \rightarrow \sim A^*$$
$$\Rightarrow\quad \mathrm{ZF} \nvdash A^*,$$

since $\mathrm{ZF} + L = \overline{1}$ is consistent by 4.7.iii.

Finally, to prove part iii, assume $\underline{K}$ satisfies

$$1 \Vdash \bigwedge_{\Box B \in S} \Delta(\Box B \rightarrow B), \quad 1 \Vdash \bigwedge_{\Box B \in S} (\Box B \rightarrow B), \quad 1 \Vdash \bigwedge_{\Delta B \in S} (\Delta B \rightarrow B), \quad 1 \not\Vdash A.$$

Again, put 1 into D. But now add 0 to $\underline{K}$; define

$$0 \Vdash p \quad \text{iff} \quad 1 \Vdash p, \text{ for } p \in S,$$

and define $*$ by

$$p^* \;=\; \bigvee\{L = \overline{x}:\; 0 \le x \le n \;\&\; x \Vdash p\}.$$

Do not put 0 into D.

Again, Lemmas 4.4, 4.5, and 4.7 and their proofs require no change. Since the interpretation $*$ differs from that used in proving Lemma 4.6, the analogue must be proven; but the details are identical for $x > 0$ and we need not repeat them. For $x = 0$ we have the following lemma.

<u>4.8. LEMMA</u>. Under the assumptions cited above, for $B \in S$,

i. $0 \Vdash B$ iff $1 \Vdash B$

ii. $0 \Vdash B \;\Rightarrow\; \mathrm{PRA} \vdash L = \overline{0} \rightarrow B^*$

iii. $0 \not\Vdash B \;\Rightarrow\; \mathrm{PRA} \vdash L = \overline{0} \rightarrow \sim B^*$.

Proof: All parts are by induction on the complexity of B.

i. Exercise.

ii & iii. The only interesting cases are $B = \Box C$ and $B = \Delta C$.

Let $B = \Box C$. Observe

$$0 \Vdash B \;\Rightarrow\; \bigwedge_{1 \le x \le n} (x \Vdash C)$$
$$\Rightarrow\; \bigwedge_{1 \le x \le n} (\mathrm{PRA} \vdash L = \overline{x} \rightarrow C^*). \qquad (*)$$

But we also have

$$0 \Vdash B \;\Rightarrow\; 1 \Vdash C$$
$$\Rightarrow\; 0 \Vdash C, \text{ by part i}$$
$$\Rightarrow\; \mathrm{PRA} \vdash L = \overline{0} \rightarrow C^*, \text{ by induction hypothesis.}$$

With (*) this yields

$$0 \Vdash \Box C \Rightarrow \bigwedge_{x \leq n} (\mathrm{PRA} \vdash L = \overline{x} \rightarrow C^*)$$

$$\Rightarrow \mathrm{PRA} \vdash \bigvee_{x \leq n} L = \overline{x} \rightarrow C^*$$

$$\Rightarrow \mathrm{PRA} \vdash C^*, \text{ since } \mathrm{PRA} \vdash \bigvee_{x \leq n} L = \overline{x}$$

$$\Rightarrow \mathrm{PRA} \vdash Pr(\ulcorner C^* \urcorner)$$

$$\Rightarrow \mathrm{PRA} \vdash L = \overline{0} \rightarrow Pr(\ulcorner C^* \urcorner).$$

Inversely,

$$0 \nVdash \Box C \Rightarrow \exists x (x \nVdash C)$$

$$\Rightarrow \exists x (\mathrm{PRA} \vdash L = \overline{x} \rightarrow \sim C^*)$$

$$\Rightarrow \exists x (\mathrm{PRA} \vdash Pr(\ulcorner C^* \urcorner) \rightarrow Pr(\ulcorner L \neq \overline{x} \urcorner))$$

$$\Rightarrow \exists x (\mathrm{PRA} \vdash Con(\mathrm{PRA} + L = \overline{x}) \rightarrow \sim Pr(\ulcorner C^* \urcorner)$$

$$\Rightarrow \mathrm{PRA} \vdash L = \overline{0} \rightarrow \sim Pr(\ulcorner C^* \urcorner),$$

since $\mathrm{PRA} \vdash L = \overline{0} \rightarrow Con(\mathrm{PRA} + L = \overline{x})$ for all $1 \leq x \leq n$.

Let $B = \Delta C$. The proof is similar. Observe

$$0 \Vdash \Delta C \Rightarrow \bigwedge_{x \in \mathrm{D}} (x \Vdash C)$$

$$\Rightarrow \bigwedge_{x \in \mathrm{D}} (\mathrm{PRA} \vdash L = \overline{x} \rightarrow C^*). \tag{1}$$

But again,

$$0 \Vdash \Delta C \Rightarrow 1 \Vdash C \Rightarrow 0 \Vdash C$$

$$\Rightarrow \mathrm{PRA} \vdash L = \overline{0} \rightarrow C^*. \tag{2}$$

Now $\quad \mathrm{ZF} \vdash L = \overline{0} \vee \bigvee_{x \in \mathrm{D}} L = \overline{x},$

while, by (1) and (2),

$$\mathrm{ZF} \vdash L = \overline{0} \vee \bigvee_{x \in \mathrm{D}} L = \overline{x} \rightarrow C^*.$$

Thus $\quad \mathrm{ZF} \vdash C^*,$

whence $\quad \mathrm{PRA} \vdash Pr_{ZF}(\ulcorner C^* \urcorner)$

$$\vdash L = \overline{0} \rightarrow Pr_{ZF}(\ulcorner C^* \urcorner).$$

Inversely,

$$0 \nVdash \Delta C \Rightarrow \exists x \in \mathrm{D}(x \nVdash C)$$

$$\Rightarrow \exists x \in \mathrm{D}(\mathrm{PRA} \vdash L = \overline{x} \rightarrow \sim C^*)$$

$$\Rightarrow \exists x \in \mathrm{D}(\mathrm{ZF} \vdash L = \overline{x} \rightarrow \sim C^*)$$

$$\Rightarrow \quad \exists x \in D(\mathsf{PRA} \vdash Pr_{ZF}(\ulcorner C^* \urcorner) \to Pr_{ZF}(\ulcorner L \neq \overline{x} \urcorner))$$

$$\Rightarrow \quad \exists x \in D(\mathsf{PRA} \vdash Con(\mathsf{ZF} + L = \overline{x}) \to \sim Pr_{ZF}(\ulcorner C^* \urcorner))$$

$$\Rightarrow \quad \mathsf{PRA} \vdash L = \overline{0} \to \sim Pr_{ZF}(\ulcorner C^* \urcorner),$$

since $\mathsf{PRA} \vdash L = \overline{0} \to Con(\mathsf{ZF} + L = \overline{x})$ for all $x \in D$. QED

With the completion of the proof of Lemma 4.8, we have essentially finished the proof of Theorem 4.2.iii. For,

$$1 \not\Vdash A \quad \Rightarrow \quad 0 \not\Vdash A, \text{ by 4.8.i}$$

$$\Rightarrow \quad \mathsf{PRA} \vdash L = \overline{0} \to \sim A^*$$

$$\Rightarrow \quad \sim A^* \text{ is true, since } L = \overline{0} \text{ is true.}$$

QED

Before exiting to the exercises, we have just a few generalities to discuss. The first of these is sufficiently important to be singled out from the rest.

4.9. REMARK. Under certain circumstances, the theories PRA and ZF can be replaced by a pair $T_0 \subseteq T_1$ of *RE* theories in which T_1 proves the reflexion schema for T_0. What are these conditions?

i. For 4.2.i and 4.2.ii, T_0 and T_1 must be Σ_1-sound. For, the proof of Lemma 4.7 (cf. Exercise 2) used the fact that PRA and ZF (here: T_0 and T_1) proved no false Σ_1-assertions of the form $\exists v(Fv = \overline{x})$.

ii. For 4.2.iii, the b $\Rightarrow$ c implication requires T_0, T_1 to be arithmetically sound: $(\Box B \to B)^*$ and $(\Delta B \to B)^*$ are assumed true, hence all arithmetic theorems B^* of T_0 and T_1 must be true. Thus, whereas 4.2.i and 4.2.ii require only Σ_1-soundness, 4.2.iii requires full arithmetic soundness.

It follows from these subremarks that PRL_{ZF} and its modifications in 4.2.ii.a-b and 4.2.iii.a-b axiomatise the T_0-provably valid, T_1-provably valid, and truly valid schemata for the following pairs:

a. $T_0 = \mathsf{PRA}$, $T_1 = \mathsf{PA}$

b. $T_0 = \mathsf{PA}$, $T_1 = \mathsf{ZF}$

c. $T_0 = \mathsf{PA}$, $T_1 = \mathsf{PA} + Reflexion(\mathsf{PA})$

d. $T_0 = \mathsf{ZF}$, $T_1 = \mathsf{ZF} + \exists\kappa(\kappa$ is an inaccessible cardinal$)$.

4.10. REMARKS. i. Carlson has generalised Theorem 4.2 to the case of a whole tower of Σ_1-sound *RE* theories $T_0 \subseteq T_1 \subseteq \ldots \subseteq T_n$, where each T_{i+1} proves the soundness of T_i.

ii. Theorems 4.2.i and 4.2.ii uniformise-- provided, once again, we replace PRA by PA: There is, for example, a single interpretation * such that, for all bi-modal sentences A, $\mathrm{PRL}_{ZF} \vdash A$ iff $\mathrm{PA} \vdash A^*$.

iii. Theorem 4.2.iii does not uniformise.

EXERCISES

1. Let $\mathrm{PRL}^{\omega}_{ZF} = \mathrm{PRL}_{ZF} + \mathit{Reflexion}_{\Box} + \mathit{Reflexion}_{\Delta}$. Show the axiom schema $\mathit{Reflexion}_{\Box}$ is redundant, i.e. show:

$$\mathrm{PRL}^{\omega}_{ZF} \vdash A \quad \text{iff} \quad \mathrm{PRL}_1 \vdash \bigwedge_{\Box B \in S} (\Delta B \to B) \wedge \bigwedge_{\Delta B \in S} (\Delta B \to B) \ \to\ A.$$

2. Prove Lemma 4.7. Note where the Σ_1-soundness of each theory is needed.

3. I did not really give proofs of Theorem 4.2 parts ii and iii so much as give recipes for such. Give a detailed proof either of 4.2.ii or of 4.2.iii.

4. Use Theorem 4.2.iii to construct sentences ϕ_i with the given properties. Which ϕ_i can be chosen Σ_1 or Π_1?

   i. ϕ_1 satisfies

      a. $\mathrm{PRA} \nvdash \phi_1, \sim\phi_1$

      b. $\mathrm{ZF} \vdash \sim Pr_{PRA}(\ulcorner \phi_1 \urcorner), \sim Pr_{PRA}(\ulcorner \sim\phi_1 \urcorner)$

      c. $\mathrm{ZF} \vdash \phi_1$

   ii. ϕ_2 satisfies

      a. $\mathrm{PRA} \nvdash \phi_2, \sim\phi_2$

      b. $\mathrm{ZF} \vdash \sim Pr_{PRA}(\ulcorner \phi_2 \urcorner), \sim Pr_{PRA}(\ulcorner \sim\phi_2 \urcorner)$

      c. $\mathrm{ZF} \nvdash \phi_2, \sim\phi_2$

   iii. ϕ_3 satisfies

      a. $\mathrm{ZF} \nvdash Con(\mathrm{PRA} + Con_{PRA} + \phi_3)$

      b. $\mathrm{ZF} \nvdash Con(\mathrm{PRA} + Con_{PRA} + \sim\phi_3)$

      c. $\mathrm{PRA} + Con_{PRA} \vdash Con(\mathrm{PRA} + \phi_3)$

      d. $\mathrm{PRA} + Con_{PRA} \vdash Con(\mathrm{PRA} + \sim\phi_3)$

      e. $\mathrm{ZF} \nvdash \phi_3, \sim\phi_3$.

5. Recall that, for every *RE* theory $T \supseteq \mathrm{PRA}$ and every sentence ϕ,

$$T \vdash \phi \quad \text{iff} \quad \mathrm{PRA} \vdash Pr_T(\ulcorner \phi \urcorner). \qquad (*)$$

i. Use (*) and Exercise 2.iv of section 2 to reduce Theorem 4.2.ii to Theorem 4.2.i.

ii. Use (*) and Exercise 2.ii of section 2 to prove: There is a Σ_1-sound *RE* theory $T \supseteq PRA$ such that, for all A,

$$PRL_1 \vdash A \quad \text{iff} \quad \forall * \text{ based on } T \ (T \vdash A^*).$$

Chapter 5
Fixed Point Algebras

Algebra rarely has anything deep to say about logic. Most applications of algebra to logic are fairly shallow, the exceptions being applications of representation theory, where one is really using the non-algebraic properties of the representations. Nonetheless, algebra does offer another perspective and a convenient language or framework in which to work. The purpose of the present chapter is the presentation of such an algebraic framework-language in which to place the results already discussed, in which goals like those of the last chapter can be expressly delineated, and in which we can state and prove a theorem that delimits, not very convincingly, the boundaries for such successful generalisations of Solovay's Completeness Theorems as were obtained in the last chapter and explains, again not convincingly, the necessarily close relation between these results and Solovay's results for PRL.

Every logical theory has associated with it a whole family of natural algebraic structures, namely its family of Lindenbaum algebras. Of these, the first is the *Lindenbaum sentence algebra* of the theory, constructed as follows: Let T be the theory in question. The elements of the Lindenbaum sentence algebra of T, which algebra we shall denote A_T, are the equivalence classes of sentences in the language of T:

$$(\phi)_T = \{\psi: \ T \vdash \phi \leftrightarrow \psi\}.$$

The algebraic operations of A_T are just those inherited from the logical operations of T:

$$(\phi)_T \wedge (\psi)_T = (\phi \wedge \psi)_T$$
$$(\phi)_T \vee (\psi)_T = (\phi \vee \psi)_T$$
$$\sim(\phi)_T = (\sim\phi)_T$$
$$(\phi)_T \rightarrow (\psi)_T = (\phi \rightarrow \psi)_T,$$

and even

$$1_T = (\phi \vee \sim\phi)_T, \quad 0_T = (\phi \wedge \sim\phi)_T.$$

The usual substitution lemma for first-order logic guarantees these operations to be well-defined.

The next obvious Lindenbaum algebra is the *Lindenbaum formula algebra* in one variable, say v_0. It consists of equivalence classes,

$$(\phi v_0)_T = \{\psi v_0 : \ T \vdash \forall v_0(\phi v_0 \leftrightarrow \psi v_0)\},$$

and inherits its algebraic operations from the logical ones in the same manner in which A_T received its operations.

Closely related to, but slightly different from the Lindenbaum formula algebra in one free variable is the algebra B_T of *extensional* formulae in one free variable. Using capital letters for formulae with v_0 free and lower case for sentences, we define Φv_0 to be extensional if, for all sentences ϕ, ψ,

$$T \vdash \phi \leftrightarrow \psi \ \Rightarrow \ T \vdash \Phi(\ulcorner\phi\urcorner) \leftrightarrow \Phi(\ulcorner\psi\urcorner).$$

The elements of B_T are then equivalence classes of extensional formulae in one free variable,

$$(\Phi v_0)_T = \{\Psi v_0 : \quad \forall \text{ sentences } \phi, \ T \vdash \Phi(\ulcorner\phi\urcorner) \leftrightarrow \Psi(\ulcorner\phi\urcorner)\},$$

and the algebraic operations on B_T are those induced as usual by the logical ones in T.

Note that B_T is not a subalgebra of the Lindenbaum formula algebra, but a quotient of such-- for we allow the identification of Φ and Ψ under a weaker assumption; instead of demanding

$$T \vdash \forall v_0(\Phi v_0 \leftrightarrow \Psi v_0),$$

or even

$$\forall x \ T \vdash \Phi\bar{x} \leftrightarrow \Psi\bar{x},$$

we assume

$$\forall \phi \ T \vdash \Phi(\ulcorner\phi\urcorner) \leftrightarrow \Psi(\ulcorner\phi\urcorner).$$

As indicated by the term "extensional", we are viewing Φ, Ψ as operators mapping A_T to itself and identifying them in B_T just in case they define one and the same function.

As just announced, we want to view the elements of B_T as functions mapping A_T to itself. However, the elements of B_T are also formulae (better: given by formulae) and the Diagonalisation Lemma applies: For each Φv_0 there is a sentence ϕ such that

$$T \vdash \Phi(\ulcorner\phi\urcorner) \leftrightarrow \phi,$$

i.e. viewing $(\Phi v_0)_T$ as a function,

$$(\Phi v_0)_T((\phi)_T) = (\phi)_T.$$

Thus, every element of B_T, viewed as a function on A_T, possesses a fixed point. For this reason, the structure (B_T,A_T) is called the *Lindenbaum fixed point algebra of* T.

In the preceding chapters we have not really looked at the full Lindenbaum fixed point algebra of any theory, but we have looked at various *subalgebras* of B_T. For example, we have looked at those extensional operators $\Phi(v_0)$ generated (in a suitable sense) by $Pr_T(v_0)$ over A_T. Thus, algebraically, the study of self-reference with respect to some operator $\Phi(v_0)$ (like $Pr_T(v_0)$ or one of the $\rho(v_0)$'s discussed in Chapter 4) over a given theory T can be viewed as the study of the use of fixed points in the subalgebra (B,A_T) of the Lindenbaum fixed point algebra (B_T,A_T) of T generated by $(\Phi v_0)_T$. Algebraically, the success of our earlier enterprises can be explained by the successful partial simulation of (B,A_T) by *finite* algebras. Such success is, in general, impossible: There are elements $(\Phi v_0)_T \in B_T$ which simply do not behave like the elements of finite fixed point algebras. The main goal of the present Chapter is to say something about the difference between such Φv_0 and $Pr(v_0)$.

Section 1, immediately following, reviews the rudiments of boolean algebra and its representation theory and also introduces the notion of a *diagonalisable algebra*, a boolean algebra equipped with an operator τ that simulates $Pr(v_0)$. The completeness of PRL with respect to Kripke models is exhibited as a representation theory for diagonalisable algebras. Section 2 formally defines the notion of a fixed point algebra, discusses finite fixed point algebras, and proves a representation theorem for finite *closed* fixed point algebras: Finite closed fixed point algebras are subalgebras of diagonalisable algebras. On the basis of this result, I lay my rather shaky claim that any extensional predicate Φv_0 admitting a successful modal analysis via finite models must reduce (somehow) to Pr or some such operator completely described by PRL. Section 3 finishes with a few miscellaneous remarks on the algebra.

Finally, before setting forth on our algebraic excursion, let me warn the reader that what he might by now be expecting is true: Mathematically, this is probably the most sophisticated chapter of the present monograph; moreover, it is the chapter that coheres least well with the rest.

1. BOOLEAN AND DIAGONALISABLE ALGEBRAS

Boolean algebras are, of course, algebras of truth values. They are also algebras of sets and we can even view them as algebraic generalisations of the Lindenbaum sentence algebras. The crucial thing is that they have two distinguished elements 0 and 1 (logically interpreted as f and t, respectively) and some basic operations, which I shall denote by $'$, $+$, and $\cdot$ (logically interpreted as $\sim$, $\vee$, and $\wedge$, respectively). The use of $+$ and $\cdot$ for $\vee$ and $\wedge$ is a little old-fashioned and not completely comfortable, but it is an established notation (i.e. it has some, however weak, force of tradition behind it) and it does separate the algebraic and logical notations. So, bearing the nearly idiosyncratic notation in mind, the reader can now confront the following definition.

1.1. DEFINITION. A *boolean algebra* is a sextuple $\underline{A} = (A,+,\cdot,',0,1)$, where

i. A is a set containing the distinguished elements 0, 1,

and ii. $'$ is a unary operation on A, and $+,\cdot$ are binary operations on A, and, moreover, these functions satisfy: For all $x,y,z \in A$:

a. $x + y = y + x$, $x \cdot y = y \cdot x$ — *Commutativity*

b. $x + (y + z) = (x + y) + z$, $x\cdot(y\cdot z) = (x\cdot y)\cdot z$ — *Associativity*

c. $x\cdot(y + z) = (x\cdot y) + (x\cdot z)$, $x + (y\cdot z) = (x + y)\cdot(x + z)$ — *Distributivity*

d. $x + 0 = x$, $x\cdot 0 = 0$ — *Zero*

e. $x + 1 = 1$, $x\cdot 1 = x$ — *Unit*

f. $x + x' = 1$, $x\cdot x' = 0$. — *Complementarity*

We often write xy for $x\cdot y$ and the precedence ranking of $',+,\cdot$ with respect to the elimination of parentheses is that of $-,+,\cdot$ in ordinary algebra.

Written in terms of $\sim$, $\vee$, and $\wedge$ instead of $'$, $+$, and $\cdot$, these identities are reasonable. The eye is jarred at seeing

$$x + y\cdot z = (x + y)\cdot(x + z) \quad \text{and} \quad x + 1 = 1,$$

but it eventually returns to its socket and learns to accept this and more.

1.2. EXAMPLES. Aside from the Lindenbaum algebras, there are two natural examples of boolean algebras:

i. $\underline{P}_X = (P(X), \cup, \cap, \sim, \emptyset, X)$, where X is a non-empty set. Here, $P(X)$ is the set of all subsets of X; $\cup$, $\cap$, and $\sim$ are the set theoretic operations of union, intersection, and complementation, respectively; and $\emptyset, X$ are the empty set and the whole set X, respectively.

ii. $\underline{V}_n = (2^n, max, min, compl, 0^n, 1^n)$, almost the n-dimensional vector space (for $n \geq 1$) over the 2-element field. For the non-algebraist, this means: 2^n is the set of n-tuples of 0's and 1's; max is the coordinatewise maximum; min the analogous minimum; $compl$ the coordinatewise switch of 0's and 1's; and $0^n, 1^n$ are the constant sequences $00\ldots0$, $11\ldots1$, respectively.

These are typical in that every finite boolean algebra is isomorphic to a powerset algebra $\underline{P}_X$ and a vector space algebra $\underline{V}_n$. Moreover, each representation has its uses. For example, among the sets there is the obvious relation of inclusion. We should look at this. If $\underline{V}_n$ is really to be a vector space over a field, it should have a genuine group theoretic addition $\dot{+}$ induced by addition modulo 2. We should look at this. Before doing these or discussing the representations, let me collect a few familiar boolean laws which the reader might consider missing from Definition 1.1.

<u>1.3. LEMMA.</u> Let $\underline{A} = (A, +, \cdot, ', 0, 1)$ be a boolean algebra. Let $x, y \in A$. Then:

i. if $x + y = 1$ and $x \cdot y = 0$, then $y = x'$

ii. *(Involution)* $x'' = x$

iii. $0' = 1, \quad 1' = 0$

iv. *(Idempotence)* $x + x = x, \quad x \cdot x = x$

v. *(De Morgan's Laws)* $(x + y)' = x' \cdot y', \quad (x \cdot y)' = x' + y'$.

The reader is invited to prove this himself (cf. Exercise 1, below).

Now, about the inclusion relation on $P(X)$: Observing that $Y \subseteq Z$ iff $Y \cap Z = Y$, we get the following definition.

<u>1.4. DEFINITION.</u> Let $\underline{A} = (A, +, \cdot, ', 0, 1)$ be a boolean algebra. We define a relation $\leq$ on A by $x \leq y$ iff $xy = x$.

<u>1.5. LEMMA.</u> Let $\underline{A} = (A, +, \cdot, ', 0, 1)$ be a boolean algebra.

i. $\leq$ is a weak partial ordering of A, i.e.

a. *(Reflexiveness)* for all $x \in A$, $x \leq x$

b. *(Antisymmetry)* for all $x,y \in A$, $x \leq y$ & $y \leq x \Rightarrow x = y$

c. *(Transitivity)* for all $x,y,z \in A$, $x \leq y$ & $y \leq z \Rightarrow x \leq z$

ii. for all $x,y \in A$, $x \leq y$ iff $x + y = y$

iii. for all $x,y \in A$, $x \leq y$ iff $y' \leq x'$

iv. for all $x,y,z \in A$,

$x \leq y \Rightarrow x \cdot z \leq y \cdot z$, $\quad x \leq y \Rightarrow x + z \leq y + z$.

Again I leave the proof to the reader as a collection of exercises. Slightly more interesting than this Lemma is the order-theoretic characterisation of the operations and distinguished elements of a boolean algbera.

1.6. LEMMA. Let $\underline{A} = (A,+,\cdot,',0,1)$ be a boolean algebra.

i. $0 = inf(A)$ and $1 = sup(A)$, i.e.

a. $\forall x \in A(0 \leq x)$, b. $\forall x \in A(x \leq 1)$

ii. for all $x,y \in A$, $x \cdot y$ is the greatest lower bound of x,y, i.e.

a. $xy \leq x$ and $xy \leq y$

b. if $z \leq x$ and $z \leq y$ then $z \leq xy$

iii. for all $x,y \in A$, $x + y$ is the least upper bound of x,y, i.e.

a. $x \leq x + y$ and $y \leq x + y$

b. if $x \leq z$ and $y \leq z$ then $x + y \leq z$

iv. for all $x \in A$, $x' = sup\ \{y: xy = 0\}$, i.e.

a. $xx' = 0$

b. if $xz = 0$ then $z \leq x'$

v. for all $x,y \in A$, $x \rightarrow y = sup\ \{z: xz \leq y\}$, i.e.

a. $x \cdot (x \rightarrow y) \leq y$

b. if $xz \leq y$ then $z \leq (x \rightarrow y)$,

where $x \rightarrow y$ is $x' + y$ (defined in analogy to $(p \rightarrow q) \leftrightarrow (\sim p \vee q)$).

Partial proof: I will illustrate the proof by handling a couple of the cases.

ii.a. Idempotence yields $xy \cdot x = xy$, i.e. $xy \leq x$, and $xy \cdot y = xy$, i.e. $xy \leq y$.

b. Suppose $z \leq x$ and $z \leq y$. Look at $z(xy)$:

$$z(xy) = (zx)y = zy, \quad \text{since } zx = z$$
$$= z, \quad \text{since } zy = z$$

whence $z \leq xy$.

iv.a. This follows by Complementarity.

b. Suppose $xz = 0$. Observe

$$z = z(x + x') = zx + zx' = 0 + zx' = zx',$$

whence $z \leq x'$.

As remarked, the rest of the proof is left to the reader. Only part v is non-trivial (cf. Exercise 3). QED

Lemmas 1.5 and 1.6 give basic properties of the order relation. Indeed, if we think of $\leq$ as deducibility, Lemma 1.6 states standard rules of inference (e.g. ii.a: $A \wedge B \rightarrow A$, $A \wedge B \rightarrow B$; ii.b: if $A \rightarrow C$ and $B \rightarrow C$, then $A \vee B \rightarrow C$; v.a: modus ponens; v.b: the Deduction Theorem). But, precisely because of this basicness and familiarity, I can afford to be cavalier in not presenting the details of the proofs, which in mathematical practice amounts to dismissing the results as insignificant-- in the present case because they are a sort of logic review. Now, unless the reader is already familiar with boolean algebras, we come upon something new, i.e. something that does not occur in Lindenbaum algebras: atoms.

1.7. DEFINITION. Let $\underline{A} = (A,+,\cdot,',0,1)$ be a boolean algebra. An element $a \in A$ is an *atom* if a is a minimal non-zero element of A, i.e. a is an atom if $a \neq 0$ and, for any $x \in A$, if $x \leq a$ then $x = 0$ or $x = a$.

The prototypical atoms are the singletons in a powerset algebra: A minimal nonempty set obviously contains exactly one element. The atoms of the vector algebras $\underline{V}_n$ are also easy to describe: They are the strings with exactly one 1. For example, in $\underline{V}_3$, the atoms are 100, 010, and 001. As already noted, the Lindenbaum algebras of consistent RE theories $T \supseteq PRA$ have no atoms-- this follows by Rosser's Theorem (a full discussion of which is coming in the next Chapter). Thus, not every boolean algebra possesses atoms. Every finite boolean algebra, however, has plenty of atoms:

1.8. LEMMA. Let $\underline{A} = (A,+,\cdot,',0,1)$ be a finite boolean algebra and let $x \in A$ be a non-zero element. There is an atom $a \in A$ such that $a \leq x$. Moreover, $x = sup\ \{a: a \leq x \text{ is an atom}\}$.

Proof: Let $x \neq 0$. Let $X = \{y \in A: \ 0 < y \leq x\}$. Since X is finite and nonempty, it possesses a minimal element, say, a. Now, a is an atom below x: $a \leq x$ and $a > 0$ by definition of X. If $y \leq a$, either $y = 0$ or $y \in X$ and, in the latter case, $y = a$ by the minimality of a.

Let $Y = \{a \in X: a \text{ is an atom}\} = \{a \leq x: \ a \text{ is an atom}\}$ and let $y = sup(Y)$. ($sup(Y)$ exists since it is $a_1 + \ldots + a_n$, where $Y = \{a_1,\ldots,a_n\}$.) Since $a \leq x$ for $a \in Y$, $y = sup(Y) \leq x$. To prove the converse inequality, assume otherwise: $y < x$. I claim that $xy' \leq x$ is non-zero and that this is a contradiction:

(1) $xy' \neq 0$: Otherwise

$$xy' = 0 \Rightarrow y' \leq x', \text{ by Lemma 1.6.iv.b}$$
$$\Rightarrow x \leq y, \text{ by Lemma 1.5.iii,}$$

which we are assuming to be false.

(2) To get the contradiction, let $a \leq xy' \leq x$ be an atom (as supplied by the already proven first statement of the Lemma). Obviously, $a \in Y$, whence $a \leq y$. But

$$a \leq xy' \ \& \ a \leq y \ \Rightarrow \ a \leq (xy')\cdot(y) = x(y'y) = 0,$$

whence $a = 0$, a contradiction. QED

By Lemma 1.8, a finite boolean algebra not only has a lot of atoms, but every element is constructed from them. A slight variation on the argument shows more:

1.9. THEOREM. (Finite Representation Theorem). Let $\underline{A} = (A,+,\cdot,',0,1)$ be a finite boolean algebra. The map

$$F: \ x \mapsto \{a \leq x: \ a \text{ is an atom}\}$$

is an isomorphism of $\underline{A}$ with the powerset algebra $\underline{P}_X = (P(X), \cup, \cap, \sim, \emptyset, X)$ of the set X of atoms of $\underline{A}$.

1.10. COROLLARY. If $\underline{A}$ is a finite boolean algebra, then A has cardinality 2^n, where n is the number of atoms of $\underline{A}$.

Proof of Theorem 1.9: By Lemma 1.8, the map F is one-to-one. Since we are dealing with finite sets, every set of atoms $\{a_{i_1},\ldots,a_{i_k}\}$ has a supremum $x = a_{i_1} + \ldots + a_{i_k}$. This almost guarantees $F(x) = \{a_{i_1},\ldots,a_{i_k}\}$ and that F is onto. What is needed is proof that there are no new atoms below the sum. To this (and another) end, we must prove the

Claim. $F(x + y) = F(x) \cup F(y)$.

Proof: Let a be an atom. If $a \in F(x) \cup F(y)$, then $a \leq x$ or $a \leq y$ and clearly $a \leq x + y$.

Conversely, let $a \in F(x + y)$, i.e. $a \leq x + y$, and suppose $a \notin F(x)$. Observe

$$a = a(x + y) = ax + ay.$$

But, if $a \notin F(x)$, i.e. if $a \nleq x$, then, since a is an atom, and $ax \leq x$, we must have $ax = 0$. Thus, extending our equation,

$$a = ax + ay = ay,$$

whence $a \leq y$, i.e. $a \in F(y)$. QED

We can now finish the proof of the Theorem. By the Claim,

$$\begin{aligned} F(a_{i_1} + \ldots + a_{i_k}) &= F(a_{i_1}) \cup \ldots \cup F(a_{i_k}) \\ &= \{a_{i_1}\} \cup \ldots \cup \{a_{i_k}\} \\ &= \{a_{i_1}, \ldots, a_{i_k}\} \end{aligned}$$

and F is onto. Moreover, the Claim also shows F to preserve the boolean sum. We must yet show that F preserves the boolean product and complement, as well as the constants:

i. $F(x \cdot y) = F(x) \cap F(y)$

ii. $F(x') = \sim F(x) = X - F(x)$

iii. $F(0) = \emptyset$

iv. $F(1) = X$.

We don't need to prove all of these: By de Morgan's Law and the preservation of sum, i follows once we've proven ii. Similarly, iii and iv are equivalent in the face of ii. Thus, it suffices to prove ii and, say, iv. Of these, iv is easy:

$$F(1) = \{a \in X: a \leq 1\} = \{a \in X\} = X.$$

ii. First note that

$$F(x) \cup F(x') = F(x + x') = F(1) = X.$$

To conclude that $F(x') = X - F(x)$, it thus suffices to show $F(x) \cap F(x') = \emptyset$. Suppose, on the contrary, $a \in F(x) \cap F(x')$. Then, $a \leq x$ and $a \leq x'$, whence $a \leq x \cdot x' = 0$, whence $a = 0$, a contradiction. QED

Theorem 1.9 has several immediate corollaries. The first, Corollary 1.10, simply asserts that the cardinality of a finite boolean algebra $\underline{A}$ is a power of 2--namely, 2^n, where n is the number of atoms of $\underline{A}$. Another corollary is the following:

<u>1.11. COROLLARY</u>. Let $\underline{A}$, $\underline{B}$ be boolean algebras of identical finite cardinality. Then: $\underline{A}$ and $\underline{B}$ are isomorphic.

Proof: Let the common cardinality be 2^n. Then the sets X and Y of atoms of $\underline{A}$ and $\underline{B}$ have common cardinality n, whence their powerset algebras $\underline{P}_X$ and $\underline{P}_Y$ are isomorphic. But $\underline{A}$ and $\underline{B}$ are isomorphic to these, whence to each other. QED

As a consequence of Corollaries 1.10 and 1.11 and the existence of the algebras $\underline{P}_X$, it follows that there is (up to isomorphism) exactly one boolean algebra of order 2, one of order 4, one of order 8, ... We denote these algebras $\underline{2}$, $\underline{4}$, $\underline{8}$, etc. There is also a notational consequence of all of this. If we think of $\underline{8}$ as $\underline{P}_X$ for $X = \{a,b,c\}$, we can represent $\underline{8}$ graphically as

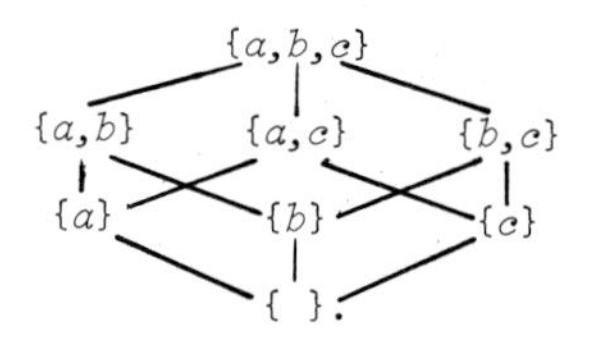

Considering a,b,c given in order, we can replace the sets by their "characteristic strings", i.e. elements of $\underline{V}_3$:

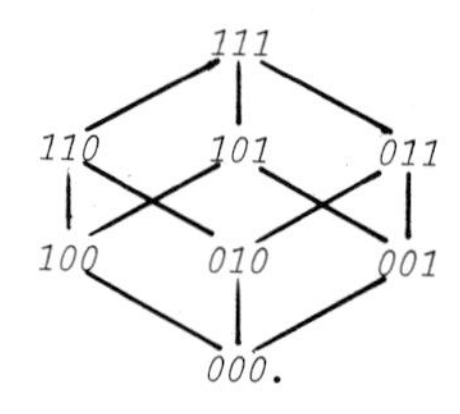

Both representations are useful.

The next major goal of this section is the discussion of the diagonalisable algebras and their representations in terms of atoms. First, however, I want to digress and discuss the group theoretic addition in a boolean algebra. We can think of it as arising by taking seriously the designation of $\underline{V}_n$ as a *vector* algebra over the 2-element field. If, indeed, *110* and *011* are vectors, we should add them:

$$011 \dot{+} 110 = (0+1, 1+1, 1+0) = (1,0,1) \quad (\text{modulo } 2)$$
$$= 101.$$

1.12. DEFINITION. Let $\underline{A} = (A,+,\cdot,',0,1)$ be a boolean algebra. We define

$$x \dot{+} y = xy' + x'y.$$

Observe that in $\underline{2}$ (= $\underline{V}_1$) this agrees with addition modulo 2:

$$0 \dot{+} 0 = 00' + 0'0 = 0 \qquad 0 \dot{+} 1 = 01' + 0'1 = 0 + 1 = 1$$
$$1 \dot{+} 0 = 10' + 1'0 = 1 \qquad 1 \dot{+} 1 = 11' + 1'1 = 0 + 0 = 0.$$

In general, a boolean algebra will not be a field (because $x \cdot x' = 0$-- any algebra other than $\underline{2}$ will have zero-divisors); but it is a ring:

1.13. LEMMA. Let $\underline{A} = (A,+,\cdot,',0,1)$ be a boolean algebra. Then $(A,\dot{+},\cdot,0,1)$ is a commutative ring with unit element, i.e.

i. $x \dot{+} y = y \dot{+} x$ *(Commutativity)*

ii. $x \dot{+} (y \dot{+} z) = (x \dot{+} y) + z$ *(Associativity)*

iii. $x \dot{+} 0 = x$ *(Zero Element)*

iv. $x \dot{+} x = 0$ *(Additive Inverse)*

v. $x \cdot y = y \cdot x$ *(Commutativity)*

vi. $x \cdot (y \cdot z) = (x \cdot y) \cdot z$ *(Associativity)*

vii. $1 \cdot x = x$ *(Unit Element)*

viii. $x \cdot (y \dot{+} z) = (x \cdot y) \dot{+} (x \cdot z)$ *(Distributivity).*

I leave the proof to the reader, noting only that v-vii have already been cited and have merely been repeated to recall the definition of a *commutative ring with unit element* (which is not really given by i-viii because the additive inverse in a ring need not be as trivial as given by iv). If we add the law of idempotence,

$$x^2 = x,$$

we get the definition of a *boolean ring*. With this, 1.13.iv is redundant, but that is not at all an exciting fact. A little more to the point is the fact that we can recover the original boolean operations from the ring structure:

1.14. LEMMA. Let $\underline{A} = (A,+,\cdot,',0,1)$ be a boolean algebra. For all $x,y \in A$:

i. $x + y = x \dot{+} y \dot{+} xy$

ii. $x' = 1 \dot{+} x.$

This is another easy exercise, which I leave to the reader. Ultimately, we will actually use this. In the next section, the group operation $\dot{+}$ will be very important to us. For now, let me simply note that, in the finite case, the group operation gives us an alternate starting point in discussing representations: If $\underline{A} = (A,+,\cdot,',0,1)$ is a finite boolean algebra, the truncated structure $\underline{A}_V = (A,\dot{+},0)$ is a finite dimensional vector space over the 2-element field. Any two such structures of dimension n are isomorphic and have cardinality 2^n (Corollary 1.10 re-derived). That $\underline{A}$ is isomorphic to $\underline{V}_n$, i.e. the extension of the uniqueness theorem to include the multiplication, requires more work.

This concludes what I wanted to say about boolean algebras. As I said, boolean algebras arise from trying to capture the algebraic essence of the Lindenbaum sentence algebra. If we think of Pr_T as an operation on this algebra, we are led to the notion of a diagonalisable algebra.

<u>1.15. DEFINITION</u>. A structure $\underline{A} = (A,+,\cdot,',\tau,0,1)$ is a *diagonalisable algebra* if $(A,+,\cdot,',0,1)$ is a boolean algebra and $\tau: A \rightarrow A$ is a function satisfying:

i. $\tau 1 = 1$

ii. for all $x,y \in A$, $\tau x \cdot \tau(x \rightarrow y) \leq \tau y$

iii. for all $x \in A$, $\tau x \leq \tau\tau x$

iv. for all $x \in A$, $\tau(\tau x \rightarrow x) \leq \tau x$,

where $x \rightarrow y$ abbreviates $x' + y$ (as in Lemma 1.6.v).

Once again, we think of τ (for *t*heorem-- one could as well use π for *p*roof, but the τ has become standard) as simulating the predicate Pr_T. Thus, i corresponds to *D1*(rule *R2* of PRL), ii to *D2*(axiom schema *A2*), iii to *D3* (schema *A3*) and iv to the Formalised Löb Theorem (*A4*).

The representation theory for diagonalisable algebras derives from that for boolean algebras: If $\underline{A}$ is a finite diagonalisable algebra, its truncation to a boolean algebra is isomorphic to a powerset algebra $\underline{P}_X$. All that remains to be done to represent $\underline{A}$ is to slap a Kripke frame structure onto X to simulate τ. This cannot be done unless we become the tiniest bit more liberal and allow frames with no distinguished minimum element, i.e. we allow an arbitrary partial ordering of the finite set X of atoms.

1.16. DEFINITIONS. Let (X,R) be a finite partially ordered set. Define the operation $\tau_R:P(X) \to P(X)$ by $\tau_R(Y) = \{a \in X: \forall b \in X(aRb \Rightarrow b \in Y)\}$, and let $\underline{P}_{X,R}$ denote $(P(X), \cup, \cap, \sim, \tau_R, \emptyset, X)$.

These definitions can easily be explained: Imagine (K,R) to be a Kripke frame and Y the set of nodes α forcing an atom p. Then $\tau_R(Y)$ is the set of nodes forcing $\Box p$. Given this, the following Lemma is fairly unremarkable:

1.17. LEMMA. Let (K,R) be a finite partially ordered set. Then: $\underline{P}_{X,R}$ is a diagonalisable algebra.

Proof: We could base a rigorous proof of this Lemma on the remark just made. The present Chapter being algebraic, however, a more algebraic proof is called for.

i. $\tau_R(X) = X$: Observe,

$$a \in \tau_R(X) \quad \text{iff} \quad \forall b \in X(aRb \Rightarrow b \in X),$$

the right-hand side of which is obviously true. Hence, all $a \in X$ are in $\tau_R(X)$, i.e. $\tau_R(X) = X$.

ii. $\tau_R(Y) \cap \tau_R(Y \to Z) \subseteq \tau_R(Z)$: Let $a \in \tau_R(Y) \cap \tau_R(Y \to Z)$. Then, for any $b \in X$,

$$aRb \Rightarrow b \in Y \qquad (1)$$

$$aRb \Rightarrow b \in Y \to Z$$

$$\Rightarrow b \notin Y \text{ or } b \in Z. \qquad (2)$$

By (1) and (2), for any $b \in X$,

$$aRb \Rightarrow b \in Z,$$

whence $a \in \tau_R(Z)$.

iii. $\tau_R(Y) \subseteq \tau_R(\tau_R(Y))$:

$$\begin{aligned} a \in \tau_R(Y) &\iff \forall b \in X(aRb \Rightarrow b \in Y) \\ &\Rightarrow \forall b,c \in X(aRbRc \Rightarrow c \in Y) \\ &\Rightarrow \forall b \in X(aRb \Rightarrow \forall c \in X(bRc \Rightarrow c \in Y)) \\ &\Rightarrow \forall b \in X(aRb \Rightarrow b \in \tau_R(Y)) \\ &\Rightarrow a \in \tau_R(\tau_R(Y)). \end{aligned}$$

(Note the use of transitivity.)

iv. $\tau_R(\tau_R(Y) \to Y) \subseteq \tau_R(Y)$: Assume this fails. Since X is finite, there is an R-maximal $a \in \tau_R(\tau_R(Y) \to Y)$ such that $a \notin \tau_R(Y)$. As we saw in proving iii,

$$a \in \tau_R(\tau_R(Y) \to Y) \quad \& \quad aRb \quad \Rightarrow \quad b \in \tau_R(\tau_R(Y) \to Y),$$

whence the R-maximality of a entails

$$aRb \quad \Rightarrow \quad b \in \tau_R(Y).$$

But $\quad a \in \tau_R(\tau_R(Y) \to Y) \quad \& \quad aRb \quad \Rightarrow \quad b \in \tau_R(Y) \to Y,$

whence $\quad a \in \tau_R(\tau_R(Y) \to Y) \quad \& \quad aRb \quad \Rightarrow \quad b \in \tau_R(Y) \to Y \quad \& \quad b \in \tau_R(Y)$

$$\Rightarrow \quad b \in Y.$$

But this puts $a \in \tau_R(Y)$, contrary to assumption. QED

The reader should, of course, feel a sense of familiarity toward the above proof. It is merely a disguised version of an earlier proof that PRL is valid in Kripke frames.

Heuristically, the Lemma says that if we have a Kripke frame and collect the set of nodes at which given atoms can be forced, we obtain a diagonalisable algebra. The Finite Representation Theorem for diagonalisable algebras says all finite diagonalisable algebras look like this.

1.18. THEOREM. (Finite Representation Theorem for Diagonalisable Algebras). Let $\underline{A} = (A,+,\cdot,',\tau,0,1)$ be a finite diagonalisable algebra. The map

$$F: \quad x \mapsto \{a \leq x: \quad a \text{ is an atom}\}$$

is an isomorphism of $\underline{A}$ with the Kripke frame algebra $\underline{P}_{X,R}$ on the set X of atoms of $\underline{A}$, where R is defined by: aRb iff $\forall x \in A(a \leq \tau x \Rightarrow b \leq x)$.

Proof: By the Finite Representation Theorem for Boolean Algebras (Theorem 1.9), we know that the isomorphism holds when we drop τ from $\underline{A}$ and τ_R from $\underline{P}_{X,R}$. Thus, it suffices to verify that

$$F(\tau x) \quad = \quad \tau_R(Fx).$$

This proof resembles that of the Completeness Theorem for PRL (Chapter 2, above), which resemblance ought to explain (or, at least, motivate) the definition of R.

Observe, for any atom a,

$$a \in F(\tau x) \quad \Rightarrow \quad a \leq \tau x$$

$$\Rightarrow \quad \forall b \in X(aRb \Rightarrow b \leq x), \text{ by definition of } R$$

$$\Rightarrow \quad \forall b \in X(aRb \Rightarrow b \in Fx), \text{by definition of } F$$

$\Rightarrow \quad a \in \tau_R(Fx)$, by definition of τ_R.

Hence $F(\tau x) \subseteq \tau_R(Fx)$.

The converse implication is the painful part of the proof; it corresponds to the part of the completeness proofs for PRL where we had $\Box C$ not belonging to some theory and we had to construct some accessible theory not containing C. We must construct an atom b such that aRb and yet $b \not\leq x$ if $a \not\leq \tau x$. To this end, assume $a \notin F(\tau x)$, i.e. $a \not\leq \tau x$. Consider $Y = \{y\colon\ a \leq \tau y\}$. This is nonempty because $1 \in Y$. Moreover, if $Y = \{y_1,\ldots,y_k\}$, then $y_0 = y_1\cdot\ldots\cdot y_k \in Y$ since, as one may easily show,

$$a \leq (\tau y_1)\cdot\ldots\cdot(\tau y_k) = \tau(y_1\cdot\ldots\cdot y_k) = \tau(y_0).$$

Now, $y_0 \not\leq x$ since otherwise $y_0 \to x = 1$

$$\tau y_0 = (\tau y_0)\cdot\tau(y_0 \to x) \leq \tau x,$$

which would imply $y_0 \notin Y$. Thus, $y_0 \not\leq x$ and $y_0\cdot x' \neq 0$, whence there is some atom $b \leq y_0\cdot x'$. This atom witnesses the fact that $a \notin \tau_R(Fx)$:

(i) aRb: For,

$$a \leq \tau y \ \Rightarrow\ y_0 \leq y, \quad \text{by choice of } y_0$$
$$\Rightarrow\ b \leq y, \quad \text{since } b \leq y_0.$$

(ii) $b \not\leq x$: For, $b \leq x'$. QED

As before, the Representation Theorem yields, as one type of application, a complete catalogue of diagonalisable algebras:

1.19. EXAMPLES. i. The only diagonalisable algebra on $\underline{2} = (\{0,1\},+,\cdot,',0,1)$ is given by: $\tau 0 = \tau 1 = 1$.

ii. There are only three diagonalisable algebras on $\underline{4} = \underline{V}_2$. These are given by:

x	$\tau_1 x$	x	$\tau_2 x$	x	$\tau_3 x$
00	11	00	01	00	10
01	11	01	11	01	10
10	11	10	01	10	11
11	11	11	11	11	11,

and correspond to the respective partial orderings:

$R1$: 01 10 $\qquad$ $R2$: 01 above 10 (01 | 10) $\qquad$ $R3$: 10 above 01 (10 | 01), of the atoms *01*, *10*. Observe that $(\underline{4},\tau_2)$ and $(\underline{4},\tau_3)$ are isomorphic.

iii. There are *19* diagonalisable algebras on $\underline{8} = \underline{V}_3$ corresponding to the *19* distinct orderings of the atoms *001*, *010*, *100*. These fall into *5* isomorphism types:

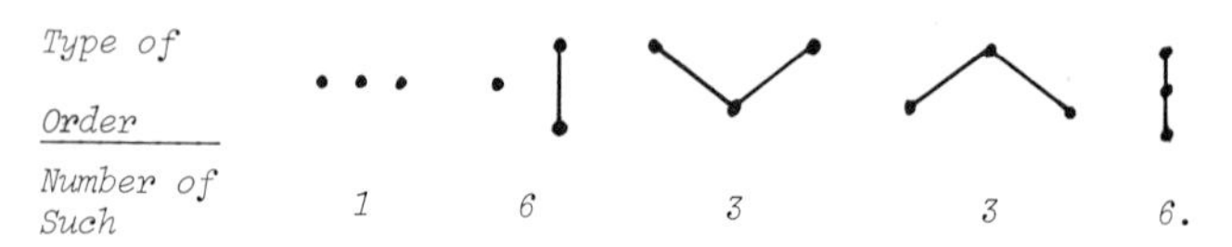

A second sort of application is the following:

1.20. APPLICATION. Let $\underline{A}$ be a diagonalisable algebra. For any $x \in A$, $\tau x \neq 0$.

For, let a be an R-maximal atom in the representation of Theorem 1.18. Since aRb vacuously implies $b \leq x$, it follows that $a \leq \tau x$.

It turns out that the R-minimal atoms also have their significance, as we shall see in the next section.

EXERCISES

1. Prove Lemma 1.3. (Hint: iv. First show $x = x\cdot(1 + 1)$ and $x = x + (0\cdot 0)$.)

2. A *lattice* is a partially ordered set $(A,\leq)$ in which every pair x,y of elements has a supremum $x + y$ and an infinum $x\cdot y$. A lattice $(A,\leq)$ is *distributive* if its suprema and infima satisfy the distributive laws,
$$x\cdot(y + z) = (x\cdot y) + (x\cdot z), \qquad x + (y\cdot z) = (x + y)\cdot(x + z).$$
A lattice $(A,\leq)$ is *complemented* if it has a top element 1, a bottom element 0, and, for every $x \in A$, an element x' such that
$$x + x' = 1, \qquad x\cdot x' = 0.$$
Show: But for the different choices of primitive operations, boolean algebras and complemented distributive lattices are the same objects.

3. Show: $x \to y = \sup\{z\colon\ xz \leq y\}$. (Hint: To show $z(x \to y) = z$ when $xz \leq y$, observe that the latter yields $xz + yz = yz$.)

4. Let $\underline{A} = (A,+,\cdot,',0,1)$ be a boolean algebra and let $a,b \in A$ with $a < b$. Define the interval $[a,b] = \{x \in A\colon\ a \leq x \leq b\}$ and show that $([a,b],+,\cdot,\nu,a,b)$ is a boolean algebra, where $+,\cdot$ are the restrictions to $[a,b]$ of the corresponding

operations on A and ν is defined by $\nu x = a + bx'$.

5. Let $\underline{A}$ be the Lindenbaum algebra of PRA.

   i. Show: $\underline{A}$ has no atoms

   ii. Show: For all $a,b \in A$, if $a < b$ there is some $c \in A$ with $a < c < b$.

   Conclude that $[a,b]$ is infinite.

6. (Duality for Homomorphisms, I). Let $\underline{A}$, $\underline{B}$ be boolean algebras. A *homomorphism* $F:\underline{A} \to \underline{B}$ is a map $F:A \to B$ preserving $+,\cdot,',0,1$. Identify $\underline{A}$, $\underline{B}$ with their respective representations $\underline{P}_X$, $\underline{P}_Y$.

   i. Let $G:Y \to X$ be a function. Show: $F_G(x) = \{a \in Y: \ Ga \in x\}$ is a homomorphism $F:\underline{P}_X \to \underline{P}_Y$.

   ii. Show: Every homomorphism $F:\underline{P}_X \to \underline{P}_Y$ is of the form F_G for some $G:Y \to X$.

   iii. Let $G_1:Z \to X$ and $G_2:Y \to Z$ and $G = G_1 \circ G_2:Y \to X$. Show: $F_G = F_{G_2} \circ F_{G_1}:\underline{P}_X \to \underline{P}_Y$.

   iv. Show: F_G is one-to-one iff G is onto.

   v. Show: F_G is onto iff G is one-to-one.

   (Hint: ii. Define $G(b)$ = the unique atom in $\bigcap\{x: \ b \in F(x)\}$.)

7. (Duality for Homomorphisms, II). A *homomorphism* $F:\underline{A} \to \underline{B}$ of diagonalisable algebras is a boolean homomorphism that preserves τ. Let $\underline{P}_{X,R_1}$ and $\underline{P}_{Y,R_2}$ be given.

   i. Let $G:Y \to X$ preserve order: $aR_2b \Rightarrow GaR_1Gb$. Show: F_G preserves τ: $F_G(\tau_{R_1}x) = \tau_{R_2}(F_G(x))$.

   ii. Show: If $F:\underline{P}_{X,R_1} \to \underline{P}_{Y,R_2}$ is a homomorphism and, as a boolean homomorphism we have $F = F_G$, then $G:Y \to X$ preserves order.

2. FIXED POINT ALGEBRAS

It is now time to consider the most general algebraic models of extensional self-reference-- the fixed point algebras.

2.1. DEFINITION. A pair of boolean algebras $(\underline{A},\underline{B})$ ((A,B) if the operations are understood) constitutes a *fixed point algebra* if B consists of functions $\alpha:A \to A$ and the following hold:

i. B contains the constant functions $\lambda x.a$ for each $a \in A$

ii. the boolean operations on B are pointwise on A:

$$(\alpha + \beta)a = \alpha(a) + \beta(a), \quad (\alpha\cdot\beta)a = (\alpha a)\cdot(\beta a), \quad \alpha'(a) = (\alpha a)'$$

$$1_B(a) = 1_A, \quad 0_B(a) = 0_A$$

iii. B is closed under composition

and iv. each $\alpha \in B$ has a fixed point $a \in A$: $\alpha a = a$.

To explain i (and, shortly, the definition of a *closed* fixed point algebra), let me quickly recall a notational convention:

2.2. λ-NOTATION. To distinguish between a function f given by an expression $f(x)$ and the value of the function at x, we write $\lambda x.f(x)$ to denote the function and $f(x)$ the value. Thus, e.g., $1_B = \lambda x.1_A$ and $' = \lambda x.(x')$.

Somewhat less formally, let me also note that we use lower case roman letters $(a,b,c,\ldots,x,y,z)$ to denote elements of A and lower case greek letters $(\alpha,\beta,\gamma,\ldots)$ to denote elements of B.

The obvious example of a fixed point algebra is the Lindenbaum fixed point algebra of a theory T-- say, PRA. As before, but for a different reason-- to be discussed in the next section, we are again interested in finite algebras. Here, the obvious examples arise from diagonalisable algebras.

2.3. DEFINITIONS. Let $\underline{A} = (A,+,\cdot,',\tau,0,1)$ be a diagonalisable algebra. We define the *diagonalisable fixed point algebra* $(\underline{B},\underline{A})$ corresponding to $\underline{A}$ by letting B be the set of all functions $\alpha: A \to A$ definable by a "polynomial" in which the free variable x lies always in the scopes of τ's. More carefully: A *polynomial* $\overline{\alpha}$ over $\underline{A}$ defining a function α over A is defined by:

i. for $a \in A$, the constant term $\overline{a}$ is a polynomial defining the function $\lambda x.a$

ii. the variable $\overline{x}$ is a polynomial defining the function $\lambda x.x$

iii. if $\overline{\alpha}$, $\overline{\beta}$ are polynomials in $\overline{x}$ defining functions α, β, respectively, then $\overline{\alpha} + \overline{\beta}$ and $\overline{\alpha}\cdot\overline{\beta}$ are polynomials defining $\alpha + \beta$ and $\alpha\cdot\beta$, respectively

iv. if $\overline{\alpha}$ is a polynomial defining α, then $\overline{\alpha}'$ and $\overline{\tau}\overline{\alpha}$ are polynomials defining $\lambda x.(\alpha x)'$ and $\lambda x.\tau\alpha x$, respectively.

The polynomials in which $\overline{x}$ lies only in the scopes of $\overline{\tau}$'s, which we shall call *diag-*

onalisable polynomials, are defined analogously by:

i. for $a \in A$, the constant $\overline{a}$ is a diagonalisable polynomial

ii. if $\overline{\alpha}$ is a polynomial, then $\overline{\tau}\circ\overline{\alpha}$ is a diagonalisable polynomial

iii-iv. as before, but with "polynomial" replaced by "diagonalisable polynomial".

With all this, $B = \{\alpha : A \to A:\ \alpha$ is defined by a diagonalisable polynomial$\}$. Since we shall only be interested in diagonalisable polynomials and their functions, we shall occasionally simply refer to such functions as τ-polynomials.

2.4. LEMMA. Let $\underline{A}$ be a diagonalisable algebra and (B,A) its corresponding diagonalisable fixed point algebra. Then: (B,A) is, in fact, a fixed point algebra.

Proof: By the de Jongh-Sambin Theorem. Let $\overline{\alpha}$ be any polynomial defining a function $\lambda x.\alpha x \in B$. We can think of $\overline{\alpha}$ as a modal formula in the variables $\overline{x}, \overline{a}_1, \ldots, \overline{a}_n$, where $\overline{a}_1, \ldots, \overline{a}_n$ are the constants occurring in $\overline{\alpha}$, and where $\overline{\tau}$ is read as $\Box$. By the de Jongh-Sambin Theorem, there is another modal formula $\overline{\beta}(\overline{a}_1, \ldots, \overline{a}_n)$ such that

$$\mathrm{PRL} \vdash \overline{\beta} \leftrightarrow \overline{\alpha}(\overline{\beta})$$

Since PRL proves this, and since the axioms on τ in the definition of a diagonalisable algebra were precisely the axioms of PRL, it follows that

$$\overline{\beta}(\overline{a}_1, \ldots, \overline{a}_n) = \overline{\alpha}(\overline{\beta}(\overline{a}_1, \ldots, \overline{a}_n), \overline{a}_1, \ldots, \overline{a}_n)$$

in any diagonalisable algebra in which $a_1, \ldots, a_n$ have been interpreted. In particular, this holds in $(\underline{A}, \tau)$, whence, for $b = \beta(a_1, \ldots, a_n)$, we have $b = \alpha(b)$, i.e. α has a fixed point in A. QED

Lest the reader find himself confused by the polynomials $\overline{\alpha}$, functions α, and modal formulae-- here denoted $\overline{\alpha}$, let me quickly illustrate the proof of the Lemma: Suppose α is given by the polynomial (and let me cease the overlining)

$$\tau(x + a_1)' \cdot a_2.$$

We consider the modal formula,

$$\sim\Box(p \vee q_1) \wedge q_2$$

with diagonal

$$D(q_1, q_2) = \sim\Box(\sim\Box(f \vee q_1) \wedge q_2) \wedge q_2$$

(most quickly found by Example 4.17 with $\nabla p = \Box(p \vee q_1) \vee \sim q_2$). Letting $b = \tau(\tau(a_1)' \cdot a_2)' \cdot a_2$, we have $\alpha b = b$ as desired.

Since we have a lot of finite diagonalisable algebras, we also have a lot of diagonalisable fixed point algebras, whence a lot of fixed point algebras. There is a non-obvious quantitative assertion here: Distinct finite diagonalisable algebras yield distinct diagonalisable fixed point algebras. Having published an example purporting to show this not to be the case, I consider this simple truth to be non-obvious. The simple proof of this result will be left as an exercise (Exercise 3, below). For all that, there is something new: There are non-diagonalisable fixed point algebras arising as proper subalgebras of the diagonalisable ones. In the finite case, there are, as my colleague David Hobby has shown, no other ones: Every finite fixed point algebra is a subalgebra of a finite diagonalisable fixed point algebra. The proof of this requires a bit too much universal algebra to be presented here and I shall present my earlier partial result instead. But, first... some preliminaries:

2.5. EXAMPLES. i. On $\underline{2}$, there is only the trivial fixed point algebra:

$$A = \{0,1\}, \qquad B = \{\lambda x.0, \lambda x.1\}$$

ii. On $\underline{4}$, the fixed point algebras are exactly the diagonalisable ones.

The assertion of Example i is fairly obvious; that of Example ii is not and will be proven in the Exercises (Exercise 6, below).

The interesting new behaviour occurs on $\underline{8}$. For example, some nontrivial diagonalisable algebras are contained in others:

2.6. EXAMPLE. Represent $\underline{8}$ as $\underline{V}_3$ with atoms $a = 001$, $b = 010$, and $c = 100$. Let τ_1, τ_2, τ_3 be the τ-operators induced by the respective partial orderings,

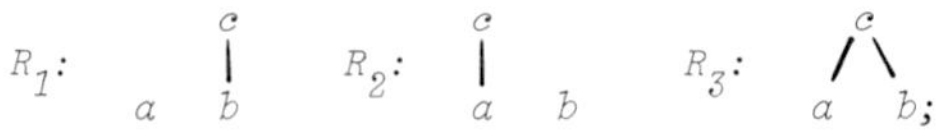

i.e. τ_1, τ_2, τ_3 are defined by:

x	$000, 001, 010, 011$	$100, 101, 110, 111$
$\tau_1 x$	101	111
$\tau_2 x$	110	111
$\tau_3 x$	100	111

Then: τ_1, τ_2 are definable in terms of τ_3 and are in the diagonalised fixed point algebra of τ_3. For,

$$\tau_1 = \lambda x.(001 + \tau_3 x) = \lambda x.(a + \tau_3 x)$$
$$\tau_2 = \lambda x.(010 + \tau_3 x) = \lambda x.(b + \tau_3 x).$$

That neither τ_1 nor τ_2 is definable from the other is readily established by enumerating all the elements of the given algebras; in addition to the constant functions, each algebra possesses 8 other functions, giving a total of 16 functions. The algebra generated by τ_3 has, of course, more such functions. Exercise 3 has more to say on this.

On $\underline{8}$, nondiagonalisable fixed point algebras appear. To illustrate this, we need a tiny bit of notation.

2.7. NOTATION. Let (B,A) be a fixed point algebra and let α B. By $A(\alpha)$, we mean the set of all functions generated over A (i.e. the constant functions $\lambda x.a$) by composition and the boolean operations. $(A(\alpha),A)$ is, obviously, the smallest fixed point algebra over A containing α. We also let $A(\alpha)$ denote the entire algebra $(A(\alpha),A)$.

Now for some non-diagonalisable fixed point algebras:

2.8. EXAMPLES. i. Let τ_1 be the τ-operator

x	$000,001,010,011$	$100,101$	$110,111$
$\tau_1 x$	100	110	111

dual to the ordering

where a, b, and c are 001, 010, and 100, respectively, as in Example 2.6. $A(\tau_1)$ is a proper subalgebra of the diagonalisable fixed point algebra determined by τ_1, as is the algebra $(\lambda x.\tau_1(x'))$. Each of these algebras contains non-constant τ-operators.

ii. Let the operator α_0 be defined by

x	$000,001$	everything else
$\alpha_0 x$	111	110

α_0 is of the form $\lambda x.\tau_2(x')$, where τ_2 is the τ-operator dual to $b \diagdown a \diagup c$.

$A(\alpha_0)$ possesses no non-constant τ-operator; in particular, $\tau_2 \notin A(\alpha_0)$.

We can give a quick proof of the first non-definability result: A simple induction shows that every $\alpha \in A(\tau_1)$ maps 000, 001, 010, 011 onto the identical image. But $\alpha = \lambda x.\tau_1(x')$ maps 001 and 010 onto 111 and 110, respectively. Thus, $\alpha = \lambda x.\tau_1(x') \notin A(\tau_1)$. A similar argument shows $\tau_1 \notin A(\lambda x.\tau_1(x'))$. The other assertions are verified by inspection: In ii, for example, a bit of paperwork reveals $A(\tau_2)$ to possess 16 elements, only two of which map 111 to 111. One of these is the constant τ-operator; the other fails to preserve modus ponens.

These non-diagonalisable fixed point algebras are subalgebras of diagonalisable ones. As remarked earlier, this is true of all finite fixed point algebras and, indeed, the main goal of this section is to prove that every finite fixed point algebra satisfying an additional closure property is embeddable in a finite diagonalisable fixed point algebra. First, however, we will need to develop a little machinery.

There are two technical preliminaries. The first is a sort of restriction or retraction construction.

2.9. DEFINITIONS. Let (B,A) be a fixed point algebra and $a < b$ elements of A. The *interval algebra* $[a,b]$ is defined by i. taking as domain the interval

$$[a,b] = \{c \in A:\ a \le c \le b\},$$

ii. restricting the boolean sum $+$ and product $\cdot$ to $[a,b]$, iii. taking as complement the operation

$$\nu x = b\cdot(a + x'),$$

and iv. taking a,b as the $0,1$ elements, respectively. The interval algebra, so defined, is a boolean algebra (as the reader proved in Exercise 4 of the preceding section). For $\alpha \in B$, define $\alpha^r : B \to [a,b]$ by

$$\alpha^r(x) = b\cdot(a + \alpha(x)).$$

Letting α^r ambiguously denote the function just defined and its restriction to $[a,b]$, we define

$$B^r = \{\alpha^r : [a,b] \to [a,b]:\ \alpha \in B\}.$$

Defining boolean operations on B^r by their pointwise behaviour on $[a,b]$, we obtain an algebra $(B^r,[a,b])$, which we call the *induced algebra*.

Those familiar with universal algebra will recognise that, were we to ignore the designated elements and the complement, the induced algebra would be a *retraction*. The non-lattice information is not preserved under the return embedding of $[a,b]$ into A. However, like a retraction, the induced algebra is a homomorphic image of the original algebra and, almost like a retraction, it is simultaneously almost also a subalgebra. In the finite case, it is smaller as well and this smaller size, along with the close relation to the original, makes the induced algebra construction a tool for a mathematical induction. Before using it for such, we must, of course, verify one little fact:

2.10. LEMMA. Let (B,A) be a fixed point algebra, $a,b \in A$ with $a < b$. The induced algebra $(B^r,[a,b])$ is a fixed point algebra.

Proof: As already remarked, the reader verified in the last Exercise collection that $[a,b]$ is a boolean algebra under its assigned operations. Moreover, as is implicit in the proof of this Exercise, the map

$$x \mapsto b\cdot(a + x)$$

is a homomorphism of A onto $[a,b]$ (i.e. it preserves $+,\cdot$, complement and the distinguished elements). From this immediately follows the closure of B^r under the boolean operations, i.e. the fact that B^r is a boolean algebra.

If $c \in [a,b]$, then $b\cdot(a + c) = c$ and $(\lambda x.c)^r$ is the constant function c on $[a,b]$.

Further, B^r is closed under composition: Let $\alpha^r,\beta^r \in B^r$ and observe

$$\lambda x.\beta^r(\alpha^r x) \quad = \quad \lambda x.(b\cdot(a + \beta(\alpha^r x)) \quad = \quad (\lambda x.(\beta(\alpha^r x)))^r \in B^r,$$

this latter because $\alpha^r \in B$ and B is closed under composition.

It only remains to verify that each $\alpha^r \in B^r$ possesses a fixed point in $[a,b]$. But α^r can be viewed as an element of B, whence it has a fixed point $c \in A$. However,

$$c \;=\; \alpha^r c \in range(\alpha^r) \;\subseteq\; [a,b].$$

QED

[Remark: Exercise 8, below, will clarify some of the above. In particular, $\beta^r\alpha^r = (\beta\alpha)^r$ holds under special circumstances.]

As an example of the use of induced algebras in applications, we have the following theorem:

2.11. THEOREM. (Uniqueness of Fixed Points). Let (B,A) be a finite fixed point algebra. Then: For any $\alpha \in B$, there is a unique fixed point $a \in A$.

Proof: By induction on the cardinality of A, $|A|$.

Basis. $|A| = 2$. By inspection: B consists of two constant functions, whence each function has only one fixed point.

Induction step. Suppose $\alpha \in B$ had two distinct fixed points $a,b \in A$. There are two cases to dispose of.

Case 1. $a \neq b'$. We consider the induced algebra on $(ab, a+b)$, which is a proper subalgebra of A, whence of lower cardinality than A. The induction hypothesis applies and every $\beta^r \in B^r$ has only one fixed point in $(ab, a+b)$. However, this assertion contradicts the observations that

$$\alpha^r a = (a+b)\cdot(ab + \alpha(a)) = (a+b)\cdot(ab+a) = (a+b)a = a$$

$$\alpha^r b = (a+b)\cdot(ab + \alpha(b)) = (a+b)\cdot(ab+b) = (a+b)b = b.$$

Case 2. $a = b'$. Because $|A|$ is finite, the successive ranges of $\alpha, \alpha^2, \alpha^3, \ldots$ eventually settle down to some set X of cardinality, say, m. If, say, $range(\alpha^n) = X$, then $\alpha^n \restriction X$ is a permutation and a basic result of group theory tells us then that $(\alpha^n)^{m!}$ is the identity on X. If $X = \{a,b\}$, then $\lambda x.(\alpha^{n\cdot m!}(x))'$ has no fixed point, a contradiction. Thus there is some third element $c \in X$. But a,c are non-complementary fixed points of $\alpha^{n\cdot m!}$, another contradiction, as we saw in Case 1. QED

<u>2.12. COROLLARY</u>. Let (B,A) be a finite fixed point algebra.

i. $\forall \alpha \in B \, \exists n \in \omega (\alpha^n \text{ is constant})$

ii. $\forall \alpha \in B \; \alpha$ is not a homomorphism of A.

For the proof, cf. Exercise 4, below.

These results show there to be a tremendous difference between the finite fixed point algebras and the infinite ones. In the Lindenbaum fixed point algebras (B_T, A_T) for theories T only a little stronger than PRA (perhaps also for PRA?), there are many extensional formulae without unique fixed points; indeed, some of these define homomorphisms of the Lindenbaum sentence algebras. It follows that finite algebras will not adequately model all interesting arithmetic self-reference.

A word about the proof of the Uniqueness Theorem: The fleeting reference to group theory in the last part of the proof is a bit disappointing, the more so as there is a simple completely group theoretic proof of the result. The reader will find this in the exercises (specifically, Exercise 5); our more official application of the group theory (i.e. of $\dot{+}$) will be to produce the right interval $[a,b]$ on which to induce an algebra.

Two considerations must govern the choice of an interval $[a,b]$ with which to work in an induction like that of the proof of Theorem 2.11. First, of course, the cardinality of $[a,b]$ must be less than that of A, hence $[a,b]$ must omit something. Second, for delicate work, $[a,b]$ must be very closely related to A; in particular, it must be as large as possible. Obvious candidates are $[a,1]$, where a is an atom. Unfortunately, not every atom represents a good choice. With diagonalisable algebras, however, there are partial orderings on the atoms and these lead to natural choices:

2.13. LEMMA. Let (B,A) be a finite diagonalisable fixed point algebra based on the τ-operator τ, i.e. B consists of all appropriate polynomials in τ. Then: If an atom $a \in A$ is minimal in the dual ordering R of the atoms of A, then, for all $x \in A$ and all $\alpha \in B$, $\alpha x = \alpha(a + x)$.

Proof: First note that, for any α, the conclusion is equivalent to the assertion: For all $x \in A$,

$$\alpha x = \alpha(a \dot{+} x)$$

(Why is this the case?) We shall prove this by induction on the generation of α.

Basis. α has the form $\lambda x.\tau\beta x$, where β is a purely boolean expression, i.e. β is constructed without the use of a τ. Regardless of how complicated an expression we might have for β, once we write it as a polynomial in $\dot{+}$ and $\cdot$ and simplify it, it takes on the form

$$\beta x = b_0 \dot{+} b_1 x,$$

for some $b_0, b_1 \in A$. (By idempotence, there are no terms of higher degree.) Now:

$$\beta(a \dot{+} x) = b_0 \dot{+} b_1 a \dot{+} b_1 x$$

$$= \begin{cases} b_0 \dot{+} b_1 x, & \text{if } b_1 a = 0 \\ a \dot{+} (b_0 \dot{+} b_1 x), & \text{otherwise} \end{cases}$$

since a is atomic. But this means

$$\beta(a \dot{+} x) = \begin{cases} \beta x, & \text{if } b_1 a = 0 \\ a \dot{+} \beta x, & \text{otherwise.} \end{cases}$$

Hence: To prove the basis it suffices to show, for all $y \in A$,

$$\tau y = \tau(a \dot{+} y),$$

or even, for all $y \in A$, $\tau y = \tau(a + y)$.

Let y be given. For any atom b, we have

$$b \le \tau y \quad \text{iff} \quad \forall \text{ atoms } c\big(bRc \Rightarrow c \le y\big), \quad \text{by 1.17 \& 1.18}$$

$$\text{iff} \quad \forall \text{ atoms } c\big(bRc \Rightarrow c \le a + y\big),$$

by the R-minimality of a. But 1.17 and 1.18 again apply to this last to yield:

$$b \le \tau y \quad \text{iff} \quad b \le \tau(a + y),$$

for any atom b, whence $\tau y = \tau(a + y)$.

Induction step. α is built up from expressions of the form $\lambda x.\tau\beta x$ and constants by boolean operations and composition. Given the basis, this step is trivial. QED

The atoms satisfying the conclusion of this Lemma are precisely the ones to apply the induced algebra construction to. Because of this important role they will play, we give them a fancy name.

2.14. DEFINITION. Let (B,A) be a fixed point algebra. An atom $a \in A$ is called a *fundamental atom* if a satisfies one of the following equivalent conditions:

i. $\forall \alpha \in B \, \forall x \in A\big(\alpha x = \alpha(x + a)\big)$

ii. $\forall \alpha \in B \, \forall x \in A\big(\alpha x = \alpha(x \dot{+} a)\big)$.

It can be shown that every finite fixed point algebra possesses a fundamental atom. The proof, due to David Hobby, requires too much universal algebra to be presented here and I will instead present my earlier proof for finite *closed* fixed point algebras, that is, finite fixed point algebras possessing some extra closure properties. What can easily be proven for all finite fixed point algebras is our next lemma, which must be preceded by a definition.

2.15. DEFINITION. Let (B,A) be a fixed point algebra. The equivalence relation $\simeq$ on A is defined by: For $a,b \in A$,

$$a \simeq b \quad \text{iff} \quad \forall \alpha \in B\big(\alpha a = \alpha b\big).$$

2.16. LEMMA. Let (B,A) be a finite fixed point algebra. For each $a \in A$ there is an element $b \in A$ with $a \neq b$, but $a \simeq b$; i.e. the $\simeq$-equivalence classes all have cardinality at least 2.

Proof: By contradiction. Suppose $a \in A$ is such that, for each $b \in A$, there is an $\alpha_b \in B$ with $\alpha_b a \neq \alpha_b b$.

Claim. We can assume $\alpha_b a = 0$, $\alpha_b b > 0$.

To prove the claim, let $\alpha_b a = c$, and replace α_b by $\beta = \lambda x.(c \dotplus \alpha_b x)$. For, $\beta a = 0$ and $\beta b \neq 0$.

To complete the proof of the Lemma, define

$$\beta = \lambda x.\,sup\,\{\alpha x: \ \alpha \in B \ \&\ \alpha a = 0\}.$$

The *sup* being a finite sum, β is in B. Now, $\beta a = 0$ and, by the Claim, $\beta b \neq 0$ for all $b \neq a$. If we now define

$$\gamma = \lambda x.(a \dotplus \beta x),$$

we obtain an element $\gamma \in B$ with the peculiar property that $\gamma a = a$ and $\gamma b \neq a$ for any $a \neq b \in A$, which curious property contradicts Corollary 2.12.i, whereby $\lim_{n\to\infty} \gamma^n b = a$ for all $b \in A$. QED

By this Lemma, we have $0 \simeq a$ for some non-zero a, but we do not know that a is atomic nor, even assuming a to be atomic, that it would be fundamental: $x \simeq x + a$, for all $x \in A$. To draw this conclusion we need the extra closure conditions already hinted at. Before citing them, let me quickly digress to refer the reader to Exercise 6, wherein the present Lemma distinguishes itself by enabling the reader to verify the assertion of Example 2.5.ii that the only fixed point algebras on $\underline{4}$ are the diagonalisable ones.

The digression being over, we finally arrive at the desired definition:

2.17. DEFINITION. A fixed point algebra (B,A) is *closed* if, in addition to properties i-iv of Definition 2.1 of a fixed point algebra, it satisfies

v. for all $a \in A$ and $\alpha \in B$, $\lambda x.\alpha(x \dotplus a) \in B$ and $\lambda x.\alpha(a \cdot x) \in B$.

Note that condition v is equivalent to the more general

v'. for all $\alpha \in B$ and any boolean polynomial $p(x)$ over A, $\lambda x.\alpha p(x) \in B$.

I have chosen to emphasise v over v' because a. it would appear easier to verify in practice and b. it is the form that is used in the following proofs.

2.18. LEMMA. Let (B,A) be a fixed point algebra closed under

$$\alpha \in B \Rightarrow \lambda x.\alpha(x \dot{+} c) \in B$$

for all $c \in A$. Then: For any $a,b \in A$,

$$a \simeq b \iff a \dot{+} b \simeq 0.$$

Proof: Suppose $a \simeq b$. Let $\alpha \in B$ be given and $\beta = \lambda x.\alpha(x \dot{+} b)$. Observe:

$$a \simeq b \Rightarrow \beta a = \beta b \Rightarrow \alpha(a \dot{+} b) = \alpha(b \dot{+} b) \Rightarrow \alpha(a \dot{+} b) = \alpha 0.$$

As α is arbitrary, this entails $a \dot{+} b \simeq 0$.

Conversely, suppose $a \dot{+} b \simeq 0$. Let α be given and define $\beta = \lambda x.\alpha(x \dot{+} b)$ as before. Observe:

$$a \dot{+} b \simeq 0 \Rightarrow \beta(a \dot{+} b) = \beta 0 \Rightarrow \alpha(a \dot{+} b \dot{+} b) = \alpha(0 \dot{+} b) \Rightarrow \alpha a = \alpha b.$$

As before, we conclude $a \simeq b$. QED

It might be worth noting both that not every finite fixed point algebra satisfies the closure condition used in this proof and that such algebras can fail to satisfy the conclusion-- cf. Exercise 7, below.

Lemma 2.18 is very important. For an atom $a \in A$ to be fundamental, it must satisfy $x \simeq x \dot{+} a$, for all $x \in A$. By the Lemma, to do this under the stated closure condition-- in particular if (B,A) is closed-- it will suffice to have $0 \simeq a$. By Lemma 2.16, we know for finite A that $0 \simeq b$ for some non-zero b. To get an atom, we appeal to the other extra closure property of a closed fixed point algebra.

2.19. LEMMA. Let (B,A) be a fixed point algebra closed under

$$\alpha \in B \Rightarrow \lambda x.\alpha(c \cdot x) \in B$$

for all $c \in A$. Then: For any $a,b \in A$,

$$a \simeq 0 \Rightarrow a \cdot b \simeq 0.$$

Proof: Let $a \simeq 0$ and $b \in A$. For any given $\alpha \in B$, let $\beta = \lambda x.\alpha(b \cdot x)$ and observe,

$$a \simeq 0 \Rightarrow \beta a = \beta 0 \Rightarrow \alpha(a \cdot b) = \alpha(0 \cdot b) \Rightarrow \alpha(a \cdot b) = \alpha 0.$$ QED

2.20. COROLLARY. Let (B,A) be a finite closed fixed point algebra. Then: (B,A) possesses a fundamental atom.

This is an immediate consequence of Lemmas 2.16, 2.18, and 2.19.

We are almost ready to prove that every finite closed fixed point algebra (B,A) is a subalgebra of a diagonalisable algebra over A. The proof will be an induction like that of the proof of the Uniqueness of Fixed Points. To carry out such an induction, we need one technical lemma:

2.21. LEMMA. Let (B,A) be a closed fixed point algebra, $a,b \in A$ with $a < b$. Then the induced algebra $(B^r,(a,b))$ is closed.

The proof is fairly trivial and I omit it.

We now have all the tools needed to prove the representation theorem.

2.22. THEOREM. (Representation Theorem for Finite Closed Fixed Point Algebras). Let (B,A) be a finite closed fixed point algebra. Then: There is some τ-operator τ on A such that every element of B is a τ-polynomial, i.e. (B,A) is a subalgebra of the diagonalisable fixed point algebra induced by τ.

Theorem 2.22 is a consequence of the following more technical result.

2.23. THEOREM. Let (B,A) be a closed fixed point algebra with $|A| = 2^n$. Then:

$$|B| \leq 2^{2^n - 1} = 2^{|A| - 1}.$$

Moreover, every such B is a subalgebra of one of exactly $n!$ (diagonalisable) algebras for which equality obtains.

Theorem 2.23 is proven by induction on $|A| = 2^n$, i.e. by induction on n. The cases $n = 1$ and $n = 2$ follow readily from the assertions of Examples 2.5 and require no assumption of closedness. The first nontrivial case, in which the forthcoming proof needs this assumption, occurs when $n = 3$. I remark that either case $n = 1$ or $n = 2$ can be taken as the basis of the induction.

The proof of Theorem 2.23 is miserably non-canonical. It consists, firstly, of taking a closed fixed point algebra (B,A), where A has cardinality 2^{n+1}, and showing B to be a subset of one of at most $(n + 1)!$ boolean algebras B_0 such that (B_0,A) is a closed fixed point algebra with $|B_0| = 2^{2^{n+1} - 1}$, and, secondly, of exhibiting $(n + 1)!$ distinct such diagonalisable fixed point algebras (B_0,A) of the given cardinality. It will follow that (B,A) has been embedded in a diagonalisable algebra and the induction step of the induction will have been completed.

The first step requires a relatively simple lemma possessing a long, but routine proof.

2.24. LEMMA. Let (B,A) be a finite closed fixed point algebra with fundamental atom $a \in A$. Let $(B^r,(a,1))$ be the induced algebra and $C \supseteq B^r$ an algebra such that $(C,(a,1))$ is a closed fixed point algebra. Define

$$B_0 = \{\alpha : A \to A:\ \alpha^r \in C \ \&\ \forall x \in A(\alpha x = \alpha(x \dot{+} a))\}.$$

Then: i. $B \subseteq B_0$

ii. (B_0,A) is a closed fixed point algebra with fundamental atom a

iii. $|B| \leq |B_0| = |C|\cdot 2^{|A|/2}$.

But for the relativisation to C, which is a technical matter, the Lemma makes sense: It is exactly the sort of use of the induced algebra construction we would have to make: Go down from (B,A) to $(B^r,(a,1))$ and come back up to the largest (B_0,A) inducing the same algebra and having a as a fundamental atom. However, natural or not, the result still requires a proof.

Proof of Lemma 2.24: Lemma 2.21 gives us the existence of some C, namely B^r, for which $(C,(a,1))$ is a closed fixed point algebra. Although we only referred previously to α^r for $\alpha \in B$, it is clear that α^r can be defined for any $\alpha : A \to A$ and the definition of B_0 makes sense.

We have three things to prove. Of these, assertion i, that $B \subseteq B_0$, is trivial. Assertion iii on the cardinality of B_0 is a mere calculation: There are $|C|$ choices of $\alpha^r \in C$ (the first factor). For each such α^r, there are $2^{|A|/2}$ possible $\beta \in B_0$ (the second factor) for which $\beta^r = \alpha^r$. To see this latter, note that a. for any given value $\alpha^r x$, there are two possible values of βx-- namely, $\alpha^r x$ and $a \dot{+} \alpha^r x$, and b. by the assumption that $\alpha x = \alpha(x \dot{+} a)$, this choice is made only once for each pair x, $x \dot{+} a$, and there are $|A|/2$ such pairs.

The heart of the proof is assertion ii: (B_0,A) is a closed fixed point algebra with fundamental atom a. Given the rest, the fact that a is a fundamental atom is trivial.

B_0 is a boolean algebra. Let $\alpha,\beta \in B_0$ and let $\gamma = \alpha + \beta$. Note that

$$\alpha^r \in C \ \&\ \beta^r \in C \ \Rightarrow\ (\alpha + \beta)^r = \alpha^r + \beta^r \in C$$

$$\forall x \in A\big(\alpha x = \alpha(x + a) \ \&\ \beta x = \beta(x + a)\big) \ \Rightarrow\ \forall x \in A\big(\gamma x = \gamma(x + a)\big).$$

Thus, $\gamma \in B_0$. Similarly, the reader can show that, if $\alpha \in B_0$, then $\alpha' \in B_0$.

B_0 is closed under composition. Let $\alpha,\beta \in B$ and let $\gamma = \lambda x.\alpha\beta x$. For all $x \in A$,

$$\gamma^r x \ = \ a + \alpha\beta(x) \ = \ a + \alpha(a + \beta x), \text{ since } \forall y \in A\big(\alpha y = \alpha(a + y)\big)$$
$$= \ \alpha^r\beta^r(x).$$

Since C is closed under composition, it follows that $\gamma^r \in C$. But also, for any $x \in A$,

$$\gamma x \ = \ \alpha\beta x \ = \ \alpha\beta(a + x) \ = \ \gamma(a + x),$$

since $\beta x = \beta(a + x)$. Hence $\gamma \in B_0$.

B_0 has fixed points. Let $\alpha \in B_0$ and let $b \in (a,1)$ be a fixed point for $\alpha^r \in C$. Either $\alpha b = \alpha^r b = b$ and b is a fixed point or $\alpha(a \dot{+} b) = \alpha b = a \dot{+} \alpha^r b = a \dot{+} b$ and $a \dot{+} b$ is a fixed point.

(B_0,A) is closed. Let $\alpha \in B_0$, $b \in A$ and consider $\beta = \lambda x.\alpha(x \dot{+} b)$. A little algebra shows that, for all $x \in A$,

$$\alpha(x \dot{+} b) \ = \ \alpha(x \dot{+} (b + a)),$$

whence we can assume without loss of generality that $b \in (a,1)$. Since C is closed, $\beta^r = \lambda x.\alpha^r(x \dot{+} b) \in C$. But, for any $x \in A$,

$$\beta x \ = \ \alpha(x \dot{+} b) \ = \ \alpha((x \dot{+} b) \dot{+} a) \ = \ \alpha((x \dot{+} a) \dot{+} b) \ = \ \beta(x \dot{+} a).$$

However, $x + a$ is one of x and $x \dot{+} a$ and it follows that $\beta x = \beta(x + a)$, whence $\beta \in B_0$. Similarly, $\lambda x.\alpha(b\cdot x) \in B_0$.

This completes the proof. QED

Now, assume half of Theorem 2.23 as an induction hypothesis on n: Each closed fixed point algebra (B,A) with $|B| = 2^n$ can be embedded in one of (exactly) $n!$ closed fixed point algebras (B_1,A), where $|B_1| = 2^{|A| - 1}$. So assume we are given a closed fixed point algebra (B,A) with A having the next cardinality: $|A| = 2^{n+1}$. Choose a fundamental atom $a \in A$ and look at $(B^r,(a,1))$. By induction hypothesis, $(B^r,(a,1))$ can be embedded in one of exactly $n!$ closed fixed point algebras $(C,(a,1))$ with $|C| = 2^{2^n - 1}$. By the Lemma, this lifts to an embedding of (B,A) into (B_0,A), where

$$|B_0| \ = \ |C|\cdot 2^{|A|/2} \ = \ 2^{2^n - 1}\cdot 2^{2^n} \ = \ 2^{2^{n+1} - 1}.$$

But how many choices of B_0 are there? A has $n + 1$ atoms for the first choice and there are $n!$ C's for the next, yielding an *upper* bound of $(n + 1)\cdot n! \ = \ (n + 1)!$

(This is only an upper bound because not all atoms need to be fundamental. Also, we have not verified that distinct choices of atoms will yield distinct B_0's.)

This gives us half of Theorem 2.23 for $n+1$. To get the other half, and along with it Theorem 2.22, for $n+1$, it will suffice to obtain exactly $(n+1)!$ distinct *diagonalisable* fixed point algebras (B,A) with $|B| = 2^{|A|} - 1$ and $|A| = 2^{n+1}$. This is actually a simple task.

2.25. LEMMA. Let $k \geq 2$ be given and A a boolean algebra of cardinality 2^k with set of atoms $P = \{a_0,\ldots,a_{k-1}\}$. For each total ordering $\prec$ of P, the corresponding diagonalisable fixed point algebra (B,A) satisfies

$$|B| = 2^{|A|} - 1.$$

Moreover, the $k!$ distinct total orderings of P yield distinct fixed point algebras.

Proof: The easiest thing to show is that distinct orderings of P yield distinct algebras B. The most revealing proof, which obviates our choice of diagonalisable algebras, uses Exercise 3, referred to back in Example 2.6: The incompatibility of the distinct total orderings entails the non-interdefinability of their dual τ-operators, whence the distinctness of the diagonalisable fixed point algebras they determine. Nonetheless, an *ad hoc* proof is readily accessible and I give it here before launching into the big cardinality calculation.

Let $\prec_1$, $\prec_2$ be distinct total orderings of P, say:

$$a_{i_{k-1}} \prec_1 \ldots \prec_1 a_{i_0} \qquad a_{j_{k-1}} \prec_2 \ldots \prec_2 a_{j_0}.$$

Let m be minimum so that $a_{i_m} \neq a_{j_m}$ and let $b = a_{i_0} + \ldots + a_{i_{m-1}} = a_{j_0} + \ldots + a_{j_{m-1}}$ (= 0, if $m = 0$). Letting τ_1,τ_2 denote the τ-operators dual to $\prec_1$, $\prec_2$, respectively, we have

$$\tau_1 b = b + a_{i_m}, \qquad \tau_2 b = b + a_{j_m},$$

whence

$$\tau_1 b \cdot \tau_2 b = (b + a_{i_m})\cdot(b + a_{j_m}) = b,$$

since a_{i_m}, a_{j_m} are distinct atoms incomparable to b. But we also have $\tau_1 1 \cdot \tau_2 1 = 1$, whence $\tau_1 \cdot \tau_2$ has two fixed points and it is contradictory to assume both τ_1 and τ_2 to belong to one and the same fixed point algebra.

Now for the real work of the proof! We will prove by induction on k that the diagonalisable fixed point algebra (B,A) determined by the τ-operator dual to a total ordering $\prec$ of the atoms $P = \{a_0,\ldots,a_{k-1}\}$ satisfies $|B| = 2^{2^k-1}$. The basis of the induction, i.e. the case $k = 2$, is verified by inspection and is left to the reader.

So let $|A| = 2^{k+1}$ and let (B,A) be the diagonalisable fixed point algebra determined by a total ordering $\prec$ of the (now slightly larger) set $P = \{a_0,\ldots,a_k\}$ of atoms of A. Let, further, a be the $\prec$-minimum (hence: fundamental) atom of A and consider the induced algebra $(B^r,(a,1))$. By Lemma 2.24,

$$|B_0| = |B^r|\cdot 2^{2^k},$$

where B_0 is the maximum algebra inducing B^r. First, we must show $B = B_0$, so that

$$|B| = |B^r|\cdot 2^{2^k},$$

and then that $(B^r,(a,1))$ is the diagonalisable fixed point algebra on $(a,1)$ determined by the "restriction" of the total ordering $\prec$ of P to $P - \{a\}$. For, then, the induction hypothesis yields

$$|B^r| = 2^{2^k-1},$$

whence
$$|B| = 2^{2^k-1}\cdot 2^{2^k} = 2^{2^{k+1}-1},$$

and we will have completed the proof.

First, $B = B_0$: Since $B \subseteq B_0$, we need only show $B_0 \subseteq B$. Let $\alpha \in B_0$. By the definition of B_0, there is some $\alpha_0 \in B$ such that $\alpha_0^r = \alpha^r$. Now $\alpha_1 = \lambda x.(a'\cdot\alpha_0 x) \in B$ is the minimum function inducing α^r. Toward defining α explicitly from α_0, observe that, for any $b \in A$,

$$a\cdot\tau b = \begin{cases} a, & b = 1 \text{ or } b = 1 \dot{+} a \quad (\text{i.e. } b \simeq 1) \\ 0, & \text{otherwise.} \end{cases}$$

Letting $b_0,\ldots,b_m \in A$ enumerate all $b \in A$ such that $\alpha b = a + \alpha_1 b$, we see

$$\alpha = \lambda x.\left(\alpha_1 x + a\cdot sup\ \{\tau(x \dot{+} b_i \dot{+} 1):\ i \le m\}\right) \in B.$$

Since $\alpha \in B_0$ was arbitrary, we see that $B = B_0$.

Next, we show that $(B^r,(a,1))$ is diagonalisable with the promised ordering. There are three steps to the proof of this.

First, note that $\tau^r = \lambda x.(a + \tau x)$ is a τ-operator on $(a,1)$: For all $x,y \in (a,1)$,

i. $\tau^r 1 = a + \tau 1 = a + 1 = 1$

ii. $\tau^r x \cdot \tau^r (x \to y) = (a + \tau x)(a + \tau(x \to y)) = a + \tau x \cdot \tau(x \to y)$

$\leq a + \tau y = \tau^r y$

iii. $\tau^r x = a + \tau x \leq a + \tau\tau x = a + \tau(a + \tau x) = \tau^r \tau^r x$

iv. $\tau^r x \leq x \Rightarrow a + \tau x \leq x$

$\Rightarrow \tau x \leq x$

$\Rightarrow x = 1.$

(Note: Assertion iv is redundant: By the existence of fixed points, the proof of Löb's Theorem applies and iv is derivable from i-iii.)

Second, the set P^r of atoms of $(a,1)$ consists of those elements of $(a,1)$ of the form $a + a_i$, where $a \neq a_i \in P$ is an atom of the original algebra. What is the ordering $\prec^r$ on P^r dual to τ^r? Note, for $a_i, a_j \in P - \{a\}$,

$$a_i \prec a_j \iff \forall x \in A(a_i \leq \tau x \Rightarrow a_j \leq x)$$

$$\iff \forall x \in A(a + a_i \leq a + \tau x \Rightarrow a + a_j \leq a + x) \quad (*)$$

$$\iff \forall x \in (a,1)(a + a_i \leq \tau^r x \Rightarrow a + a_j \leq x) \quad (**)$$

$$\iff a + a_i \prec^r a + a_j,$$

where (*) follows since a is disjoint from a_i, a_j and (**) follows since a is fundamental.

Third, we must see that B^r consists of precisely the appropriate τ-polynomials in τ^r. But B^r includes τ^r and is closed, whence it contains all of these polynomials. The simplest proof of the converse is had by taking $\alpha \in B$ and noting that its definition in terms of τ relativises to one of α^r in terms of τ^r. An alternate approach is to interweave the induction of this Lemma with that of the proof of Theorem 2.23: By the current induction hypothesis, B^r contains a diagonalisable fixed point subalgebra of cardinality $2^{2^k - 1}$; by that of Theorem 2.23 ($k = n$), B^r has cardinality at most $2^{2^k - 1}$. Hence $(B^r, (a,1))$ is a diagonalisable fixed point algebra.

This completes the proof. QED

With the proof of Lemma 2.25 completed, we have finished the proof of Theorem 2.23 and, with it, that of Theorem 2.22. We now know that every finite closed fixed point algebra is a subalgebra of a diagonalisable fixed point algebra. What does this do for us? I shall attempt an explanation in the next section.

EXERCISES

1. Let (B,A) be a fixed point algebra with A finite. Show: B is finite.

2. Verify the assertions of Example 2.8: Generate $A(\tau_1), A(\lambda x.\tau_1(x'))$, and $A(\alpha_0)$ and find all τ-operators on A in these algebras.

3. Let $\underline{A}$ be a finite boolean algebra and suppose τ_1,τ_2 are distinct τ-operators on A with dual orderings R_1,R_2, respectively, of the set P of atoms of A.

   i. Show: If τ_1 is definable in terms of τ_2, i.e. if τ_1 is a polynomial in τ_2, then $R_1 \subseteq R_2$.

   ii. Show: The diagonalisable fixed point algebras determined by τ_1,τ_2 are distinct.

   (Hint: i. Let a be a fundamental atom (cf. Lemma 2.13) of the diagonalisable fixed point algebra (B,A) based on either τ. Let R be the dual of τ and R^r the dual of τ^r defined on $(B^r,(a,1))$. Show: For any atoms b,c other than a, $b\,R\,c$ iff $a + b\ R^r\ a + c$. Then apply induction.)

4. Prove Corollary 2.12.

5. (Alternate proof of the Uniqueness of Fixed Points). Let (B,A) be a finite fixed point algebra. For any $a \in A$, define

   $$F_a = \{\alpha \in B:\ \ \alpha a = a\}.$$

   i. Show: For $a,b \in A$, $|F_a| = |F_b| = |B|/|A|$.

   ii. Show: For any $\alpha \in B$ there is a unique $a \in A$ such that $\alpha a = a$.

   (Hints: i. For $c,d \in A$, define $P_{cd} = \{\alpha \in B:\ \ \alpha c = d\}$ and use $\dot{+}$ to show $|P_{cd}| = |P_{cc}|$; ii. show: $|B| = \sum_{a \in A} |F_a|$ and that this implies F_a,F_b are disjoint for $a \neq b$.)

6. Prove Example 2.5.ii: The only fixed point algebras on $\underline{4}$ are the diagonalisable ones. (Hint: Use Lemma 2.16 to show there are only three possibilities for $\approx$: one with all four elements equivalent, and two in which 0 is equivalent to one atom and 1 to the other.)

7. Let (B,A) be the algebra of Example 2.8.ii. Show that it satisfies neither the

premise nor the conclusion of Lemma 2.18.

8. Let (B,A) be a fixed point algebra, $a,b \in A$ with $a < b$.

i. Show that the map $\alpha \mapsto \alpha^r$ of B to B^r is a boolean homomorphism:

$$(\alpha + \beta)^r = \alpha^r + \beta^r, \qquad (\alpha \cdot \beta)^r = \alpha^r \cdot \beta^r, \qquad (\alpha')^r = (\alpha^r)'$$

$$(0_B)^r = 0_{B^r} = \lambda x.a, \qquad (1_B)^r = 1_{B^r} = \lambda x.b.$$

ii. Show that, if a is a fundamental atom and $b = 1$, the map also respects compositions: $(\alpha \circ \beta)^r = \alpha^r \circ \beta^r$.

3. DISCUSSION

The significance, as I see it, of the Representation Theorem for Finite Closed Fixed Point Algebras is that it gives something of an upper bound on the success available in analysing extensional self-reference via finite structures and that it offers a theoretical explanation for the close relation between the Δ or ∇ operators and the $\square$ that we saw in the last Chapter. The fixed point algebras seem to be the most general algebraic modelling of extensional self-reference possible. The additional closure restriction is *logically* harmless and, in any event, David Hobby has removed it from the list of hypotheses necessary for the representation theorem. Thus, if we have any natural extensional operator $\Phi(v_0)$ that we can model adequately with *finite* (which finiteness is necessary for many applications) algebras, the representation theorem tells us that these finite models are sub-diagonalisable. Hence, Φ must have some close relation to a natural τ-operator, like $Pr(v_0)$ or $Pr_T(v_0)$. This was indeed the case with, e.g., the Mostowski operator studied in the last Chapter.

As already emphasised, this vague argument is not totally convincing. One thing that is missing is any canonicality of the representation. The representation of finite boolean algebras as powerset algebras is canonical and extends to infinite boolean algebras under the proper generalisation of the notion of atom and the introduction of topological considerations. The representation of diagonalisable algebras is also moderately canonical and extends to the infinite case if the right topological restrictions are made, and a duality is even attained if the duals are padded properly. As we see by Theorem 2.11 and Corollary 2.12, there is no extension to the infinite case of the representation theorem for finite closed fixed point algebras. A canonical representation theorem would, presumably, yield a more uniform reduction of finite fixed point algebras to finite diagonalisable algebras. This lack of uniformity takes the bite out of my argument.

All the same, my argument is not totally *un*convincing. Each success we had *was*-- somehow-- closely related to the operators $Pr_T(v)$. Moreover, by Theorem 2.11 and Corollary 2.12, there is no hope of using finite algebras to study extensional operators Φv_0, like homomorphisms of Lindenbaum sentence algebras, which have more than one fixed point. Pressed to make a definitive statement, I would say that the prospective researcher who wants to use modal logic to study extensional operators is best advised to stick close to Pr or other known arithmetic τ-operators; to do otherwise would seem to require a breakthrough comparable to Solovay's Completeness Theorems.

There does yet remain one possibility, namely the use of well-behaved infinite algebras. Bernardi's original proof of the existence of explicitly defined fixed points for parameter-free modal formulae took this form: The free diagonalisable algebra over the empty set is just the algebra of finite and cofinite sets of natural numbers. A natural topology exists on this structure and Bernardi showed, for appropriate $A(p)$, that $\lim_{n \to \infty} A^n(t)$ exists and is a fixed point D to $A(p)$,

$$\mathsf{PRL} \vdash D \leftrightarrow A(D).$$

However, this result was neither as general as the de Jongh-Sambin result nor as effective as my own, these latter results both having been established by means of finite structures-- as we saw in Chapter 2, above. Moreover, there is the problem of finding good infinite fixed point algebras. At present, the only concrete examples I know of are the diagonalisable ones and the Lindenbaum fixed point algebras (and, of course, subalgebras generated by particular operators).

With respect to the Lindenbaum fixed point algebras, there is one result which may, or may not, look encouraging: Solovay has shown that, if T is any *RE*-theory in which Peano arithmetic, PA, can be interpreted, then its Lindenbaum fixed point algebra (B_T, A_T), is isomorphic to that of PA, (B_{PA}, A_{PA}). Encouragement comes from the thought that this might mean some overall uniformity allowing a global understanding of extensional self-reference. Discouragement comes of realising that there are no non-constant distinguished elements in these algebras: PA and PA + $\sim Con_{PA}$, for example, cease to have isomorphic such algebras when their respective provability predicates are distinguished or even when one adds a predicate $P(\ulcorner\phi\urcorner)$ asserting ϕ to be (equivalent to) a Σ_1-sentence. This latter fact is particularly annoying in view of the traditional interest in the behaviour of Σ_1-sentences.

I have just explained virtually everything I know about the infinite fixed point algebras. I have no intuition on whether or not they will prove useful in the study of self-reference, but I think I can say that their own study ought to turn out interesting.

For the present, we have reached the summit of our knowledge of extensional self-reference and it is time to turn to the study of non-extensional self-reference.

Part III
Non-Extensional Self-Reference

Chapter 6

Rosser Sentences

Gödel's Incompleteness Theorem asserts that the self-referential sentence ϕ defined by

$$T \vdash \phi \leftrightarrow \sim Pr_T(\ulcorner \phi \urcorner)$$

is underivable and that, if T is sufficiently sound, it is not refutable. Indeed, if $T = PRA + \sim Con_{PRA}$, then $T \vdash \sim\phi$ and the sufficient soundness condition is seen to be necessary. In 1936, J. Barkley Rosser showed how to get around this by proving the following:

ROSSER'S THEOREM. Let T be a consistent RE extension of PRA. Then: There is a Σ_1-sentence ϕ such that

i. $T \nvdash \phi$

ii. $T \nvdash \sim\phi$

iii. $T \vdash Con_T \rightarrow \sim Pr_T(\ulcorner \phi \urcorner)$

iv. $T \vdash Con_T \rightarrow \sim Pr_T(\ulcorner \sim\phi \urcorner)$.

The truth of this theorem has already been established twice in Chapter 3-- in Exercise 4 of section 1 and Examples 2.4 and 2.5 of section 2 of that Chapter. As remarked in Example 2.4 of that Chapter and as the reader was supposed to prove in an ensuing exercise, Rosser's Theorem cannot be proven by the kind of self-referential sentence available in PRL; the proofs of Rosser's Theorem given depended on Solovay's Completeness Theorems and their far more sophisticated self-referential construction. Rosser's original proof of his Theorem used a form of self-reference that was much simpler than, yet is related to and even a prototype of Solovay's. It is Rosser's self-referential sentence construction that we wish to study in this Chapter.

Rosser's construction can be explained as follows: Like Gödel, he wants a sentence asserting its own unprovability, but he first tampers with the notion of provability by defining:

$$Pr_T^R(\ulcorner\phi\urcorner): \quad \exists v_0\big(Prov_T(v_0,\ulcorner\phi\urcorner) \wedge \forall v_1 < v_0 \sim Prov_T(v_1,\ulcorner\sim\phi\urcorner)\big).$$

The Rosser sentence is just the Gödel sentence for this representation of provability:

$$\phi \leftrightarrow \sim Pr_T^R(\ulcorner\phi\urcorner).$$

Ignoring the complexity (this ϕ is Π_1, that in the statement of the Theorem is Σ_1), it is now an easy matter to prove Rosser's Theorem.

Proof of Rosser's Theorem: Let T be a consistent *RE* theory containing PRA and let ϕ satisfy

$$T\vdash \phi \leftrightarrow \sim Pr_T^R(\ulcorner\phi\urcorner). \qquad (*)$$

i. $T\nvdash \phi$: Suppose on the contrary that $T\vdash \phi$. Let x be the smallest number such that $Prov_T(\overline{x},\ulcorner\phi\urcorner)$ is true. By consistency, $\sim\phi$ is not provable, whence $Prov_T(\overline{y},\ulcorner\sim\phi\urcorner)$ is always false. Hence

$$Prov_T(\overline{x},\ulcorner\phi\urcorner) \wedge \forall v_1 < \overline{x} \sim Prov_T(v_1,\ulcorner\sim\phi\urcorner)$$

is true and $Pr_T^R(\ulcorner\phi\urcorner)$ is a true Σ_1-sentence. Thus,

$$T\vdash Pr_T^R(\ulcorner\phi\urcorner).$$

But also $T\vdash \sim Pr_T^R(\ulcorner\phi\urcorner)$ by (*), a contradiction.

ii. $T\nvdash \sim\phi$: Let $T\vdash \sim\phi$ and let x be minimum such that $Prov_T(\overline{x},\ulcorner\sim\phi\urcorner)$. Again it follows that

$$T\vdash Prov_T(\overline{x},\ulcorner\sim\phi\urcorner). \qquad (1)$$

But, $\sim\phi \leftrightarrow Pr_T^R(\ulcorner\phi\urcorner)$ and, since $T\vdash \sim\phi$, we have

$$T\vdash \exists v_0\big(Prov_T(v_0,\ulcorner\phi\urcorner) \wedge \forall v_1 < v_0 \sim Prov_T(v_1,\ulcorner\sim\phi\urcorner)\big). \qquad (2)$$

In the presence of (1), (2) becomes

$$T\vdash \exists v_0 < \overline{x}\big(Prov_T(v_0,\ulcorner\phi\urcorner) \wedge \forall v_1 < v_0 \sim Prov_T(v_1,\ulcorner\sim\phi\urcorner)\big).$$

But this assertion is *PR*, whence true (for, otherwise, its true negation is provable--contrary to consistency). But this means that $Prov_T(\overline{y},\ulcorner\phi\urcorner)$ is true for some y, i.e. that $T\vdash \phi$, again a contradiction.

iii-iv: That these underivability results can be formally derived on the assumption of consistency ought not to be too surprising given the outline above. As this formalisation will follow directly from results of this and the next Chapter, I will not give the details here. QED

The reference to the *least* x in each of parts i and ii of the proof is an informal induction; obviously the induction will have to be formal in the proofs of parts iii and iv. For the mere independence result, however, we do not need the full power of the assumption that T contains the induction of PRA: Rosser's Theorem, parts i and ii, applies to much weaker theories. Our earlier proofs, via Solovay's Completeness Theorems, only apply to stronger theories because of the need to prove the function defined by recursion to be total. Hence, Rosser's original proof is somewhat more general. Indeed, for most of the applications we gave of Solovay's theorems, if we drop the formalisability assertions, we can apply Rosseresque tricks and obtain results for weaker theories. We shall not do this, however, for, the crucial thing *here* is not really the incompleteness phenomenon we call Rosser's Theorem, but, rather, the instance of self-reference. Rosser's sentence is the simplest interesting example of non-extensional self-reference. By its non-extensionality, our earlier analysis of self-reference does not apply. The goal of the present Chapter is to supply some such analysis; the goal of the next Chapter is to analyse more sophisticated types of non-extensional self-reference.

The insight that a Rosser sentence is a Gödel sentence for a modified provability predicate, though offering a heuristic mnemonic, is not much use in this analysis. What we must do is look closely at the comparison of witnesses of the provability assertions in Pr_T^R.

DEFINITION. Let $\exists v_0 \phi v_0$, $\exists v_1 \psi v_1$ be formulae (in practice: usually Σ_1-sentences). By their *witness comparison formulae* we mean

$$\exists v_0 \phi \preccurlyeq \exists v_1 \psi: \quad \exists v_0 (\phi v_0 \wedge \forall v_1 < v_0 \sim\psi v_1)$$
$$\exists v_0 \phi \prec \exists v_1 \psi: \quad \exists v_0 (\phi v_0 \wedge \forall v_1 \le v_0 \sim\psi v_1).$$

Among Σ_1-formulae (or, indeed, any formulae with leading existential quantifiers) ϕ, ψ, we write $\phi \preccurlyeq \psi$ and $\phi \prec \psi$ without exhibiting the quantifiers. Note that $\preccurlyeq$ and $\prec$ are a weak and associated strict pre-ordering on the set of witnessed (i.e. true) existentially quantified assertions, but not necessarily provably so: If $\phi = \exists v \psi v$, the assertion $\phi \rightarrow (\phi \preccurlyeq \phi)$, i.e.

$$\exists v\psi v \rightarrow \exists v(\psi v \wedge \forall v_0 < v \sim\psi v_0),$$

is an instance of the Least Number Principle. The schema, $\phi \rightarrow (\phi \preccurlyeq \phi)$, for existentially quantified ϕ is thus equivalent to induction and not all instances are provable in PRA. However, in the important Σ_1-case, this is a theorem schema of PRA.

Getting back to Rosser sentences, the relevance of the witness comparison formulae is that the modified proof predicate is a witness comparison:

$$Pr^R(\ulcorner\phi\urcorner): \quad Pr(\ulcorner\phi\urcorner) \preccurlyeq Pr(\ulcorner\sim\phi\urcorner).$$

Hence, Rosser's sentence,

$$\phi \leftrightarrow \sim(Pr(\ulcorner\phi\urcorner) \preccurlyeq Pr(\ulcorner\sim\phi\urcorner)),$$

is a negated witness comparison. Since the negation merely complicates the issue, it has become customary to look at the dual Σ_1-sentence,

$$\phi \leftrightarrow .Pr(\ulcorner\sim\phi\urcorner) \preccurlyeq Pr(\ulcorner\phi\urcorner).$$

DEFINITION. A *Rosser sentence* is a sentence ϕ satisfying

$$\text{PRA} \vdash \phi \leftrightarrow .Pr(\ulcorner\sim\phi\urcorner) \preccurlyeq Pr(\ulcorner\phi\urcorner).$$

REMARK. Our officially designated Σ_1-Rosser sentences are not quite the negations of the traditional Π_1-versions. If

$$\phi \leftrightarrow \sim(Pr(\ulcorner\phi\urcorner) \preccurlyeq Pr(\ulcorner\sim\phi\urcorner)),$$

then

$$\sim\phi \leftrightarrow .Pr(\ulcorner\phi\urcorner) \preccurlyeq Pr(\ulcorner\sim\phi\urcorner) \qquad (1)$$

is not necessarily equivalent to

$$\sim\phi \leftrightarrow .Pr(\ulcorner\sim\sim\phi\urcorner) \preccurlyeq Pr(\ulcorner\sim\phi\urcorner), \qquad (2)$$

which would make $\sim\phi$ a Rosser sentence. For, $\preccurlyeq$ is not extensional: If, say, $\sim\sim\phi$ has a shorter proof than ϕ, (2) will not say the same thing as (1). With this *caveat* in mind, however, the reader may wish to think of our Σ_1-Rosser sentences as almost satisfying the equivalence,

$$\phi \leftrightarrow Pr^R(\ulcorner\sim\phi\urcorner),$$

dual to the traditional Π_1-Rosser equivalence,

$$\phi \leftrightarrow \sim Pr^R(\ulcorner\phi\urcorner).$$

The ensuing analysis of Rosser sentences carries over for all variants with mere notational changes.

And what about this analysis? It so happens that all one needs to add to our modal understanding of *Pr* to understand Rosser sentences is to add an analysis of the witness comparison. In section 1, therefore, we define three modal systems, R^-, R, and R^ω and prove their completeness with respect to appropriate Kripke models. Both R^- and R are analogues to PRL and R^ω is an analogue to PRL^ω. The splitting in the first case is one of the two notions of semantics: R^- is the modal logic best fitting the Kripke semantics and R is the one complete with respect to arithmetic interpretations; unlike the familiar case, these two roles do not conflate. The arithmetic completeness just referred to is established in section 2. It differs a little from the familiar in that the interpretation of $\Box$, although always equivaltent to *Pr*, is not fixed to being *Pr*. In the face of non-extensionality, the reader may or may not regard this as serious. It has the effect of changing the ordering (or, representation thereof) of proofs. As I consider the exact ordering unimportant, I consider the situation entirely satisfactory; there are those, however, who do not, but who have yet to offer substantive reasons for their displeasure. The reader will, however, have to judge for himself the subsequent importance of the applications in section 3 to the problem of the uniqueness and explicit definability of Rosser sentences. It turns out that Rosser's sentences are occasionally unique and occasionally not.

I may as well mention at this point that, but for some of the exercises, the results of the present Chapter are due jointly to David Guaspari and Robert Solovay.

1. MODAL SYSTEMS FOR ROSSER SENTENCES

Our modal systems for studying Rosser sentences will have the following language:

PROPOSITIONAL VARIABLES: $p, q, r, \ldots$

TRUTH VALUES: t, f

PROPOSITIONAL CONNECTIVES: $\sim, \wedge, \vee, \rightarrow$

MODAL OPERATOR: $\Box$

WITNESS COMPARISONS: $\preccurlyeq$, $\prec$.

There is one restriction on the rules of formula construction: $\preccurlyeq, \prec$ can only be applied to pairs of *boxed* formulae: If $\Box A$, $\Box B$ are formulae, so are $\Box A \preccurlyeq \Box B$ and $\Box A \prec \Box B$. The other connectives ($\sim, \wedge, \vee, \rightarrow, \Box$) are total.

The necessity of some restriction is obvious: Witness comparisons are supposed to imitate the witness comparisons of the arithmetic language. Thus, e.g., $(\sim\Box A) \preccurlyeq (\sim\Box B)$ is meaningless. Our restriction is, with respect to this objection, too restrictive. The assertions $(\Box A \wedge \Box B) \prec \Box C$, $\Box A \preccurlyeq (\Box B \vee \Box C)$, $\Box A \preccurlyeq (\Box B \preccurlyeq \Box C)$ and their like have natural meanings: With conjunctions on the left side of a comparison, one maximises so that, e.g.,

$$(\Box A \wedge \Box B) \prec \Box C. \leftrightarrow .(\Box A \prec \Box C) \wedge (\Box B \prec \Box C).$$

Conjunctions on the right are minimised; and disjunctions on the left (right) are minimised (maximised), so that, e.g.,

$$\Box A \preccurlyeq (\Box B \vee \Box C). \leftrightarrow .(\Box A \preccurlyeq \Box B) \wedge (\Box A \preccurlyeq \Box C).$$

Nestings of comparisons are perfectly natural since a comparison of Σ_1-sentences is itself Σ_1. However, they too are unnecessary since their meanings are also easily determined. For example,

$$\Box A \preccurlyeq (\Box B \preccurlyeq \Box C). \leftrightarrow .\big((\Box B \preccurlyeq \Box C) \wedge (\Box A \preccurlyeq \Box B)\big) \vee \big(\sim(\Box B \preccurlyeq \Box C) \wedge \Box A\big).$$

Hence, the obvious legitimate extensions of our applications of $\preccurlyeq$ and $\prec$ are redundant and we can avoid them for simplicity's sake. Later, should we decide to extend our definition to allow more complexity inside comparisons, we can do so without difficulty.

The witness comparisons, $\Box A \preccurlyeq \Box B$ and $\Box C \prec \Box D$ model Σ_1-formulae, which have a special relation to $\Box$, namely Demonstrable Σ_1-Completeness (0.6.22):

$$\mathsf{PRA} \vdash \phi \rightarrow Pr(\ulcorner \phi \urcorner)$$

for all $\phi \in \Sigma_1$. Obviously, this will have to be mirrored modally by a special axiom schema. To this end, we have the following definition.

1.1. DEFINITION. The class of Σ-*formulae* is defined inductively by:

i. $\Box A$, $\Box A \preccurlyeq \Box B$, and $\Box A \prec \Box B$ are Σ-formulae for any A, B

ii. if A, B are Σ-formulae, so are $A \wedge B$, $A \vee B$.

The Σ-formulae given by clause i are called *strict* Σ-*formulae* and their class is written $s\Sigma$.

That finishes our description of the language. The next step is axiomatisation and is almost straightforward: One takes the axioms of PRL, a schema of Σ-completeness, and the obvious order axioms and obtains a theory that will match the appropriate Kripke models perfectly. Unfortunately, it does not match arithmetic: Consider a node α forcing $\Box f$. Will α force $\Box t \prec \Box f$ or $\Box f \preccurlyeq \Box t$? Obviously, we want the former, but the theory with the nice Kripke models will not decide the issue. However, it will prove $\Box(\Box t \prec \Box f)$ and the additional soundness rule, $\Box A \;/\; A$, fills the gap. Thus, we will have two theories, R^- and R. R^- is complete with respect to Kripke models and R with respect to arithmetic interpretations. The bridge between Kripke models and arithmetic interpretations is provided by R^ω-- the analogue to PRL^ω: It happens that R^ω relates to *both* R^- and R as PRL^ω relates to PRL. Hence, everything works: We connect R^- with the Kripke models, use A-sound such models to get completeness of R^ω with respect to arithmetic interpretations and read off results for R. There are still a few novelties to these proofs, but I do not need to elaborate on them here.

1.2. DEFINITION. The modal system R^- in the language just described has the following axioms and rules of inference:

AXIOMS. $A1$. All tautologies

$A2$. $\Box A \wedge \Box(A \to B) \to \Box B$

$A3$. $\Box A \to \Box\Box A$

$A4$. $\Box(\Box A \to A) \to \Box A$

$A5$. $A \to \Box A$, all $A \in \Sigma$

$A6$. Order axioms:

$A6.i$. ($\preccurlyeq$ pre-orders the true boxed formulae). For all A,B,C with principal connective $\Box$,

$A \to .A \preccurlyeq A, \qquad A \preccurlyeq B. \to A$

$(A \preccurlyeq B) \wedge (B \preccurlyeq C) \to .A \preccurlyeq C$

$A \vee B \to .(A \preccurlyeq B) \vee (B \prec A)$

$A6.ii$. ($\prec$ is the associated strict pre-ordering). For appropriate A,B,

$A \prec B. \to .A \preccurlyeq B, \qquad A \preccurlyeq B. \to \sim(B \prec A)$

$A6.iii$. (True sentences are witnessed earlier than false ones.) For appropriate A,B,

$A \wedge \sim B \to .A \prec B.$

RULES. $R1$. $A,\ A \to B \;/\; B$

$R2$. $A \;/\; \Box A$.

1.3. REMARKS. i. Axiom schema *A3* is a subschema of *A5* and can be dropped.

ii. Axiom schema *A5* is equivalent to the subschema obtained by restricting A to being $s\Sigma$, i.e. a strictly Σ-formula.

iii. In *A6*, I have not drawn the boxes. Obviously, $A \to .A \preccurlyeq A$ indicates the schema $\Box A \to .\Box A \preccurlyeq \Box A$. By not exhibiting the boxes and only referring to A's being of the "appropriate form", we can later modify these axioms by simply redefining the adjective "appropriate". This will happen when one allows $A \preccurlyeq B$ to exist for all $A,B \in \Sigma$ or when one adds new modal operators.

The theory R^- is strong enough to prove Rosser's Theorem, but is not complete with respect to arithmetic interpretations. To obtain such completeness, a new rule of inference is needed.

1.4. DEFINITION. R is the theory obtained by adding to the axioms and rules of R^- the additional rule of inference:

R3. $\Box A \,/\, A$.

Recall that PRL is closed under *R3* and we did not have this dichotomy of theories.

Parallel to PRL^ω we will have a theory R^ω. The theory R^ω will not be closed under *R2*-- for the same reason that PRL^ω was not. To define R^ω and discuss the Kripke models for R^-, we once again invoke our convention proscribing the use of *R2*: R^- is now thought of as given by the *R2*-free axiomatisation of PRL and further axioms $\boxed{s}$ *A5* and $\boxed{s}$ *A6*, i.e.

$\boxed{s}$ *A5*. $\boxed{s}\,(A \to \Box A)$, all $A \in \Sigma$

$\boxed{s}$ *A6*. $\boxed{s}\,(A \to .A \preccurlyeq A)$, for A boxed

etc.

With this in mind, we can define R^ω.

1.5. DEFINITION. R^ω is the theory extending (the *R2*-free formulation of) R^- by the additional schema of soundness:

A7. $\Box A \to A$, all A.

By rights, we should define R^ω as an extension of R-- since we claim R to axiomatise provable schemata and R^- to be of mere technical use. This doctrinaire objection is readily disposed of: R^ω is trivially closed under the rule *R3* and, hence, if we choose an $R2$-free formulation of R, we see R^{ω} to be the extension of R by the soundness schema. Technically, however, we will need R^{ω} to be defined in terms of R^{-} because R^{-} will have the good Kripke model theory.

Except for warning the reader once again against attempting substitution within the brave new context of witness comparison, we have nothing of a syntactic nature to discuss and so move on to the model theory.

The Kripke model theory for R^{-} offers no major surprises. One minor novelty will be our restriction to models whose frames are finite irreflexive *trees*. The reason for this restriction is the later necessity (in the proof of Lemma 1.10) of performing a certain construction on a model by induction from the bottom up. If there are distinct α_1, α_2 in the frame with a common extension β, the work done at α_1 and α_2 might not compatibly extend to β. (The restriction to trees is not essential for completeness, merely for constructions of this sort.)

1.6. DEFINITION. A *Kripke pseudo-model for* R^{-} is a quadruple $\underline{K} = (K,<,\alpha_0,\Vdash)$, where $(K,<,\alpha_0)$ is a finite irreflexive tree with root α_0 and $\Vdash$ is a forcing relation satisfying the usual conditions on the connectives: For $\alpha \in K$,

ii-iii. $\alpha \Vdash t$, $\alpha \nVdash f$

iv. $\alpha \Vdash \sim A$ iff $\alpha \nVdash A$

v-vii. $\alpha \Vdash A \circ B$ iff $(\alpha \Vdash A) \circ (\alpha \Vdash B)$, for $\circ \in \{\wedge, \vee, \rightarrow\}$

viii. $\alpha \Vdash \Box A$ iff $\forall \beta > \alpha (\beta \Vdash A)$.

In effect, a Kripke pseudomodel for R^{-} is simply a Kripke model (for PRL) in which witness comparisons are treated as new atoms. A Kripke model for R^{-} will simply be a Kripke pseudomodel in which the axioms of R^{-} are valid.

1.7. DEFINITION. A Kripke pseudo-model $\underline{K}$ for R^{-} is a *Kripke model for* R^{-} iff, for all appropriate A,B,D,E,

i. $\alpha \Vdash A \preccurlyeq B \Rightarrow \forall \beta > \alpha (\beta \Vdash A \preccurlyeq B)$

ii. $\alpha \Vdash A \prec B \Rightarrow \forall \beta > \alpha (\beta \Vdash A \prec B)$

iii. if $D \rightarrow E$ is an instance of $A6$, $\alpha \Vdash D \Rightarrow \alpha \Vdash E$.

This last definition is admittedly a bit artificial, but it makes the following assertion obvious:

1.8. LEMMA. (Soundness Theorem). Let $\underline{K}$ be a Kripke model for R^-. Then R^- is valid (true) in $\underline{K}$, i.e. for any sentence A,

$$R^- \vdash A \Rightarrow \forall \alpha \in K(\alpha \Vdash A)$$
$$\Rightarrow \alpha_0 \Vdash A.$$

We will prove completeness for R^- by reduction to completeness for PRL. That is, we consider the witness comparisons simply to be new, strangely labelled, propositional atoms of the ordinary modal language, but which obey new axioms. We can prove the completeness of R^- with respect to its models by reduction to the completeness of PRL with respect to its models. Allowing infinite models and recalling the strong completeness of PRL with respect to *models* in which it is valid, such a completeness result is trivial. However, we want completeness with respect to *frames*-- and *finite* ones at that. Thus, as in reducing PRL^ω to PRL, we must reduce R^- to PRL in finite pieces. The next definition and lemma will allow us to do this.

1.9. DEFINITIONS. A set S of formulae in the language of R^- is *adequate* iff i. S is closed under subformulae and ii. S is closed under witness comparison, i.e. S is adequate iff

i. if $A \in S$ and B is a subformula of A, then $B \in S$

and ii. if $\Box A, \Box B \in S$, then $\Box A \preccurlyeq \Box B$ and $\Box A \prec \Box B$ are in S.

In analogy with the notation,

$$S(A) = \{B:\ B \text{ is a subformula of } A\},$$

we write,

$$S^+(A) = S(A) \cup \{\Box B \preccurlyeq \Box C,\ \Box B \prec \Box C:\ \Box B, \Box C \in S(A)\},$$

for the smallest adequate set containing A.

1.10. EXTENSION LEMMA. Let S be an adequate set of formulae and let $\underline{K}_S = (K,<,\alpha_0,\Vdash_S)$ be a Kripke pseudo-model for R^- which forces all axioms of R^- involving only formulae in S. Then, there is a forcing relation $\Vdash$ on $(K,<,\alpha_0)$ for which the resulting structure $\underline{K} = (K,<,\alpha_0,\Vdash)$ is a Kripke model for R^- and, moreover, for all $\alpha \in K$ and $A \in S$,

$$\alpha \Vdash_S A \text{ iff } \alpha \Vdash A.$$

Intuitively, the Extension Lemma asserts that, to define a Kripke model, we need only define the forcing relation on an adequate set of formulae in such a way as to satisfy the appropriate requirements. For, then there will be some extension of what we've done that will yield a genuine Kripke model. Note that, in discussing PRL we did this implicitly-- a forcing relation had no extra requirements to satisfy and was completely determined by any action on the remaining atoms. Here, the relation is *not* determined by its action on atoms, but by its action on atoms *and* witness comparisons, and, moreover, the latter actions are constrained by their persistence (*A5*) and ordering (*A6*) requirements.

Proof of Lemma 1.10: The proof is a routine but hideous nested induction. The outer induction is on the complexity of nestings of witness comparison connectives in a given formula (*N.B.* While we have not allowed, e.g., $\Box A \preccurlyeq (\Box B \prec \Box C)$, we have allowed $\Box A \preccurlyeq \Box(\Box B \prec \Box C)$.); and the inner induction is on the ordering of the model (but this time going from the bottom up instead of from the top down).

We begin with a curious definition of complexity: For $A = p$ atomic or $A \in S$, we define $c(A) = 0$. For other A, we proceed by induction on the usual complexity of A: For A not atomic and not in S, we define

$$c(A) = c(B), \quad \text{if } A = \sim B \text{ or } \Box B$$

$$c(A) = \max\{c(B), c(C)\}, \text{ if } A = B \circ C \text{ for } \circ \in \{\wedge, \vee, \rightarrow\}$$

$$c(A) = 1 + \max\{c(B), c(C)\}, \text{ if } A = B \circ C \text{ for } \circ \in \{\preccurlyeq, \prec\}.$$

In words: For $A \notin S$, $c(A)$ counts the number of nestings of $\preccurlyeq$ or $\prec$ in A.

Given $\underline{K}_S = (K, <, \alpha_0, \Vdash_S)$, we define a new relation $\Vdash$ in a sequence of stages by induction on $c(A)$. That is, we shall define a sequence of relations $\Vdash_n$ such that, if $c(A) = n$,

$$\alpha \Vdash A \quad \text{iff} \quad \alpha \Vdash_n A$$

and, if $A \in S$,

$$\alpha \Vdash A \quad \text{iff} \quad \alpha \Vdash_S A.$$

<u>Stage 0</u>. For A of the form p, $\Box B \preccurlyeq \Box C$, or $\Box B \prec \Box C$, with $c(A) = 0$, simply define

$$\alpha \Vdash_0 A \quad \text{iff} \quad \alpha \Vdash_S A,$$

for all $\alpha \in K$. Using the traditional truth conditions (i.e. 1.6.ii-1.6.viii), $\Vdash_0$ automatically extends to all A with $c(A) = 0$. Moreover, an easy induction on the length of such sentences A shows, for $c(A) = 0$ and all $\alpha \in K$,

$$\alpha \Vdash_0 A \quad \text{iff} \quad \alpha \Vdash_S A.$$

In particular,

i. $\Vdash_0$ agrees with $\Vdash_S$ on S

ii. the axioms of R^- mentioning only formulae in S are $\Vdash_0$-forced at all nodes.

Stage $n + 1$. We assume $\Vdash_n$ ($\supseteq \Vdash_{n-1} \supseteq \cdots \supseteq \Vdash_0$) has been nicely defined on $\{A\colon\ c(A) \le n\}$. By "nicely defined" we mean two things: First, we assume that, for each α, the relations,

$$\alpha \Vdash_n \Box B \preccurlyeq \Box C \quad \text{and} \quad \alpha \Vdash_n \Box B \prec \Box C$$

form a weak and associated strict pre-partial ordering on the *appropriate formulae*, which, when $n = 0$ are the boxed formulae in S and when $n > 0$ are the boxed formulae of complexity $< n$ (for, these are the formulae $\Box B, \Box C$ for which $c(\Box B \preccurlyeq \Box C) = c(\Box B \prec \Box C) \le n$). For convenience, let us denote the set of these appropriate formulae by X^n. These partial orderings are assumed to satisfy:

P1. they are linear on $\{\Box B \in X^n\colon\ \alpha \Vdash_n \Box B\}$

P2. they place all true boxed formulae in X^n before all false ones, i.e. if $\Box B, \Box C \in X^n$, then: $\alpha \Vdash_n \Box B \wedge \sim\Box C \Rightarrow \alpha \Vdash_n \Box B \prec \Box C$

P3. they yield the empty relation on the set of false boxed formulae.

The second assumption of niceness of definition is persistence: For $\alpha < \beta$ and appropriate $\Box B, \Box C$,

$$\alpha \Vdash_n \Box B \preccurlyeq \Box C \Rightarrow \beta \Vdash_n \Box B \preccurlyeq \Box C$$
$$\alpha \Vdash_n \Box B \prec \Box C \Rightarrow \beta \Vdash_n \Box B \prec \Box C.$$

Define

$$X^{n+1} = \{\Box B\colon\ c(\Box B) \le n\}.$$

The extension of $\Vdash_n$ to $\Vdash_{n+1}$ must be such as to i. extend the pre-partial ordering already given on X^n, ii. satisfy properties *P1*-*P3* with respect to $n + 1$, and iii. be upwards persistent. Fortunately, this is not a difficult task to perform. The most manifest point is that, to satisfy iii, we must start at the bottom of the tree and carry the positive information upwards-- i.e. we will induct on the ordering of the tree.

Let $\alpha \in K$. We define $\Vdash_{n+1}$ on X^{n+1} by substages:

Substage 0. If $\Box B, \Box C \in X^n$, define

$$\alpha \Vdash_{n+1} \Box B \circ \Box C \quad \text{iff} \quad \alpha \Vdash_n \Box B \circ \Box C,$$

for $\circ \in \{\preccurlyeq, \prec\}$.

Substage 1. If α is the origin, proceed to substage 2. Otherwise, let β be the unique predecessor of α. (*N.B.* Here is where we use the fact that $(K,<,\alpha_0)$ is a tree.) By our tree-induction, $\Vdash_{n+1}$ has already been completely determined on X^{n+1} at β. For $\Box B, \Box C \in X^{n+1}$ and $\circ \in \{\preccurlyeq, \prec\}$, if $\beta \Vdash_{n+1} \Box B \circ \Box C$, then define $\alpha \Vdash_{n+1} \Box B \circ \Box C$.

Substage 2. Set $X_\alpha^{n+1} = \{\Box B \in X^{n+1}: \alpha \Vdash_n \Box B \ \&\ \beta \nVdash_n \Box B\}$, where β is again the immediate predecessor to α, unless α is the origin, in which case we drop the condition on β from the definition. Choose any linear pre-ordering of X_α^{n+1} extending that of X_α^n (and consistent with what has been done so far) and define

$$\alpha \Vdash_{n+1} \Box B \preccurlyeq \Box C \ (\Box B \prec \Box C) \quad \text{for } \Box B, \Box C \in X^{n+1}$$

iff $\Box B$ weakly (strongly) precedes $\Box C$ in this pre-ordering.

Substage 3. For $\Box B, \Box C \in X^{n+1}$ such that $\alpha \Vdash_n \Box B$, but $\alpha \nVdash_n \Box C$, define

$$\alpha \Vdash_{n+1} \Box B \prec \Box C \quad \text{and} \quad \alpha \Vdash_{n+1} \Box B \preccurlyeq \Box C.$$

Substages 1 and 2 guarantee *P1*; substage 3 guarantees *P2*; and our failure to order false boxed sentences guarantees *P3*. Finally, substage 1 guarantees persistence. (Of these, only *P1* requires comment. The verification is by induction on the tree ordering: Suppose $\alpha \Vdash_n \Box B, \Box C$. Either both are forced at β, only one is forced at β, or neither are forced at β. In the first case, β decides $\Box B \preccurlyeq \Box C$ or $\Box C \prec \Box B$ by induction hypothesis; in the second case, β decided the ordering during its own substage 3; and in the third case α decides it in substage 2.)

Next use 1.6.ii-1.6.viii to define $\Vdash_{n+1}$ on all A with $c(A) = n+1$.

To complete the proof of the Lemma, define, for all $\alpha \in K$ and all A,

$$\alpha \Vdash A \quad \text{iff} \quad \exists n(\alpha \Vdash_n A).$$

It is not hard to see that $\underline{K} = (K,<,\alpha_0,\Vdash)$ is indeed a Kripke model for R^- and that, for $A \in S$ and $\alpha \in K$,

$$\alpha \Vdash A \quad \text{iff} \quad \alpha \Vdash_S A.$$

I leave these details to the reader. QED

The proof of the Extension Lemma, though routine, is so long that the unwary reader may easily have forgotten our immediate goal, which is to prove the completeness of R^- with respect to its Kripke models. The Extension Lemma is, however, more than merely a lemma towards the proof of the Completeness Theorem; it is an important tool in constructing models. The following example provides an illustration of this use as well as a brief pause between the Lemma and the Theorem.

1.11. EXAMPLE. $R^- \nvdash \Box t \prec \Box f$.

Proof: Let $A = \Box t \prec \Box f$ and note that the smallest adequate set containing A is

$$S = S^+(A) = \{\Box t \prec \Box f,\ \Box t \preccurlyeq \Box f,\ \Box f \prec \Box t,\ \Box f \preccurlyeq \Box t,\ \Box t \prec \Box t,\ \Box t \preccurlyeq \Box t,\ \Box f \prec \Box f,\ \Box f \preccurlyeq \Box f,\ \Box t,\ \Box f,\ t,\ f\}.$$

Let $\underline{K}_S$ be the following one-point pseudo-model (with atoms and witness comparisons outside S not forced at all):

$$\alpha_0 \cdot \Box f \prec \Box t \quad \text{(and: } \Box f \preccurlyeq \Box t,\ \Box t \preccurlyeq \Box t,\ \Box f \preccurlyeq \Box f).$$

$\underline{K}_S$ satisfies all the requirements of the hypothesis of the Extension Lemma, i.e. it is a Kripke pseudo-model satisfying all appropriate instances of *A5* (vacuously) and *A6* (the order axioms on $\Box f, \Box t$). Thus, the Lemma tells us that $\Vdash_S$ can be suitably extended to all formulae to yield a model of R^-. Since $\alpha_0 \Vdash \Box f \prec \Box t$ in this model, it follows that $R^- \nvdash \Box t \prec \Box f$. QED

This example can be multiplied: For instance, a two-point model will show

$$R^- + \Box t \prec \Box f \nvdash \Box^2 t \prec \Box f.$$

The addition of the rule *R3* yields all sentences $\Box^n t \prec \Box^k f$ for $n,k > 0$. (Cf. Exercise 4, below.)

The actual production of models will be necessary in section 3, below, when we apply the arithmetic simulation of such models, i.e. the arithmetic completeness result of the next section, to questions about Rosser sentences. For now, Example 1.11 will have to suffice and we must establish completeness.

1.12. COMPLETENESS THEOREM FOR R^-. Let A be given. The following are equivalent:

i. $R^- \vdash A$

ii. A is valid in all Kripke models $\underline{K} = (K, <, \alpha_0, \Vdash)$, i.e. for all $\alpha \in K$, $\alpha \Vdash A$

iii. A is true in all Kripke models $\underline{K} = (K, <, \alpha_0, \Vdash)$, i.e. $\alpha_0 \Vdash A$.

Proof: The implications i $\Rightarrow$ ii $\Rightarrow$ iii are just the assertions of the Soundness Lemma and I have no intention of repeating here the disclaimer of any need to include a detailed proof. The crucial thing here is the proof of the implication iii $\Rightarrow$ i, which, as usual, is done contrapositively: If $R^- \nvdash A$, we show how to construct a model in which A is not true.

Assume $R^- \nvdash A$. By the Extension Lemma, this amounts to a simple implication not being provable in PRL, which in turn implies the existence of a countermodel to this implication, which will in turn give the countermodel to A. The details follow.

Let $\underline{L}$ be the usual modal language and $\underline{R}$ the extension of $\underline{L}$ by the addition of $\preccurlyeq, \prec$. We can embedd $\underline{R}$ into $\underline{L}$ by simply adding to $\underline{L}$ the new atoms

p_{BC} for each sentence $\Box B \preccurlyeq \Box C$ of $\underline{R}$

and q_{BC} for each sentence $\Box B \prec \Box C$ of $\underline{R}$.

The addition being minor, we still call the language $\underline{L}$. Using these new atoms, we can interpret every sentence D of $\underline{R}$ by a new sentence, say D^r, of $\underline{L}$ simply by replacing all *maximal* occurrences in D of sentences $\Box B \preccurlyeq \Box C$ ($\Box B \prec \Box C$) by p_{BC} (q_{BC}, respectively). The inner structure of $\Box B \preccurlyeq \Box C$ gets lost, but what is important is not what is inside $\Box B \preccurlyeq \Box C$, but rather how $\Box B \preccurlyeq \Box C$ relates to the things inside it-- and these logical relations can be recovered.

Let

$$S^+\text{-}Comp = \{\Box B \circ \Box C . \rightarrow \Box(\Box B \circ \Box C): \ \Box B, \Box C \in S(A) \ \& \ \circ \in \{\preccurlyeq, \prec\}\}$$

be the set of instances of $s\Sigma$-Completeness for comparisons in $S^+(A)$;

and let

$$S^+\text{-}Ord = \{\text{instances of order axioms for } \Box B, \Box C \in S(A)\},$$

e.g. $\Box B \rightarrow . \Box B \preccurlyeq \Box B$

$(\Box B \preccurlyeq \Box C) \wedge (\Box C \preccurlyeq \Box D) \rightarrow (\Box B \preccurlyeq \Box D)$

etc.

for $\Box B, \Box C, \Box D \in S(A)$.

Let $(S^+\text{-}Comp)^r$ be the translation into $\underline{L}$ of $\boxed{s} \bigwedge S^+\text{-}Comp$; i.e., ignoring the strong box, $(S^+\text{-}Comp)^r$ is a finite conjunction of sentences,

$$p_{BC} \rightarrow \Box p_{BC}, \quad q_{BC} \rightarrow \Box q_{BC}, \quad \text{for } \Box B, \Box C \in S(A).$$

Similarly, $(S^+\text{-}Ord)^r$ denotes the strongly boxed finite conjunction of translations of the appropriate order axioms:

$$\Box(B^r) \to p_{BB}, \quad q_{BC} \to p_{BC}, \quad \text{etc.}$$

for $\Box B, \Box C \in S(A)$.

Now, it obviously is the case that

$$\mathsf{PRL} \vdash \boxed{s}(\bigwedge S^+\text{-}Comp) \wedge \boxed{s}(\bigwedge S^+\text{-}Ord) \to A \quad \text{in } \underline{R}$$

(i.e. if we drop $A5$, $A6$ from R^-) iff

$$\mathsf{PRL} \vdash (S^+\text{-}Comp)^r \wedge (S^+\text{-}Ord)^r \to A^r \quad \text{in } \underline{L}$$

For, the only difference between, e.g., $\Box B \preccurlyeq \Box C$ and p_{BC} in either sort of proof is notational.

We are now in position to prove the result. Assume $\mathsf{R}^- \nvdash A$ and observe:

$$\mathsf{R}^- \nvdash A \Rightarrow \mathsf{PRL} \nvdash \boxed{s}(\bigwedge S^+\text{-}Comp) \wedge \boxed{s}(\bigwedge S^+\text{-}Ord) \to A \quad \text{in } \underline{R}$$

$$\Rightarrow \mathsf{PRL} \nvdash (S^+\text{-}Comp)^r \wedge (S^+\text{-}Ord)^r \to A^r \quad \text{in } \underline{L}$$

$$\Rightarrow \exists \underline{K}^L(\alpha_0^L \Vdash^L (S^+\text{-}Comp)^r \wedge (S^+\text{-}Ord)^r \wedge \sim A^r),$$

where $\underline{K}^L$ is a finite tree Kripke model for PRL and the superscript "$\underline{L}$" indicates the language forced and that $\underline{K}^L$ is a Kripke model for PRL (following Chapter 2) rather than one for R^- (following Definition 1.6). This existence assertion follows from the Completeness of PRL with respect to models on finite trees. Using $\underline{K}^L$, we define a new Kripke pseudo-model $\underline{K}^R$ for R^- by taking

$$K^R = K^L, \qquad <^R = <^L, \qquad \alpha_0^R = \alpha_0^L$$

(henceforth dropping the superscripts from these coinciding components of $\underline{K}^L$ and $\underline{K}^R$), and defining, for each $\alpha \in K$,

$$\alpha \Vdash^R p \quad \text{iff} \quad \alpha \Vdash^L p, \quad \text{for } p \in \underline{R}$$

$$\left.\begin{array}{l} \alpha \Vdash^R \Box B \preccurlyeq \Box C \quad \text{iff} \quad \alpha \Vdash^L p_{BC} \\ \alpha \Vdash^R \Box B \prec \Box C \quad \text{iff} \quad \alpha \Vdash^L q_{BC} \end{array}\right\} \text{ for } \Box B, \Box C \in \underline{R}.$$

A straightforward induction on the length of D in $\underline{R}$ shows

$$\alpha \Vdash^R D \quad \text{iff} \quad \alpha \Vdash^L D^r, \qquad (*)$$

for all $\alpha \in K$.

Since, for any $\alpha \in K$, $\alpha \Vdash^L (S^+\text{-}Comp)^r \wedge (S^+\text{-}Ord)^r$ (because these strongly boxed assertions were forced at α_0), it follows that

$$\alpha \Vdash^R \bigwedge S^+\text{-}Comp \quad \text{and} \quad \alpha \Vdash^R \bigwedge S^+\text{-}Ord$$

for all $\alpha \in K$. This means that $\underline{K}^R$ is a Kripke pseudo-model for R^- satisfying all the persistence and order axioms naming only boxed sentences of the adequate set $S^+(A)$. Hence, the Extension Lemma applies and yields $\underline{K} = (K,<,\alpha_0,\Vdash)$ such that

$$\alpha_0 \Vdash A \quad \text{iff} \quad \alpha_0 \Vdash^{R} A.$$

But: $$\alpha_0 \nVdash^{L} A \;\Rightarrow\; \alpha_0 \nVdash^{R} A$$
$$\Rightarrow\; \alpha_0 \nVdash A,$$

and A is false in a Kripke model for R^-. QED

The completeness result for R^ω is, by analogy with that for PRL^ω, easily conjectured; it is not as easily proven.

1.13. DEFINITION. Let A be given. A Kripke model $\underline{K} = (K,<,\alpha_0,\Vdash)$ for R^- is A-*sound* iff, for all $\Box B \in S(A)$, $\alpha_0 \Vdash \Box B \rightarrow B$.

1.14. COMPLETENESS THEOREM FOR R^ω. For all A, the following are equivalent:

i. $R^\omega \vdash A$

ii. $R^- \vdash \bigwedge\!\!\bigwedge_{\Box B \in S(A)} (\Box B \rightarrow B) \rightarrow A$

iii. A is true in all A-sound Kripke models, i.e. if $\underline{K}$ is A-sound, then $\alpha_0 \Vdash A$.

The equivalence ii $\Leftrightarrow$ iii is an immediate consequence of the Completeness Theorem for R^-, and the implication ii $\Rightarrow$ i is obvious. The tricky thing is to show i $\Rightarrow$ ii or i $\Rightarrow$ iii. In proving the corresponding implication for PRL and PRL^ω (Chapter 2, section 4), we argued contrapositively by assuming an A-sound countermodel to $\bigwedge\!\!\bigwedge(\Box B \rightarrow B) \rightarrow A$ and constructing an even sounder countermodel by tacking on successive nodes below the origin. We wish to do so now, but will have a little more trouble: We must mix the construction given in the proof of the Extension Lemma with that of Chapter 2, section 4; moreover, since the construction of the Extension Lemma proceeds from the bottom up, we cannot add one new node at a time, but must add all the new ones at once. As this construction might be useful for more than just proving the Completeness Theorem for R^ω, we shall isolate it as a lemma.

1.15. LEMMA. Let A be a sentence, X a set of sentences, and $\underline{K} = (K,<,\alpha_0,\Vdash)$ an A-sound Kripke model for R^-. Then: There is another model $\underline{K}' = (K',<',\alpha_0',\Vdash')$ which is $\bigwedge\!\!\bigwedge X$-sound and which satisfies, for all $B \in S(A)$,

$$\alpha_0' \Vdash B \quad \text{iff} \quad \alpha_0 \Vdash B.$$

(In particular, $\underline{K}'$ is A-sound.)

Proof sketch: Let $\underline{K}$ be given. The new model will be constructed by adding new nodes $\alpha_1,\ldots,\alpha_{N+1}$ below α_0: $\alpha_{N+1} <' \alpha_N <' \ldots <' \alpha_0$.
Thus, $<'$ will simply extend $<$ on $K' = K \cup \{\alpha_1,\ldots,\alpha_{N+1}\}$ and $\alpha_0' = \alpha_{N+1}$. The number N will be determined by X, which we assume to be adequate and, for the sake of convenience, to contain $S = S^+(A)$. The new forcing relation, $\Vdash'$, will be defined to i. agree with $\Vdash$ on S, ii. validate all instances of $A5$ and $A6$, and iii. make the new model $\mathfrak{M}$ X-sound. The first part is readily done, the second is done in analogy to the construction of the Extension Lemma, and the third will be explained shortly.

First, for all $\alpha \in K$ and $B \in S$, define

$$\alpha \Vdash' B \quad \text{iff} \quad \alpha \Vdash B;$$

for α_i ($i = 1,\ldots,N+1$), we will define

$$\alpha_i \Vdash' B \quad \text{iff} \quad \alpha_0 \Vdash B.$$

Without knowing anything else about N, we can prove by induction that $\Vdash'$ restricted to S satisfies all the defining clauses of a forcing relation, e.g.

$$\alpha \Vdash' \Box B \quad \text{iff} \quad \forall \beta > \alpha (\beta \Vdash' B)$$

for $\Box B \in S$.

For the rest of the construction, let $d(B)$ denote the number of nestings of in B, i.e.

$$d(p) = 0$$

$$d(B \circ C) = \max\{d(B), d(C)\}, \quad \text{for } \circ \in \{\wedge, \vee, \rightarrow, \preccurlyeq, \prec\}$$

$$d(\sim B) = d(B)$$

$$d(\Box B) = 1 + d(B).$$

Let $N = \max\{d(B):\ B \in X\}$.

The decision on whether to have $\alpha \Vdash' B$ or not, for $B \notin S$, will be made by induction on $d(B)$. In this, the construction resembles that underlying the Extension Lemma. Now, however, we will also want to satisfy

$$\text{if } i,j \geq d(B), \text{ then } \alpha_i \Vdash' B \Longleftrightarrow \alpha_j \Vdash' B \qquad (*)$$

for all $B \in X$. (In fact, the construction will yield this for all B, though vacuously for $d(B)$ too large.) Since $N \geq d(B)$ for all such B, we will have $\alpha_N \Vdash' B$ iff

$\alpha_{N+1} \Vdash' B$ for $B \in X$, whence

$$\alpha_{N+1} \Vdash' \Box B \to B$$

and the new model will be $\bigwedge X$-sound.

To avoid circumlocutions, let us say that as much of $\Vdash'$ as has been defined at any stage is Y-satisfactory for an adequate set Y if every instance of $A5$ and $A6$ naming only elements of Y is everywhere forced. Thus, $\Vdash'$ is S-satisfactory by the definition of $\Vdash'$ given for elements of S and the S-satisfactoriness of $\Vdash$.

Let $D_m = \{B\colon\ d(B) \le m\}$. We extend $\Vdash'$ by induction on m in such a way that $\Vdash'$ will be $S \cup D_m$-satisfactory and (*) shall hold for $d(B) \le m$.

Basis. For each $\alpha \in K$ and atom $p \in X - S$, let $\alpha \Vdash' p$. The relation $\Vdash'$ extends automatically to all sentences in D_0. The result is $S \cup D_0$-satisfactory since it is S-satisfactory and D_0 contains no boxed formulae. The restriction of (*) to formulae of depth ≤ 0 holds trivially.

Induction step. Assume the definition of $\Vdash'$ to have been made on $S \cup D_m$, to be $S \cup D_m$-satisfactory, and that (*) holds for $d(B) \le m$. We extend the definition of $\Vdash'$ to D_{m+1} in three stages for the sets

$Y_1 = \{\Box B\colon\ d(B) = m\}$

$Y_2 = \{B\colon\ B$ is $C \preccurlyeq C'$ or $C \prec C'$; $C, C' \in S \cup D_m$; and at least one of C, C' is in $Y_1\}$

$Y_3 = \{B\colon\ B$ is a boolean combination of formulae from D_m with one or more formulae from $Y_1 \cup Y_2\}$.

Stage 1. The extension is automatic: for $\Box B \in Y_1$ and $\alpha \in K$,

$$\alpha \Vdash' \Box B \quad \text{iff} \quad \forall \beta > \alpha(\beta \Vdash' B).$$

The crucial thing-- if $m + 1 \le N$-- is to check (*): Let $i, j \ge m + 1$, with, say, $i > j$. On the one hand,

$$\begin{aligned} \alpha_i \Vdash' \Box B &\Rightarrow \forall \beta > \alpha_i(\beta \Vdash' B) \\ &\Rightarrow \forall \beta > \alpha_j(\beta \Vdash' B), \quad \text{since } \alpha_i <' \alpha_j \\ &\Rightarrow \alpha_j \Vdash' \Box B. \end{aligned}$$

Conversely,

$$\alpha_j \Vdash' \Box B \Rightarrow \forall \beta > \alpha_i(\beta \Vdash' B)$$

and, since the only β's greater than α_i, not greater than α_j, are $\alpha_j, \alpha_{j+1}, \ldots, \alpha_{i-1}$, it suffices to show

$$\alpha_j \Vdash' \Box B \;\Rightarrow\; \alpha_j, \ldots, \alpha_{i-1} \Vdash' B$$

But: $\quad \alpha_j \Vdash' \Box B \;\Rightarrow\; \alpha_{j-1} \Vdash' B$ (since $j \geq m+1$, α_{j-1} exists)

$$\Rightarrow\; \forall k \in \{m, \ldots, N+1\}\ (\alpha_k \Vdash' B),$$

by the fact that (*) holds for $d(B) = m$. Hence,

$$\alpha_j \Vdash' \Box B \;\Rightarrow\; \alpha_i \Vdash' \Box B,$$

and (*) holds for $\Box B$.

Stage 2. Starting from the bottom of the model and working one's way up, one can construct (as in the proof of the Extension Lemma) a satisfactory extension of $\Vdash'$ to Y_2. Any such extension will continue to satisfy (*). I leave this verification as an easy exercise to the reader.

Stage 3. Everything extends automatically.

This completes the construction-- and the proof: Applying (*) to $m = N$, we saw

$$\alpha_{N+1} \Vdash' \Box B \to B$$

for all $\Box B \in X$, whence $\underline{K}$ is $\bigwedge X$-sound; we arranged to have

$$\alpha_{N+1} \Vdash' B \quad \text{iff} \quad \alpha_0 \Vdash B$$

for $B \in S$; and since $\Vdash'$ is $S \cup D_m$-satisfactory for all m, $\underline{K}$ is a model of R^-. QED

As I said, Lemma 1.15 may have other applications. However, its immediate importance to us is in proving the Completeness Theorem for R^ω.

Proof of Theorem 1.14: As already remarked, we need only show that, if

$$\mathsf{R}^- \nvdash \bigwedge_{\Box B \in S(A)} (\Box B \to B) \to A,$$

then $\quad \mathsf{R}^\omega \nvdash A.$

Assume, by way of contradiction, the failure of the implication: Let $\underline{K}$ be a model of R^- satisfying

$$\alpha_0 \Vdash \bigwedge_{\Box B \in S(A)} (\Box B \to B) \wedge \sim A$$

and let $\mathsf{R}^\omega \vdash A$. Let X be the smallest adequate set containing $\sim A$ and those sentences $\Box C \to C$ used in a proof of A. Applying Lemma 1.15 to $\underline{K}$ and X, we see that there is a $\bigwedge X$-sound countermodel to A, contradicting the assumption that

$$\mathsf{R}^- \vdash \bigwedge_{\Box C \in X} (\Box C \to C) \to A. \qquad \text{QED}$$

<u>1.16. COROLLARY</u>. (Completeness Theorem for R). Let A be given. The following are equivalent:

i. $\mathrm{R} \vdash A$

ii. $\mathrm{R}^{\omega} \vdash \Box A$

iii. A is valid in all A-sound Kripke models, i.e. if $\underline{K}$ is A-sound, then, for all $\alpha \in K$, $\alpha \Vdash A$.

Proof: i $\Rightarrow$ ii. Simply note

$$\mathrm{R} \vdash A \;\Rightarrow\; \mathrm{R} \vdash \Box A \;\Rightarrow\; \mathrm{R}^{\omega} \vdash A.$$

ii $\Leftrightarrow$ iii. Trivial by the Completeness Theorem for R^{ω}.

iii $\Rightarrow$ i. By contraposition. Note that

$$\mathrm{R} \nvdash A \;\Rightarrow\; \forall n (\mathrm{R} \nvdash \Box^{n} A)$$

$$\Rightarrow\; \forall n (\mathrm{R}^{-} \nvdash \Box^{n} A).$$

Pick N to be the number of distinct boxed subformulae of A and let $\underline{K}$ be a Kripke model of $\sim\Box^{N+1} A$. There is in $\underline{K}$ a sequence $\alpha_0 < \alpha_1 < \ldots < \alpha_{N+1}$ such that, for each i,

$$\alpha_i \Vdash \sim\Box^{N+1-i} A$$

(where $\sim\Box^{0} A = \sim A$).

For $i = 0,\ldots,N$, let

$$X_i = \{\Box B \in S(A): \ \alpha_i \Vdash \Box B\}$$

and observe that the $N + 1$ nested sets $X_N \supseteq \ldots \supseteq X_0$ contain at most N formulae. Thus, for some k, $X_{k+1} = X_k$ and the truncation $\underline{K}_{\alpha_{k+1}}$ is an A-sound model in which A is not valid (since $\alpha_{k+1} < \alpha_{N+1} \Vdash \sim A$). QED

<u>*EXERCISES*</u>

1. i. Using

$$\exists v_1 \phi \wedge \exists v_2 \psi. \leftrightarrow . \exists v \exists v_1 \le v \exists v_2 \le v (\phi \wedge \psi)$$

$$\exists v_1 \phi \vee \exists v_2 \psi. \leftrightarrow . \exists v (\phi v \vee \psi v),$$

show that witness comparisons involving conjunctions and disjunctions are redundant, e.g. $(\phi \wedge \psi) \preccurlyeq \chi. \leftrightarrow .(\phi \preccurlyeq \chi) \wedge (\psi \preccurlyeq \chi)$.

ii. Show the redundancy of nested comparisons.

2. Prove that the usual trick of adding $\Box A$ as an axiom for each axiom A allows one

to drop $R2$ as a rule of inference from the theories R^- and R.

3. Show semantically that R^- proves Rosser's Theorem:

$$R^- \vdash \Box(p \leftrightarrow .\Box\sim p \preccurlyeq \Box p) \wedge \sim\Box f. \rightarrow .\sim\Box p \wedge \sim\Box\sim p.$$

4. i. Show: $R \vdash \Box t \prec \Box f$

ii. Show: $R^- + \Box t \prec \Box f \nvdash \Box t \prec \Box^2 f$

iii. Show: For all $n,k > 0$, $R \vdash \Box^n t \prec \Box^k f$.

5. Complete the proof of Lemma 1.15.

6. Show: For any Σ-sentence A, $R^\omega \vdash A \Rightarrow R \vdash A$.

7. (Smoryński). Let p be a new variable and define A^p by:

$q^p = q$, for all atoms q

$(A \circ B)^p = A^p \circ B^p$, for $\circ \in \{\wedge, \vee, \rightarrow, \preccurlyeq, \prec\}$

$(\sim A)^p = \sim(A^p)$, $(\Box A)^p = \Box(p \rightarrow A^p)$.

Show: $R^- \vdash A$ iff $R \vdash A^p$.

(Hint: $\Leftarrow$. By contraposition: Transform a model $\underline{K}$ with $\alpha_0 \nVdash A$ into an A^p-sound countermodel to A^p by defining $\alpha \Vdash p$ iff $\alpha > \alpha_0$.)

8. (Švejdar). Define the modal system Z^- by replacing axiom schema $A5$ of R^- by

$A5'$. $\Box A \rightarrow \Box(\sim B \rightarrow (\Box A \prec \Box B))$.

i. Making the corresponding change in the definition of a Kripke model, prove completeness: For any A, $Z^- \vdash A$ iff A is true in all Kripke models for Z^-.

ii. Prove: $Z^- \vdash \Box(\Box A \preccurlyeq \Box B) \wedge \Box B \rightarrow \Box A$

iii. Prove: $Z^- \vdash \Box(p \leftrightarrow .\Box\sim p \preccurlyeq \Box p) \wedge \sim\Box f. \rightarrow .\sim\Box p \wedge \sim\Box\sim p.$

(Remark: This Exercise is more than mere axiomatics: If ϕ,ψ are Σ_1, but have several leading existential quantifiers rather than one and the comparison only involves the leading quantifier, then $\phi \preccurlyeq \psi$ will not be Σ_1 and persistence can fail for the comparison. We will encounter this behaviour in Chapter 7, below.)

9. (Smoryński). Recall Chapter 4. Let R_1^- be the theory extending R^- by the addition of one extra "proof predicate" Δ, i.e. add to the language of R^- the modal operator Δ, declare ΔA to be $s\Sigma$, and allow it to enter into witness comparisons with boxed as well as delta-ed sentences and declare these comparisons to

be $s\Sigma$. Axiomatise R_1^- by taking all schemata (in the new big language) of R^- and PRL_1 as axioms and $R1$,$R2$ as rules of inference.

i. Define Carlson pseudo-models and Carlson models for R_1^- and prove the Extension Lemma and Completeness Theorem for R_1^-.

ii. Define R_1^{ω} by adding the soundness schemata,

S1. $\square A \to A$

S2. $\Delta A \to A$,

to the *R2*-free formulation of R_1^-. Prove Completeness: For any A, the following are equivalent:

a. $R_1^{\omega} \vdash A$

b. $R_1^- \vdash \bigwedge\!\!\!\bigwedge_{S} (\Delta B \to B) \to A$

c. A is true in all A-sound Carlson models for R_1^-,

where "S" and "A-sound" are defined appropriately.

iii. Define R_1 by adding to the *R2*-free formulation of R_1^- the rule of inference,

R3'. $\Delta A \;/\; A$.

State and prove a Completeness Theorem.

iv. Prove Rosser's Theorem for the stronger theory in the weaker:

$$R_1^- \vdash \square(p \leftrightarrow .\Delta{\sim}p \preccurlyeq \Delta p) \wedge {\sim}\Delta f. \to .{\sim}\Delta p \wedge {\sim}\Delta{\sim}p.$$

10. (Smoryński). Repeat Exercise 9 for the operator ∇ of MOS, i.e. define theories $RMOS^-$, $RMOS^{\omega}$, and RMOS and:

i-iii. Prove Completeness Theorems for them

iv. Prove Mostowski's Theorem:

$$RMOS^- \vdash \square(p \leftrightarrow .\nabla{\sim}p \preccurlyeq \nabla p) \wedge {\sim}\nabla f. \to .{\sim}\nabla p \wedge {\sim}\nabla{\sim}p.$$

(Remark: Recall that, although we could use Solovay's Second Completeness Theorem to prove Rosser's Theorem in varying degrees of generality, the arithmetic interpretations underlying the arithmetic completeness theorems for PRL_1 and MOS did not admit of such applications. For, these interpretations constructed the theories to which the applications could be made. That R_1^- and $RMOS^-$ prove the results intended to follow as corollaries means we now get the desired generaliy-- once we've discussed the arithmetic interpretations. We do this in Chapter 7, below.)

2. ARITHMETIC INTERPRETATIONS

The obvious (indeed, the intended) interpretation of R within arithmetic is given by interpreting $\square$ as Pr and the modal witness comparisons by the arithmetic ones. Under such an interpretation,

A1 and *R1* are logical phenomena

A2-*A3* and *R2* become the Derivability Conditions

A4 becomes the Formalised Löb's Theorem

A5 becomes a special case of Demonstrable Σ_1-Completeness

A6 becomes a bunch of order properties of the arithmetic witness comparisons, and

R3 becomes a special case of the Σ_1-soundness of PRA:

$$\mathrm{PRA} \vdash Pr(\ulcorner\phi\urcorner) \;\Rightarrow\; Pr(\ulcorner\phi\urcorner) \text{ is true}$$
$$\Rightarrow\; \mathrm{PRA} \vdash \phi.$$

Clearly R is *sound* with respect to this interpretation. Is it complete? Should it be?

The answer to both questions is, "No." One thing we have not taken into account is the fact that, for any two distinct theorems ϕ,ψ, either

$$Pr(\ulcorner\phi\urcorner) \prec Pr(\ulcorner\psi\urcorner) \quad \text{or} \quad Pr(\ulcorner\psi\urcorner) \prec Pr(\ulcorner\phi\urcorner)$$

is true. Axioms can be added to R to handle this, but which of

$$\square t \prec \square\square t \quad \text{and} \quad \square\square t \prec \square t$$

should be true? If we allow the interpretation of T to vary, we might not have to worry about this; if we fix the interpretation, we must settle every possibility,

$$\square^n t \prec \square^k t \quad \text{or} \quad \square^k t \prec \square^n t$$

for $k,n > 0$. Obviously, we really don't want to bother with this-- and this isn't the worst: The predicate $Prov(v_1,v_0)$ asserts that v_1 is, among other things, a finite sequence, say $(x_0,\ldots,x_{k-1})$. Now, which of

$$(x_0,\ldots,x_{k-1}) \quad \text{and} \quad (y_0,\ldots,y_{m-1})$$

is smaller will depend on the exact choice of coding of finite sequences we made. Even the values $x_0,\ldots,x_{k-1}$ and $y_0,\ldots,y_{m-1}$ depend hereditarily on such a choice. In short, even so simple a question of deciding between

$$\square t \prec \square\square t \quad \text{and} \quad \square\square t \prec \square t$$

could well depend on *irrelevant* facts.

What is relevant to Rosser's Theorem is the mode of reasoning-- some facts about

Pr (*A1*-*A4*,*R1*-*R3*) and some about the witness comparisons (*A5*,*A6*). The exact order used in the comparisons doesn't much matter. Moreover, this is true for all of the traditional incompleteness phenomena involving witness comparisons (cf. the next Chapter). Thus, any decent completeness result ought to be with respect to a variety of orderings. Unfortunately, such a completeness result remains an open problem. What we can do is obtain completeness if, in addition to tampering with the ordering of the witnesses, we tamper with the witnesses themselves-- or, what amounts to the same thing, we vary the interpretation of $\Box$. Following this, it will turn out that we can obtain a partial-- but not insignificant-- result in which $\Box$ is held fixed and the ordering is varied.

2.1. DEFINITION. Let T be a (consistent) *RE* extension of PRA with proof predicate Pr_T. A Σ_1-formula $Th(v)$ is called a *standard proof predicate for* T if

$$\mathrm{PRA} \vdash \forall v\big(Th(v) \leftrightarrow Pr_T(v)\big).$$

(The pointwise tradition that has so far been upheld in this monograph would prefer the condition:

$$\text{for all } \phi,\ \mathrm{PRA} \vdash Th(\ulcorner\phi\urcorner) \leftrightarrow Pr_T(\ulcorner\phi\urcorner).$$

The soundness result requires only this, but the completeness result establishes the stronger equivalence of the Definition and, moreover, this extra strength is needed for the refinement to be discussed at the end of this section.)

2.2. DEFINITION. Let T be an *RE* extension of PRA and $Th(v)$ a standard proof predicate for T. An assignment * of arithmetic sentences p^* to modal atoms p extends to an *arithmetic interpretation* based on T as follows:

$$(\sim A)^* = \sim A^*$$

$$(A \circ B)^* = A^* \circ B^*, \quad \text{for } \circ \in \{\wedge, \vee, \rightarrow, \preccurlyeq, \prec\}$$

$$(\Box A)^* = Th(\ulcorner A^* \urcorner).$$

Because *Th* is equivalent to Pr_T, the Derivability Conditions and Löb's Theorem still hold, whence *A2*-*A4* and *R2* are valid under arithmetic interpretations. *A1* and *R1*, being logical matters, remain valid; as we haven't changed the ordering, the truths of *A5* and *A6* are not affected; and *R3* is still a matter of Σ_1-soundness. Hence, we can regard the following as settled:

2.3. THEOREM. (Soundness of Arithmetic Interpretations). Let T be a (consistent) *RE* extension of PRA and let $*$ range over arithmetic interpretations based on standard proof predicates. Then: For any modal sentence A,

i. $R^- \vdash A \Rightarrow \forall *(\mathrm{PRA} \vdash A^*)$
$\Rightarrow \forall *(T \vdash A^*)$

ii. if T is Σ_1-sound,
$R \vdash A \Rightarrow \forall *(\mathrm{PRA} \vdash A^*)$
$\Rightarrow \forall *(T \vdash A^*)$

iii. if T is sound,
$R^\omega \vdash A \Rightarrow \forall *(A^* \text{ is true})$.

Obviously, the goal of this section is to prove the converses to these implications. The first two will be left as exercises to the reader and only the third-- in the case T = PRA (a purely notational convenience)-- will be proven in the text. Specifically, we will prove the following:

2.4. THEOREM. (Guaspari-Solovay Arithmetic Completeness Theorem). Let A be a modal sentence and let $\underline{K}$ be an A-sound Kripke model for R^- in which A is true. Then: There is an arithmetic interpretation $*$ such that A^* is true. Restated: $R^\omega \vdash A$ iff, for all $*$, A^* is true.

Proof: Needless to say, the proof will be a variation of of that of Solovay's Second Completeness Theorem (Chapter 3, section 2).

Let $\underline{K} = (\{1,\ldots,n\},R,1,\Vdash)$ be an A-sound model in which A is true. As before, we assume a very special set $K = \{1,\ldots,n\}$ with special origin 1. Let

$$S = S^+(A) \cup \{\sim B:\ B \in S(A)\}.$$

S is adequate, contains all subformulae of formulae appearing in it, and is closed under negation in the sense that, for any $B \in S$, either $\sim B \in S$ or $B = \sim C$ for some $C \in S$. Add a new root 0 below K and define,

for $p \in S$: $0 \Vdash p$ iff $1 \Vdash p$

for $p \notin S$: $0 \Vdash p$.

By the proof of Lemma 1.15, this extends to an A-sound Kripke model of R^- in which, for all $B \in S$,

$0 \Vdash B$ iff $1 \Vdash B$.

Again following the proofs of Solovay's Completeness Theorems, define a function F in terms of itself and its limit L:

$$F0 = 0$$

$$F(x+1) = \begin{cases} \text{least } y\,(FxRy \;\&\; Prov(\overline{x+1}, \ulcorner L \neq \overline{y} \urcorner)), & \text{if such exists} \\ F(x), & \text{otherwise.} \end{cases}$$

The key properties of F and L were proven in Chapter 3, section 2, and can be found there.

We modify slightly the arithmetic interpretations of the atoms so that we will be able to recover B from B^* for any modal sentence B: List the atoms in order, say, $p_0, p_1, \ldots$, and define,

for $p_k \in S$:
$$p_k^* = \begin{cases} (\bigvee \{L = \overline{x}:\ x \Vdash p_k\}) \wedge \overline{k} = \overline{k}, & \text{if some } x \Vdash p_k \\ \overline{0} = \overline{1} \wedge \overline{k} = \overline{k}, & \text{otherwise} \end{cases}$$

for $p_k \notin S$:
$$p_k^*:\quad L = \overline{0} \wedge \overline{k} = \overline{k}$$

For more complex formulae, we define

$$(B \circ C)^* = B^* \circ C^*, \quad \text{for } \circ \in \{\wedge, \vee, \rightarrow, \preccurlyeq, \prec\}$$

$$(\sim B)^* = \sim B^*, \qquad (\Box B)^* = Th(\ulcorner B^* \urcorner),$$

for some, as yet unspecified, standard proof predicate $Th(\cdot)$.

We define $Th(\cdot)$ by defining a function G enumerating finite sets of theorems and letting

$$Th(v):\quad \exists v_0 (v \in D_{G(v_0)})$$

where D_x is the finite set with canonical index x (as defined in Chapter 0, following 5.7). $G(x)$ is the "set of theorems proven at step x". Needless to say, G is defined by reference to $Th(\cdot)$ and to G itself, whence we must use the Recursion Theorem to define it. Intuitively, G is defined by re-ordering proofs so that they agree with the information supplied by $\underline{K}$. In actual fact, we alter the order in which the theorems are generated and do not even touch the proofs.

I shall give an informal, anthropomorphic description of G, which the energetic overachiever can formalise on his own. G will be defined in a series of stages, but, at any stage m several things may be done and the completed domain of G at the end of stage m will be of the form $\{0,\ldots,k_m\}$, with k_m not necessarily equalling m.

Stage $2m$. Suppose, if $m > 0$, that $G(x)$ has been defined for all $x < k_{2m}$. Check if $Prov(\overline{m},\ulcorner\phi\urcorner)$ for any $\ulcorner\phi\urcorner \le m$. If not, set $k_{2m+1} = k_{2m}$ and move on to the next stage. If so, check if ϕ is B^* for some $\Box B \in S$. If not, define

$$G(k_{2m}) = \{\ulcorner\phi\urcorner\},$$

(i.e. the index $2^{\ulcorner\phi\urcorner}$ of this set), and make $k_{2m+1} = k_{2m} + 1$. The crucial case is if $Prov(\overline{m},\ulcorner B^*\urcorner)$ for some $\Box B \in S$. If this happens, do nothing: Set $k_{2m+1} = k_{2m}$ and move on to the next stage.

Stage $2m + 1$. Unless $m = 0$ or $F(m) \ne F(m - 1)$ (i.e. F has moved), simply set $k_{2m+2} = k_{2m+1}$ and move on to the next stage. If $m = 0$, let

$$Y = \{B^*: \ \Box B \in S \ \& \ 0 \Vdash \Box B\};$$

if $m > 0$, let

$$Y = \{B^*: \ \Box B \in S \ \& \ F(m) \Vdash \Box B \ \& \ F(m - 1) \nVdash \Box B\}.$$

If Y is empty, there is not much to do but move on to the next stage. So, suppose Y is non-empty and split it up into equivalence classes induced by $\preccurlyeq$:

$$\ulcorner B_1^*\urcorner \simeq \ulcorner B_2^*\urcorner \quad \text{iff} \quad F(m) \Vdash (\Box B_1 \preccurlyeq \Box B_2) \wedge (\Box B_2 \preccurlyeq \Box B_1).$$

Let $E_0, E_1, \ldots, E_{s-1}$ be the equivalence classes listed in increasing order under $\prec$. For $0 \le i \le s - 1$, define

$$G(k_{2m+1} + i) = E_i,$$

set $k_{2m+2} = k_{2m+1} + s$, and move on to the next stage.

This completes the definition of G. It is easy to see that G is total and that a formalisation of the definition will yield a formal proof in PRA that G is total.

Having defined G, we have defined Th and, therewith, the interpretation $*$. It remains to prove three lemmas:

<u>2.5. LEMMA</u>. Let $B \in S$, $0 \le x \le n$. Then:

i. $x \Vdash B \Rightarrow \text{PRA} \vdash L = \overline{x} \to B^*$

ii. $x \nVdash B \Rightarrow \text{PRA} \vdash L = \overline{x} \to \sim B^*$.

<u>2.6. LEMMA</u>. For each $\Box B \in S$ and each $0 \le x \le n$,

$$\text{PRA} \vdash L = \overline{x} \to (\text{"}x \Vdash \Box B\text{"} \leftrightarrow Pr(\ulcorner B^*\urcorner)).$$

2.7. LEMMA. PRA$\vdash$ $\forall v(Pr(v) \leftrightarrow Th(v))$.

Lemma 2.5 looks a lot like Lemmas 1.10 and 2.3 of Chapter 3. It is, however, not one of these lemmas and must be proven anew, because $(\Box B)^*$ differs from the old interpretation and because the witness comparisons offer new cases in the proof. Lemma 2.6 is a bridge between Lemmas 2.5 and 2.7, the latter Lemma being the assertion that $Th(v)$ is a standard proof predicate. Thus, Lemmas 2.6 and 2.7 are not part of the construction or the proof that it works, but, rather, are proof of its meaningfulness.

The proof of Theorem 2.4 will be complete on proving these Lemmas. For, by Lemma 2.7, * is an arithmetic interpretation and:

$$1 \Vdash A \;\Rightarrow\; 0 \Vdash A$$

$$\Rightarrow\; \text{PRA} \vdash L = \overline{0} \rightarrow A^*, \text{ by Lemma 2.5}$$

$$\Rightarrow\; A^* \text{ is true,}$$

since we know from Chapter 3 that $L = \overline{0}$ is true.

Proof of Lemma 2.5: By induction on the length of a formula $B \in S$. The Lemma is trivially true for atoms, and the cases for the propositional connectives are equally trivial.

$\Box$-*Case*. Let $B = \Box C$ and $0 \le x \le n$.

Suppose $x \Vdash \Box C$. We must show

$$\text{PRA} \vdash L = \overline{x} \rightarrow Th(\ulcorner C^* \urcorner).$$

Argue in PRA: Since $L = x$, there is some y such that $x = F(y) \Vdash \Box C$. Let m be minimum such that $F(m) \Vdash \Box C$. At stage $2m + 1$, $\ulcorner C^* \urcorner$ is put into the set Y of sentences declared proven by G, i.e. $Th(\ulcorner C^* \urcorner)$ is true.

If $x \nVdash \Box C$ and $L = x$, then no $F(m)$ ever forces $\Box C$ and C^* is never declared proven by G at any stage $2m + 1$, i.e. $\sim Th(\ulcorner C^* \urcorner)$.

$\preccurlyeq, \prec$ *Case*. Let $B = \Box C \preccurlyeq \Box D$ and $0 \le x \le n$.

Suppose $x \Vdash \Box C \preccurlyeq \Box D$. Again, assume $L = x$. Since $x \Vdash \Box C \preccurlyeq \Box D. \rightarrow \Box C$, we see that $x \Vdash \Box C$ and, applying the result of the previous case, C^* is declared proven by G at some stage $2m + 1$ with $F(m)\, R\, x$ or $F(m) = x$. Since $F(m) \Vdash \Box C$, we have one of

$$F(m) \Vdash \Box C \preccurlyeq \Box D \quad \text{or} \quad F(m) \Vdash \Box D \prec \Box C.$$

By Σ-persistence and the fact that $x \Vdash \Box C \preccurlyeq \Box D$, the latter cannot be the case. Now, either $F(m) \Vdash \Box D$ and C^* was explicitly output at least as early as D^* at this stage, i.e.

$$\ulcorner C^* \urcorner \in D_{G(k_m+i)}, \quad \ulcorner D^* \urcorner \in D_{G(k_m+j)}, \quad \text{with } i \le j,$$

or $F(m) \nVdash \Box D$ and D^* could not have been output until after stage $2m + 1$-- if at all. Hence, $Th(\ulcorner C^* \urcorner) \preccurlyeq Th(\ulcorner D^* \urcorner)$.

Similarly, if $x \Vdash \Box C \prec \Box D$, then

$$\mathsf{PRA} \vdash L = \overline{x} \to .Th(\ulcorner C^* \urcorner) \prec Th(\ulcorner D^* \urcorner).$$

If $x \nVdash \Box C \preccurlyeq \Box D$, either $x \nVdash \Box C$, whence

$$\mathsf{PRA} \vdash L = \overline{x} \to \sim Th(\ulcorner C^* \urcorner)$$

$$\vdash L = \overline{x} \to \sim (Th(\ulcorner C^* \urcorner) \prec Th(\ulcorner D^* \urcorner)),$$

or $x \Vdash \Box D \prec \Box C$, whence

$$\mathsf{PRA} \vdash L = \overline{x} \to .Th(\ulcorner D^* \urcorner) \prec Th(\ulcorner C^* \urcorner)$$

$$\vdash L = \overline{x} \to \sim (Th(\ulcorner C^* \urcorner) \preccurlyeq Th(\ulcorner D^* \urcorner)).$$

Similarly, if $x \nVdash \Box C \prec \Box D$, we conclude

$$\mathsf{PRA} \vdash L = \overline{x} \to \sim (Th(\ulcorner C^* \urcorner) \prec Th(\ulcorner D^* \urcorner)).$$

QED

Proof of Lemma 2.6: This proof is delightful in that it isn't the usual induction. It merely unites what we've just proven with facts about L from Chapter 3. It is a proof by cases carried out in PRA. Assume $L = x$.

Case 1. $x \Vdash \Box B$. Observe

$$x \Vdash \Box B \to \forall y (x R y \to y \Vdash B)$$

$$\to \forall y \big(x R y \to Pr(\ulcorner L = \overline{y} \to B^* \urcorner)\big), \text{ by 2.5}$$

$$\to Pr(\ulcorner \bigvee_{x R y} L = \overline{y} \to B^* \urcorner)$$

$$\to Pr(\ulcorner B^* \urcorner),$$

since $L = x \to Pr(\ulcorner \bigvee_{x R y} L = \overline{y} \urcorner)$. (If x is terminal, one has $L = x \to Pr(\ulcorner \overline{0} = \overline{1} \urcorner)$, whence $L = x \to Pr(\ulcorner B^* \urcorner)$.)

Case 2. $x \nVdash \Box B$. Observe

$$x \nVdash \Box B \to \exists y (x R y \wedge y \nVdash B)$$

$$\to \exists y \big(x R y \wedge Pr(\ulcorner L = \overline{y} \to \sim B^* \urcorner)\big), \text{ by 2.5}$$

$$\rightarrow \ \exists y\big(x\,R\,y \wedge \ Pr(\ulcorner B^* \rightarrow L \neq \overline{y}\urcorner)\big)$$

$$\rightarrow \ \exists y\big(x\,R\,y \wedge \ (Pr(\ulcorner B^{*}\urcorner) \rightarrow Pr(\ulcorner L \neq \overline{y}\urcorner))\big). \qquad (*)$$

But, again, from Chapter 3 we know

$$L = x \rightarrow \ \forall y(x\,R\,y \rightarrow \sim Pr(\ulcorner L \neq \overline{y}\urcorner)),$$

which, with (*), yields

$$L = x \rightarrow \ \big(x \Vdash\!\!\!\!/ \ \Box B \rightarrow \sim Pr(\ulcorner B^{*}\urcorner)\big).$$

QED

(Remark: As a cute exercise, the reader might like to use the A-soundness of $\underline{K}$ to show PRA$\vdash$ $\overline{0} \Vdash \ \Box B \rightarrow Pr(\ulcorner B^{*}\urcorner)$ without using the assumption $L = 0$.)

And, finally, we have the

Proof of Lemma 2.7: This is meant to be done formally within PRA, but we shall argue informally.

The only sentences at which $Pr(v)$ and $Th(v)$ might disagree are those finitely many B^* for which $\Box B \in S$; for any other sentence ϕ, we put $\ulcorner \phi \urcorner$ into some $G(x)$ when and only when $Prov(y, \ulcorner \phi \urcorner)$ happened to hold for some y. Thus, it suffices to show, for each B^* with $\Box B \in S$, that $Pr(\ulcorner B^{*}\urcorner) \leftrightarrow Th(\ulcorner B^{*}\urcorner)$. (Reasoning outside PRA, this means PRA$\vdash$ $Pr(\ulcorner B^{*}\urcorner) \leftrightarrow Th(\ulcorner B^{*}\urcorner)$.)

Now, L exists and equals some x between 0 and n. Let $L = x$ and observe

$$L = x \rightarrow \ .Pr(\ulcorner B^{*}\urcorner) \leftrightarrow x \Vdash \ \Box B, \text{ by 2.6}$$

$$\rightarrow \ .Pr(\ulcorner B^{*}\urcorner) \leftrightarrow Th(\ulcorner B^{*}\urcorner), \text{ by 2.5.}$$

QED

As already remarked, the completion of the proof of Lemma 2.7 also completes that of Theorem 2.4, and, therewith, the main work of this section. The corresponding completeness proofs for R and R$^-$ are in the exercises (Exercises 1 and 2, respectively), leaving only one small promised discussion: In the introduction to this Chapter, I said that the choice of a standard proof predicate really amounted to no more than a change in the ordering of the proofs. This, as we saw in the proof of Theorem 2.4, is not strictly true: At stage $2m + 1$, we needed to order several proofs when we didn't know we had any, so we created them out of the air; similarly, in stage $2m$, we ruthlessly suppressed existing proofs. However, the equivalence

$$Th(v) \leftrightarrow Pr(v)$$

suggests that we simply perform a relabelling, identifying the new phoney proofs with the old real ones. The only obstacle to this is: If B^* is provable, with $\Box B \in S$, there are infinitely many x such that

$$Prov(x, \ulcorner B^* \urcorner)$$

but only finitely many x such that

$$\ulcorner B^* \urcorner \in D_{G(x)}.$$

Now, this obstacle is far from insurmountable: At each odd-numbered stage of the construction, produce new numbers x and put

$$\ulcorner B^* \urcorner \in D_{G(x)}$$

for all such B^* that have already been output by G at an earlier odd-numbered stage.

Let us define now a function, say, P which will take x to a set of proofs of sentences in $G(x)$: Look at the stage k at which $G(x)$ was defined. If $k = 2m$, then m was a proof of some sentence ϕ not of the form B^* for $\Box B \in S$. In this case, let $P(x) = \{m\}$ (i.e. $P(x)$ is the index 2^m of the set $\{m\}$). Note that every proof of ϕ gets listed this way.

If $k = 2m + 1$, then $G(x)$ consists of sentences, say, $B_1^*, \ldots, B_m^*$ with $\Box B_1, \ldots, \Box B_m \in S$. Although the number m might not be a proof of any of these, they do have proofs. For each B_i^*, let m_i be the least number y such that

i. $Prov(y, \ulcorner B_i^* \urcorner)$

and ii. $y \notin \bigcup_{j<x} D_{P(j)}$

Then set

$$P(x) = \{m_1, \ldots, m_n\}.$$

Observe that, under the modification of G cited above, every proof of B_i^* will get listed exactly once this way.

Now, what does P do for us? A few moment's thought will convince the reader that the following hold:

i. $PRA \vdash \forall v(\exists v_0 \exists v_1 \leq v_0(v_1 \in D_{P(v_0)} \wedge Prov(v_1, v)) \leftrightarrow Th(v))$

ii. $PRA \vdash \forall v v^*((\exists v_0 \exists v_1 \leq v_0(v_1 \in D_{P(v_0)} \wedge Prov(v_1, v)) \preccurlyeq$

$\preccurlyeq \exists v_0 \exists v_1 \leq v_0(v_1 \in D_{P(v_0)} \wedge Prov(v_1, v^*)). \leftrightarrow .$

$\leftrightarrow . Th(v) \preccurlyeq Th(v^*))$

and similarly for $\prec$. What has been accomplished is this: Writing

$$v_0 \leq_P v_1: \quad \exists v(v_0 \in D_{P(v)}) \preccurlyeq \exists v(v_1 \in D_{P(v)}) \tag{1}$$

$$v_0 <_P v_1: \quad \exists v(v_0 \in D_{P(v)}) \prec \exists v(v_1 \in D_{P(v)}), \tag{2}$$

and

$$\exists v\phi v \preccurlyeq_P \exists v\psi v: \quad \exists v\big(\phi v \wedge \forall v_1 <_P v \sim\psi v_1\big) \tag{3}$$

$$\exists v\phi v \prec_P \exists v\psi v: \quad \exists v\big(\phi v \wedge \forall v_1 \leq_P v \sim\psi v_1\big), \tag{4}$$

we see that

$$\text{PRA} \vdash Th(v) \preccurlyeq Th(v^*). \leftrightarrow .Pr(v) \preccurlyeq_P Pr(v^*) \tag{5}$$

$$\text{PRA} \vdash Th(v) \prec Th(v^*). \leftrightarrow .Pr(v) \prec_P Pr(v^*). \tag{6}$$

We are not done yet. The equivalences (5) and (6) reveal the comparisons involving the new proof predicates to be equivalent to comparisons involving the usual proof predicate and a new ordering of the proofs. However, these are *equivalences* and the improved formulae on the right-hand sides cannot replace the old comparisons of the left-hand sides in the *non-extensional* settings of further witness comparisons. In short, we cannot jump immediately to a completeness result like Theorem 2.4 for arithmetic interpretations under which $\Box$ becomes Pr and the modal witness comparisons are interpreted by the new arithmetic comparisons (3) and (4). What we can do, after a cosmetic improvement in P, is jump to such a conclusion for a limited class of modal formulae.

The cosmetic improvement is this: The orderings (1) and (2) are partial, defined only on the set of (codes of) proofs. We simply extend it in any convenient way, e.g. defining

$$F(x) = \begin{cases} P(x), & \text{if } x \text{ is a proof} \\ P(x) + 2^x, & \text{if } x \text{ is not a proof,} \end{cases}$$

so that

$$D_{F(x)} = \begin{cases} D_{P(x)}, & \text{if } x \text{ is a proof} \\ D_{P(x)} \quad \{x\}, & \text{otherwise.} \end{cases}$$

(If one objects to mingling proofs and non-proofs, mix the non-proofs in in some other way.) One can then define $\leq_F$, $<_F$, $\preccurlyeq_F$, and $\prec_F$ as in (1)-(4) and verify the equivalences corresponding to (5) and (6) quite easily.

The following definition captures the relevant properties of F.

<u>2.8. DEFINITION.</u> A recursive function F is a *pre-permutation* of the natural numbers

if the following hold:

i. $\mathsf{PRA} \vdash \forall v_0 \exists v_1 (v_0 \in D_{F(v_1)})$

ii. $\mathsf{PRA} \vdash \forall v_1 \exists v_0 (v_0 \in D_{F(v_1)})$

iii. $\mathsf{PRA} \vdash \forall v_0 v_1 (v_0 \neq v_1 \rightarrow D_{F(v_0)} \cap D_{F(v_1)} = \emptyset)$.

In words, F is a pre-permutation if $D_{F0}, D_{F1}, \ldots$ is a sequence of pairwise disjoint, finite, non-empty sets the union of which is the set of all natural numbers.

The name "pre-permutation", like "pre-ordering", signifies that the function F bears the same relation to a permutation as the pre-ordering bears to an ordering: Instead of having one object called $F(x)$ in position x, we have a finite set $D_{F(x)}$. When each $D_{F(x)}$ is a singleton, F is all but a permutation of the set of natural numbers.

We might as well declare (1)-(4) to be formal matters:

2.9. DEFINITIONS. Let F be a pre-permutation. The pre-orderings $\leq_F$ and $<_F$ are defined by replacing "P" by "F" in formulae (1) and (2) under discussion; the witness comparisons $\preccurlyeq_F$, $\prec_F$ are similarly defined by replacing "P" by "F" in (3) and (4).

The point of all this can be formalised in yet another fashion:

2.10. DEFINITION. Let F be a pre-permutation and $*$ an assignment of arithmetic sentences p^* to atoms p. The *pre-permutational arithmetic interpretation* $*$ based on F and $*$ is the extension of $*$ by the following rules:

$$(A \circ B)^* = A^* \circ B^*, \quad \text{for } \circ \in \{\wedge, \vee, \rightarrow\}$$

$$(\sim A)^* = \sim A^*, \qquad (\Box A)^* = Pr(\ulcorner A^* \urcorner)$$

$$(A \circ B)^* = A^* \circ_F B^*, \quad \text{for } \circ \in \{\preccurlyeq, \prec\}.$$

The soundness of pre-permutational arithmetic interpretations is an easy exercise (Exercise 3, below) as is the following:

2.11. THEOREM. Let A be a modal sentence with no nestings of witness comparisons. If $\underline{K}$ is an A-sound Kripke model for R^- in which A is true, there is a pre-permutation F and a pre-permutational arithmetic interpretation $*$ based on F under which A^* is true.

The construction of F from a $Th(v)$ and G is as discussed above. The proof of the Theorem is an easy induction on the length of A and is left to the reader (Exercise 4).

EXERCISES

1. Prove the arithmetic completeness theorem for R: Let A be a modal sentence such that $\mathsf{R} \nvdash A$. Then, there is an arithmetic interpretation * based on a standard proof predicate $Th(v)$ for PRA such that $\mathsf{PRA} \nvdash A^*$. (Hint: Reduce the result to Theorem 2.4.) (Remark: The reduction works when PRA is replaced by a sound extension T, but the result still holds for Σ_1-sound RE extensions. It is worth noting that the construction of Theorem 2.4 still works in the Σ_1-sound case, although the full completeness only holds for fully sound T.)

2. Prove the arithmetic completeness theorem for R^-: Let A be a modal sentence such that $\mathsf{R}^- \nvdash A$. Then, there is an RE extension T of PRA (not necessarily consistent) and an arithmetic interpretation * based on a standard proof predicate $Th(v)$ for T such that $\mathsf{T} \nvdash A^*$. (Hint: Apply Exercise 7 of the preceding section and Exercise 1, above.)

3. Prove soundness for pre-permutational arithmetic interpretations *: For any modal sentence A, $\mathsf{R}^\omega \vdash A \Rightarrow A^*$ is true.

4. Prove Theorem 2.11.

3. INEQUIVALENT ROSSER SENTENCES

To a large extent, the purpose of the present section is to expound on something that could have been relegated to the exercises of the preceding section. The result of the exercise is, however, of some interest and I have decided to present it with the appropriate amount of fanfare. The result is, in case the reader hasn't guessed, the non-uniqueness of Rosser sentences.

The story of Rosser sentences is actually a lot more complicated than is indicated by the title of this section. Depending on the choice of a standard proof predicate, they can be unique or non-unique, they can refute each other or not, and they can be definable or not given by any fixed explicit definition. In short, none of the smooth behaviour of self-reference encountered in Chapters 1 and 4 carries over. The purpose of the present section is to examine some of this pathology, but not to overdo it.

An undefinability result is a nice place to begin.

3.1. THEOREM. Let A be any sentence of the modal language with witness comparisons. Then:

i. $\mathsf{R} \nvdash \sim\Box f \to \sim\Box A \wedge \sim\Box\sim A$

ii. $\mathsf{R} \nvdash A \leftrightarrow .\Box\sim A \preccurlyeq \Box A$

iii. $\mathsf{R} \nvdash \boxed{s}\,(p \leftrightarrow .\Box\sim p \preccurlyeq \Box p) \to .p \leftrightarrow A.$

Proof: i. Let A be given. Applying the construction of Lemma 1.15, one can obtain a Kripke model $\underline{K}$ for R^- that

a. is B-sound, for $B = \sim\Box f \to \sim\Box A \wedge \sim\Box\sim A$

b. is linear

and c. has at least two nodes.

Let $\alpha_{N-1} < \alpha_N$ be the two *top* nodes of K. Either $\alpha_N \Vdash A$ or $\alpha_N \Vdash \sim A$. In the former case, $\alpha_{N-1} \Vdash \Box A$, and in the latter $\alpha_{N-1} \Vdash \Box\sim A$. Thus,

$$\alpha_{N-1} \nVdash \sim\Box f \to \sim\Box A \wedge \sim\Box\sim A,$$

i.e. $\alpha_{N-1} \nVdash B$. Hence $\underline{K}$ is a B-sound model of R^- in which B is not valid (i.e. $\alpha_0 \nVdash \Box B$) and $\mathsf{R} \nvdash B$ by Corollary 1.16.

ii. In R^- we can prove Rosser's Theorem,

$$\mathsf{R}^- \vdash \Box(p \leftrightarrow .\Box\sim p \preccurlyeq \Box p) \to (\sim\Box f \to \sim\Box p \wedge \sim\Box\sim p).$$

Replacing p by A in the proof yields

$$\mathsf{R}^- \vdash \Box(A \leftrightarrow .\Box\sim A \preccurlyeq \Box A) \to (\sim\Box f \to \sim\Box A \wedge \sim\Box\sim A).$$

The unprovability in the stronger R of the conclusion yields the unprovability of the hypothesis.

iii. Because R is complete with respect to arithmetic interpretations, it is closed under the Diagonalisation Rule,

$$DR. \quad \boxed{s}\,(p \leftrightarrow B(p)) \to C \;/\; C,$$

where p has only the right sort of occurrence in B and does not occur at all in C. By this rule, the assumption,

$$\mathsf{R} \vdash \boxed{s}\,(p \leftrightarrow .\Box\sim p \preccurlyeq \Box p) \to .p \leftrightarrow A,$$

together with the provability in R of Rosser's Theorem, readily yields

$$\mathsf{R} \vdash \sim\Box f \to \sim\Box A \wedge \sim\Box\sim A,$$

contradicting i. (Exercise: Why not contradict ii?) QED

Applying the Completeness Theorem for R, we see that, for any A, there is a standard proof predicate $Th(v)$ and an interpretation * based on it such that

$$\mathrm{PRA} \nvdash Con \rightarrow \sim Pr(\ulcorner A^* \urcorner) \wedge \sim Pr(\ulcorner \sim A^* \urcorner).$$

It follows that

$$\mathrm{PRA} \nvdash A^* \leftrightarrow .Pr(\ulcorner \sim A^* \urcorner) \preccurlyeq Pr(\ulcorner A^* \urcorner).$$

(Unless A is very simple, we cannot appeal to 2.11 to obtain a pre-permutational interpretation.)

By Theorem 3.1, although the self-reference available in R allows us to obtain independent sentences, no sentence explicitly defined in the language will work. Moreover, no such sentence yields a Rosser sentence under all arithmetic interpretations. Thus, we see the failure of the explicit definability of Rosser sentences. As already announced, uniqueness can also fail.

3.2. THEOREM. (Non-Uniqueness Theorem). There is a standard proof predicate $Th(v)$ and an interpretation * based on it under which Rosser sentences are not unique, i.e. there are sentences ϕ, ψ such that

$$\mathrm{PRA} \vdash \phi \leftrightarrow .Th(\ulcorner \sim\phi \urcorner) \preccurlyeq Th(\ulcorner \phi \urcorner)$$

$$\mathrm{PRA} \vdash \psi \leftrightarrow .Th(\ulcorner \sim\psi \urcorner) \preccurlyeq Th(\ulcorner \psi \urcorner),$$

but

$$\mathrm{PRA} \nvdash \phi \leftrightarrow \psi.$$

Proof: Let

$$A = \Box(p \leftrightarrow .\Box\sim p \preccurlyeq \Box p) \wedge \Box(q \leftrightarrow .\Box\sim q \preccurlyeq \Box q) \wedge \sim\Box(p \leftrightarrow q)$$

and note that A is true in the following A-sound model

$\underline{K}$: two nodes above a common root: left node $\Box\sim p \prec \Box p,\ \Box q \prec \Box\sim q,\ p$; right node $\Box\sim q \prec \Box q,\ \Box p \prec \Box\sim p,\ q$

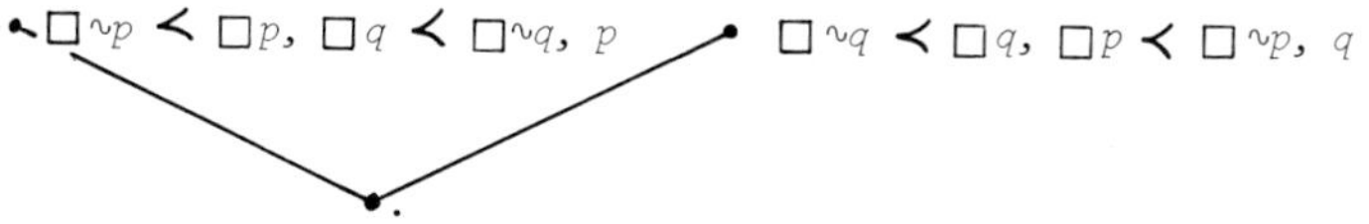

Let * be such that A^* is true and let $\phi = p^*$, $\psi = q^*$. QED

Theorem 2.11 applies also:

3.3. COROLLARY. There are a pre-permutation F and sentences ϕ, ψ such that

$$\mathrm{PRA} \vdash \phi \leftrightarrow .Pr(\ulcorner \sim\phi \urcorner) \preccurlyeq_F Pr(\ulcorner \phi \urcorner)$$

$$\mathrm{PRA} \vdash \psi \leftrightarrow .Pr(\ulcorner \sim\psi \urcorner) \preccurlyeq_F Pr(\ulcorner \psi \urcorner),$$

but $\quad PRA \nvdash \phi \leftrightarrow \psi.$

For ϕ,ψ as in either the Theorem or the Corollary, we not only have non-equivalence, but also mutual incompatibility,

$$PRA \vdash \sim(\phi \wedge \psi),$$

as follows from the fact that the model is also a B-sound model of B, where $B = \Box\sim(p \wedge q)$. Different models $\underline{K}$ will give different dependencies or even independence. Such refinements will be discussed in the Exercises.

There is another refinement-- the definability of ϕ,ψ. We will discuss this. Before doing so, however, I wish to make a polemical digression.

In their paper on Rosser sentences, Guaspari and Solovay only mentioned Theorem 3.2 and offered a few vague words about re-ordering the proofs; Corollary 3.3 was, thus, implicit rather than explicit. The talk about "standard proof predicates" together with their twice mentioning that the problem for "the usual" ordering of proofs has led to an underevaluation of this result. We must ask ourselves what is relevant here. As an analysis of methods, Theorem 3.2 is as significant as Corollary 3.3 or an answer for "the usual" proof predicate. For, what has every result using $Pr(v)$ and the witness comparisons depended on other than the Derivability Conditions and the obvious order properties codified in the axioms and rules of R^-? That the uniqueness or non-uniqueness of Rosser sentences cannot be established on this basis (There are arithmetical interpretations under which Rosser sentences are provably equivalent-- cf. Theorem 3.6, below.) is the *first* result requiring more than the usual conditions for a settlement.

If we take it to be our goal not to analyse methods, but to analyse $Pr(v)$, then I agree that the Arithmetic Completeness Theorems for R and R^ω are inadequate; but I still consider Corollary 3.3 to be convincing. Indeed, I think the devotees of "the usual" ordering have their work cut out for them, not merely in establishing the uniqueness or non-uniqueness of their Rosser sentences, but in explaining the significance of their ultimate result, whichever direction it should happen to go in. In the last section, I pointed out how the restriction to arithmetic interpretations using only $Pr(v)$ and the standard ordering would have required us, for the sake of an arithmetic completeness theorem, to make many seemingly irrelevant decisions. I should go farther here in pointing out that "the usual" ordering of proofs determined by "the usual" proof predicate is itself an artifice and somewhat irrelevant at that. As already noted, a change in the manner of coding finite sequences (e.g. using powers of 2 (as in using finite sets of ordered pairs) *vs.* using exponents on primes (and the Fundamental Theorem of Arithmetic)) will no doubt permute the proofs. A slight change in the logical setting (from, say, a Hilbert-style to a sequent calculus) will also change the proofs themselves.

All this is yet beside the point: The fact is that these formal proofs are mere idealisations of real proofs and their ordering by numerical code does not idealise any real ordering of real proofs any more than does the pre-permutational ordering of Corollary 3.3. The situation would be different if there were some good theoretical results supporting interest in the usual ordering, but all there are in the literature are a few pathologies shared equally by the pre-permutational orderings. In short, the significance of Corollary 3.3 cannot be questioned, but what is open to question is the significance of the problem of the uniqueness or non-uniqueness of Rosser sentences when "the usual" ordering is used.

But enough of this! Let us get back to work. Let us also stop calling exercises "Theorems" and label the next one an "Example".

3.4. EXAMPLE. (Definable Rosser Sentences). There is a standard proof predicate $Th(v)$ such that

$$Th(\ulcorner\overline{0} = \overline{1}\urcorner) \prec Th(\ulcorner\overline{0} = \overline{1} \wedge \overline{0} = \overline{1}\urcorner)$$

and
$$Th(\ulcorner\overline{0} = \overline{1} \wedge \overline{0} = \overline{1}\urcorner) \prec Th(\ulcorner\overline{0} = \overline{1}\urcorner)$$

are Rosser sentences for $Th(v)$.

Proof: Note that we can replace p by $\Box f \prec \Box(f \wedge f)$ and q by $\Box(f \wedge f) \prec \Box f$ in the previous model without destroying A-soundness. QED

Applying 2.11 to this yields

3.5. EXAMPLE. There are a pre-permutation F and sentences ϕ,ψ such that ϕ,ψ are F-Rosser, i.e.

$$\mathrm{PRA} \vdash \phi \leftrightarrow .Pr(\ulcorner\sim\phi\urcorner) \preccurlyeq_F Pr(\ulcorner\phi\urcorner)$$

$$\mathsf{PRA} \vdash \psi \leftrightarrow . Pr(\ulcorner \sim\psi \urcorner) \preccurlyeq_F Pr(\ulcorner \psi \urcorner)$$

and are *equivalent* to definable sentences:

$$\mathsf{PRA} \vdash \phi \leftrightarrow . Pr(\ulcorner \overline{0} = \overline{1} \urcorner) \prec_F Pr(\ulcorner \overline{0} = \overline{1} \wedge \overline{0} = \overline{1} \urcorner)$$

$$\mathsf{PRA} \vdash \psi \leftrightarrow . Pr(\ulcorner \overline{0} = \overline{1} \wedge \overline{0} = \overline{1} \urcorner) \prec_F Pr(\ulcorner \overline{0} = \overline{1} \urcorner).$$

Taking ϕ, ψ as in 3.4, 2.11 yields the conclusions of 3.5. Note that, although, say, $Pr(\ulcorner \overline{0} = \overline{1} \urcorner) \prec_F Pr(\ulcorner \overline{0} = \overline{1} \wedge \overline{0} = \overline{1} \urcorner)$ is equivalent to the F-Rosser sentence $Th(\ulcorner \overline{0} = \overline{1} \urcorner) \prec Th(\ulcorner \overline{0} = \overline{1} \wedge \overline{0} = \overline{1} \urcorner)$, our inability to substitute for equivalents in non-extensional contexts will not allow us to conclude that $Pr(\ulcorner \overline{0} = \overline{1} \urcorner) \prec_F Pr(\ulcorner \overline{0} = \overline{1} \wedge \overline{0} = \overline{1} \urcorner)$ *is* F-Rosser (cf. Exercise 4, below).

Having spent so much energy on the non-uniqueness of Rosser sentences, let us now prove their uniqueness.

3.6. THEOREM. (Equivalence of Rosser Sentences). There is a standard proof predicate $Th(v)$ such that, if ϕ, ψ are Rosser sentences for $Th(v)$, i.e. if

$$\mathsf{PRA} \vdash \phi \leftrightarrow . Th(\ulcorner \sim\phi \urcorner) \preccurlyeq Th(\ulcorner \phi \urcorner)$$

and
$$\mathsf{PRA} \vdash \psi \leftrightarrow . Th(\ulcorner \sim\psi \urcorner) \preccurlyeq Th(\ulcorner \psi \urcorner),$$

then
$$\mathsf{PRA} \vdash \phi \leftrightarrow \psi.$$

Theorem 3.6 is called a "Theorem" rather than an "Example" because it is not a quick application of the Arithmetic Completeness Theorem for R^ω, but is given its own proof via the Recursion Theorem.

Proof of Theorem 3.6: We will define a function G enumerating the theorems of PRA and define $Th(v)$ in terms of G. For convenience, we shall have G outputting a single theorem at a time, whence $Th(v)$ will be $\exists v_0 (v = G(v_0))$ instead of $\exists v_0 (v \in D_{G(v_0)})$. G will be defined by Recursion, i.e. by reference to itself and $Th(v)$.

The construction of G is again given anthropomorphically and it again proceeds by stages at which any number of values of G can be defined. Thus, we assume at the beginning of stage m the domain of G to be $\{0, \ldots, k_m - 1\}$. Unlike our earlier construction, our present one will proceed in a normal fashion until something happens--which we will call "ringing the bell"-- at which point the construction changes radically and permanently.

During the construction, we will build up a list of Rosser sentences called the "Rosser list".

Stage m. We need only state what to do before the bell rings or when it has immediately rung.

Case 1. The bell has not yet rung. Check if $Prov(m, \ulcorner\phi\urcorner)$ for some $\ulcorner\phi\urcorner \le m$. If not, move on to stage $m + 1$. If so, check if ϕ or $\sim\phi$ is on the Rosser list. If it is, ring the bell and proceed to Case 2; if not, set $G(k_m) = \ulcorner\phi\urcorner$ and look at ϕ.

If ϕ is of the form, $\psi \leftrightarrow .Pr(\ulcorner\sim\psi\urcorner) \preccurlyeq Pr(\ulcorner\psi\urcorner)$, making ψ a Rosser sentence, then put ψ on the Rosser list *unless*

a. $\sim\psi$ is already on the list

or b. $\psi = \sim\chi$ for some χ already on the list.

Go on to stage $m + 1$.

Case 2. The bell has just rung. It could have rung because ϕ or $\sim\phi$ is on the Rosser list with ϕ just proven.

Subcase 2.a. ϕ is on the Rosser list. Let this list be $\psi_0,\ldots,\psi_{k-1}$. Define

$$G(k_m + i) = \psi_i, \quad 0 \le i \le k - 1$$

$$G(k_m + k + i) = \sim\psi_i, \quad 0 \le i \le k - 1$$

and then let $G(k_m + 2k)$, $G(k_m + 2k + 1)$, ... enumerate all sentences of PRA.

Subcase 2.b. $\sim\phi$ is on the Rosser list. Again, let the list be $\psi_0,\ldots,\psi_{k-1}$, but this time let G output the negations first:

$$G(k_m + i) = \sim\psi_i, \quad 0 \le i \le k - 1$$

$$G(k_m + k + i) = \psi_i, \quad 0 \le i \le k - 1$$

and let $G(k_m + 2k)$, ... enumerate all sentences of PRA.

This gives the construction of G. The Theorem follows from a few lemmas.

3.7. LEMMA. PRA $\vdash$ If the bell rings then $\sim Con$.

3.8. LEMMA. PRA $\vdash$ $\forall v\big(Th(v) \leftrightarrow Pr(v)\big)$.

3.9. LEMMA. Let ψ be a sentence. If

$$\text{PRA} \vdash \psi \leftrightarrow .Th(\ulcorner\sim\psi\urcorner) \preccurlyeq Th(\ulcorner\psi\urcorner), \qquad (*)$$

then PRA $\vdash$ ψ is put on the Rosser list.

Lemmas 3.7 and 3.8 are easy exercises.

Proof of Lemma 3.9: Let m be a proof of (*). The only way to fail to put ψ on the Rosser list at stage m is for $\sim\psi$ or some χ for which $\psi = \sim\chi$ to be on the list already. But, by Exercise 1, below, all Rosser sentences are false and we cannot have $\psi,\sim\psi$ or $\chi,\sim\chi$ both being Rosser sentences.

QED

Proof of Theorem 3.6 continued: Let ϕ,ψ be Rosser sentences for $Th(v)$. By Lemma 3.9, they are both on the Rosser list, say, by stage m. Because all Rosser sentences are unprovable, PRA proves that they are on the list at this stage and that the bell hasn't rung by then. But,

$$Th(\ulcorner\sim\phi\urcorner) \preccurlyeq Th(\ulcorner\phi\urcorner) \quad \text{iff} \quad G \text{ outputs } \sim\phi \text{ before } \phi$$
$$\text{iff} \quad G \text{ outputs } \sim\psi \text{ before } \psi$$

this latter equivalence guaranteed by the uniform decision made when the bell rings in the construction of G. But

$$G \text{ outputs } \sim\psi \text{ before } \psi \quad \text{iff} \quad Th(\ulcorner\sim\psi\urcorner) \preccurlyeq Th(\ulcorner\psi\urcorner),$$

whence
$$\phi \leftrightarrow .Th(\ulcorner\sim\phi\urcorner) \preccurlyeq Th(\ulcorner\phi\urcorner)$$
$$\leftrightarrow .Th(\ulcorner\sim\psi\urcorner) \preccurlyeq Th(\ulcorner\psi\urcorner)$$
$$\leftrightarrow \psi$$

QED

EXERCISES

1. Show: $R^{-} \vdash \boxed{s}\,(p \leftrightarrow .\Box\sim p \preccurlyeq \Box p) \wedge \sim\Box f \rightarrow \sim p$, whence PRA $\vdash Con \rightarrow \sim\phi$ for any Rosser sentence ϕ.

2. Following 3.1, we wrote $\nvdash Con \rightarrow \sim Pr(\ulcorner A^*\urcorner) \wedge \sim Pr(\ulcorner\sim A^*\urcorner)$ and $\nvdash A^* \leftrightarrow .Pr(\ulcorner\sim A^*\urcorner) \preccurlyeq Pr(\ulcorner A^*\urcorner)$, instead of $\nvdash Th(\ulcorner\overline{0} = \overline{1}\urcorner) \rightarrow \sim Th(\ulcorner A^*\urcorner) \wedge \sim Th(\ulcorner\sim A^*\urcorner)$ and $\nvdash A^* \leftrightarrow .Th(\ulcorner\sim A^*\urcorner) \preccurlyeq Th(\ulcorner A^*\urcorner)$. Show that this was not a mistake. (Hint: Cf. the definition of $Th(v)$.)

3. Let
$$A = \Box f \prec \Box(f \wedge f) \qquad B = \Box(f \wedge f) \prec \Box(f \wedge f \wedge f).$$
Find arithmetic interpretations * under which A^*,B^* are Rosser sentences such that
   i. PRA $\vdash A^* \rightarrow B^*$, PRA $\nvdash B^* \rightarrow A^*$
   ii. A^*,B^* are independent over PRA.

4. Show that there is a standard proof predicate $Th(v)$ and sentences ϕ,ψ such that

i. $\mathrm{PRA} \vdash \psi \leftrightarrow .\mathit{Th}(\ulcorner\sim\psi\urcorner) \preccurlyeq \mathit{Th}(\ulcorner\psi\urcorner)$

ii. $\mathrm{PRA} \vdash \phi \leftrightarrow \psi$

but iii. $\mathrm{PRA} \nvdash \phi \leftrightarrow .\mathit{Th}(\ulcorner\sim\phi\urcorner) \preccurlyeq \mathit{Th}(\ulcorner\phi\urcorner)$.

5. Let $\mathit{Th}^R(\ulcorner\phi\urcorner)$ be $\mathit{Th}(\ulcorner\phi\urcorner) \preccurlyeq \mathit{Th}(\ulcorner\sim\phi\urcorner)$. Construct standard proof predicates and sentences ϕ,ψ such that

i. $\mathrm{PRA} \nvdash \mathit{Th}^R(\ulcorner\phi\urcorner) \wedge \mathit{Th}^R(\ulcorner\phi \rightarrow \psi\urcorner) \rightarrow \mathit{Th}^R(\ulcorner\psi\urcorner)$

ii. $\mathrm{PRA} \nvdash \mathit{Th}^R(\ulcorner\phi\urcorner) \rightarrow \mathit{Th}^R(\ulcorner\mathit{Th}^R(\ulcorner\phi\urcorner)\urcorner)$.

Chapter 7
An Ubiquitous Fixed Point Calculation

It was around 1976 that David Guaspari introduced the witness comparison notation. The importance of this has already been indicated by the results of Chapter 6: It provided just the right framework for a discussion of Rosser sentences. Rosser introduced the use of witness comparisons in 1936 and, from then until Guaspari introduced his notation, most applications of self-reference in arithmetic used the comparison in a surprisingly uniform manner; yet the users neither recognised the uniformity nor attempted to explain their use of self-reference other than to make a passing reference to "Rosser's trick." But when the new notation came on the scene, the uniformity was readily apparent (modulo a little standardisation).

Put into Σ_1-form, Rosser's sentence satisfies

$$\mathsf{PRA} \vdash \phi \leftrightarrow .Pr(\ulcorner\sim\phi\urcorner) \preccurlyeq Pr(\ulcorner\phi\urcorner).$$

The sentence Andrzej Mostowski used in 1960 (cf. Chapter 4), when put into Σ_1-form, reads

$$\mathsf{PRA} \vdash \phi \leftrightarrow .\mathbb{W}Pr_{T_i}(\ulcorner\sim\phi\urcorner) \preccurlyeq \mathbb{W}Pr_{T_i}(\ulcorner\phi\urcorner)$$

($\mathbb{W}Pr_{T_i}$ as in Chapter 4, section 1 (1.10) and section 3). In the same year, John Shepherdson used sentences of the form

$$\mathsf{PRA} \vdash \phi \leftrightarrow .(Pr(\ulcorner\sim\phi\urcorner) \vee \psi) \preccurlyeq (Pr(\ulcorner\phi\urcorner) \vee \chi),$$

where ψ,χ were Σ_1-sentences. (Actually, he allowed free variables in ϕ,ψ,χ.) In 1972, Petr Hájek used a sentence satisfying

$$\mathsf{PRA} \vdash \phi \leftrightarrow .Relint_{GB}(\ulcorner\sim\phi\urcorner) \preccurlyeq Relint_{GB}(\ulcorner\phi\urcorner),$$

where $Relint_{GB}(\ulcorner\psi\urcorner)$ asserts ψ to be relatively interpretable in GB, and where GB is the finitely axiomatised Gödel-Bernays set theory. The common form of these-- something implied by provability happens to $\sim\phi$ at least as early as it does to ϕ-- ought to be apparent. The witness comparisons, together with the extra modal operators introduced in Chapter 4 and some variables σ,τ for Σ_1-formulae, provide an ideal language for expressing what is common to these fixed points: They all look like

$$p \leftrightarrow {}_{\bullet}(\Delta\sim p \vee \sigma) \preccurlyeq (\nabla p \vee \tau), \qquad (*)$$

for suitable choices of $\nabla,\Delta,\sigma,\tau$-- with σ,τ possibly absent. (Actually, they look like this for identical Δ,∇; we are generalising a bit.) Once we recognise this, the next step is also fairly clear: We set up a system of modal logic in which to prove some metatheorem about p. We can then systematically apply this metatheorem to obtain *all* (not: most) the applications made of the given instances of the modal fixed point (*).

The goal of this final Chapter, thus, is to abandon our study of self-reference as an object of study and lay the foundations of a study of it as a tool for application. In the immediately following section, we define a *c*onvenient system of *m*odal *l*ogic, which we shall call CML, and prove a theorem in CML describing the behaviour of the sentence p satisfying (*). This Theorem accounts for all known applications of sentences of the form (*), and was used to obtain additional applications. To help emphasise the uniformity and fecundity of this result, we separate the Theorem from its applications and put the latter in a special section of their own, section 2.

In the 1970s, Hájek, Clark Kent, Guaspari, and Vítězslav Švejdar applied sentences ϕ that were very similar to those cited above, but did not fall under the auspices of my ubiquitous fixed point calculation. Kent, for example, relativised the provability predicate to a partial truth definition and, although Guaspari's sentences were much more complicated than this, upon simplification they too took this form. Guaspari proved that certain applications of these fixed points could not be established without the relativisation. Section 3 discusses this. Section 4 is devoted to the self-referential sentences of Hájek and Švejdar.

Both sections 3 and 4 concern PA rather than PRA. While full induction is not needed to obtain the relativisation of the proof predicates, the witness comparisons do presuppose the Least Number Principle and, as they become logically complex, need progressively more induction and we might as well assume full induction. Moreover, section 4 relies on induction in a much more subtle and much deeper manner: Whereas PRA is all but finitely axiomatised, PA is *reflexive*: It proves the consistency of each of its finite subtheories. The Hájek-Švejdar self-reference makes heavy use of this. Such results form a natural limit to the scope of this monograph. By its similarity to the ubiquitous fixed point, we can still profitably discuss the Hájek-Švejdar fixed point; to go any further would involve us in seemingly arcane considerations of the metamathematics of Peano Arithmetic.

1. AN UBIQUITOUS FIXED POINT CALCULATION

The modal system CML has the following language:

PROPOSITIONAL VARIABLES: $p, q, r, \ldots$

Σ-VARIABLES: $\sigma, \tau, \ldots$

TRUTH VALUES: t, f

PROPOSITIONAL CONNECTIVES: $\sim, \wedge, \vee, \rightarrow$

MODAL OPERATORS: $\Box, \Delta, \nabla$

WITNESS COMPARISONS: $\preccurlyeq, \prec$.

The Σ-variables are new and are intended to be interpreted by Σ_1-sentences. There are now three modal operators-- the usual box and *two* new ones to give added generality in the form of a fixed point

$$p \leftrightarrow .(\Delta{\sim}p \vee \sigma) \preccurlyeq (\nabla p \vee \tau). \qquad (*)$$

The rules of sentence formation are as one might expect-- $\preccurlyeq, \prec$ apply only to pairs of Σ-sentences; but the class of Σ-sentences is larger than before:

1.1. DEFINITION. The class of Σ-sentences is inductively generated as follows:

i. σ, τ are Σ-sentences

ii. for any A, $\Box A$, ΔA, ∇A are Σ-sentences

iii. if A,B are Σ-sentences, so are $A \circ B$ for $\circ \in \{\wedge, \vee, \preccurlyeq, \prec\}$.

I remark that we don't need to nest witness comparisons unless one is embedded in a modal operator, i.e. $A \preccurlyeq (B \prec C)$ need not be allowed in the language, although we must include $A \preccurlyeq \Box(B \prec C)$. The closure under disjunction is necessary for (*). (Actually, as remarked in the last Chapter, we could avoid this by circumlocution.)

The axiomatisation of CML is a delicate matter. Like SR, CML is not given with a single interpretation in mind; its goal is broad applicability rather than analysis of a single notion. The less it assumes, the wider will be its range of applicability.

Moreover, not all applications require the *provability* of all assertions, but need only the truth of some of these: That something is unprovable assuming the consistency of the given theory is the obvious example. Another example concerns the connective $\triangledown$ of MOS: The axiom schema

$$\Box A \rightarrow \triangledown A,$$

assumes the *uniform provability* in PRA that the theories $T_0, T_1, \ldots$ interpreting $\triangledown$ extend PRA. When we add the witness comparisons and produce the Mostowski sentence,

$$\phi \leftrightarrow .\bigvee Pr_{T_i}(\ulcorner \sim\phi \urcorner) \preccurlyeq \bigvee Pr_{T_i}(\ulcorner \phi \urcorner),$$

its independence over each T_i requires only the *truth* of all the consistencies and the *truth* that these theories are extensions of PRA. A *formalisation* of this independence result, say the implication,

$$\forall i Con(T_i) \rightarrow \forall i(\sim Pr_{T_i}(\ulcorner \phi \urcorner) \wedge \sim Pr_{T_i}(\ulcorner \sim\phi \urcorner)),$$

requires the uniform proof that each T_i extends PRA.

Another source of generality is this: If we do not require the formalisation of the result to be proven, we will not need any induction within the theory. For example, in its unformalised form, Rosser's Theorem holds for (all *RE* extensions of) much weaker theories than PRA: Over such theories one can find independent sentences. The formal assertion, that consistency implies this independence, cannot generally be proven in the weaker theories and PRA is the natural theory for such formalisations.

The point is, thus, that we have to make a choice: We are not striving for a complete axiomatisation of a single set of operators over a single theory; we are striving for an axiomatic understanding of a schema of self-reference applying in a variety of circumstances. We can aim for the greatest possible applicability or reasonable familiarity. In the present case, the price of doing the former is not any complication of the proof, but rather the acceptance of an axiom system from which many familiar axioms are dropped. The real cost here is to me-- I have to offer some rationale for accepting and remembering the reduced axiom system. Burdensome as is this task, with truly altruistic motives, I shall undertake it and aim for the generality.

Incompleteness theorems, like Rosser's or Mostowski's generalisation thereof, are generally stated for all *RE* extensions of a weaker theory than PRA. The system

R of Raphael Robinson (not to be confused with the R of Chapter 6) and the stronger, but finitely axiomatised system Q of Robinson are the two most common examples of such weak theories. Although R is weaker than Q and its mention yields more general results, Q is finitely axiomatised and, hence, more popular. We shall thus choose Q as our standard in the sequel.

The exact details of Q do not concern us; the reader not already familiar with Q can look it up in the monograph of Tarski, Mostowski, and Robinson. The crucial thing about Q, for our purposes, is that it is strong enough to prove all true Σ_1-sentences but not strong enough to reflect on this fact, and it can prove a bare minimum about the ordering. Thus, if we interpret $\Box$ by Pr_Q (actually, by $Pr_Q(\ulcorner \iota(\phi) \urcorner)$, where $\iota(\phi)$ is the more-or-less canonical interpretation (or: translation) of ϕ in the restricted language of Q (which, being finitely axiomatised, must have a finite language if it is to be Σ_1-complete)), the choice of axioms and omissions of the following theory CML makes some sense.

1.2. DEFINITION. The system CML in the language described has the following axiom schemata and rule of inference:

AXIOMS: *A1*. All tautologies

A2. Necessitations of all tautologies

A3. $\Box A \wedge \Box(A \to B) \to \Box B$

A4. $A \to \Box A$, for $A \in \Sigma$

A5. Order axioms, as in 6.1.2

A6. $\Box(A \preccurlyeq B \to \sim(B \prec A))$, for $A,B \in \Sigma$

A7. $\Box A \to \Delta A$, $\Box A \to \nabla A$

RULE: *R1*. $A,\ A \to B\ /\ B$.

Once again, if we think of $\Box$ as denoting provability in Q and we think of what PRA can prove about Q, the axioms are fairly self-explanatory: *A1*, *A5*, and *R1* require no explanation; *A2*, *A3* are built into Pr_T regardless of the choice of non-logical axioms; *A6* follows from the provability in Q of the formula

$$v_0 \leq v_1 \to \sim(v_1 < v_0);$$

and *A4* is PRA's recognition that Q is strong enough to prove all true Σ_1-sentences.

A7 requires a bit more comment: For full generality, we want to interpret Δ and ∇ by formulae stronger than, but not necessarily provably stronger than Pr_Q. Because of the finite axiomatisability of Q, *A7* will be provable when we interpret $\Box$ by Pr_Q and Δ,∇ by, say, Pr_T, $Relint_{GB}$, or even $\bigvee Pr_{T_i}$ for a *finite* sequence $T_0,\ldots,T_{n-1}$ of consistent *RE* extensions of Q; *A7* need not be provable when one of Δ,∇ is interpreted by $\bigvee Pr_{T_i}$ for an infinite *RE* sequence of consistent extensions of Q. Thus, sometimes *A7* will translate into a provable schema and sometimes it will be cited, along with consistency, as a true, but unprovable hypothesis.

As for what is missing, the glaring omissions are the Formalised Löb's Theorem, which may be false for weak theories like Q, and *R2*, which is certainly false under the present interpretation:

$$\mathrm{PRA} \vdash \phi \quad \not\Rightarrow \quad \mathrm{PRA} \vdash Pr_Q(\ulcorner \phi \urcorner)$$

The reader might also note that *A6* contains only the necessitation of one order axiom. Some of the others are false, e.g. $\Box(A \rightarrow .A \preccurlyeq A)$ is false because Q does not prove the induction implicit in $A \rightarrow .A \preccurlyeq A$. Of the other order axioms, I confess not to have bothered checking which ones are and are not derivable in Q; only *A6* is actually used in practice.

Augmented by a generalised consistency assumption, CML will suffice to prove the main Theorem of this section. Some generalisations in the Exercises require an additional axiom schema. The next two definitions list this and some consistency schemata.

1.3. DEFINITION. Axiom schema *A8* is the schema

A8. $\Box(A \rightarrow B) \wedge \Delta A \rightarrow \Delta B, \quad \Box(A \rightarrow B) \wedge \nabla A \rightarrow \nabla B.$

1.4. DEFINITIONS. The following are generalisations of consistency:

Simple Consistency. $\sim\Box f$

Extra Consistency. $\sim\Delta f, \quad \sim\nabla f$

Super Consistency. $\Box A \rightarrow \sim\Delta\sim A, \quad \Box A \rightarrow \sim\nabla\sim A$

Compatibility. $\Delta A \rightarrow \sim\nabla\sim A$.

These schemata are more-or-less self-explanatory. The various implications holding among the consistency schemata are explored in Exercise 1, below. In the absence of *A8*, super consistency is the key notion. (In fact, in a non-modal guise it has also turned up under the name mono-consistency in the work of Per Lindström.)

We already know from the last Chapter that we cannot substitute equivalents inside witness comparisons. We also saw that, to prove Rosser's Theorem, we hardly needed to make such substitutions. We barely need to do so now either, but it will still be worth mentioning the following:

1.5. LEMMA. For all sentences A, B,

i. $\mathsf{CML} \vdash \Box(A \leftrightarrow B) \rightarrow .\Box A \leftrightarrow \Box B$

ii. $\mathsf{CML} + A8 \vdash \Box(A \leftrightarrow B) \rightarrow .\nabla A \leftrightarrow \nabla B$

iii. $\mathsf{CML} + A8 \vdash \Box(A \leftrightarrow B) \rightarrow .\Delta A \leftrightarrow \Delta B.$

I leave the proof as an exercise to the reader (cf. Exercise 2, below).

A more pertinent syntactic matter concerns the role of $\prec$ as a strong negation of $\preccurlyeq$. A consequence of Demonstrable Σ_1-Completeness and, say, Rosser's Theorem is the fact that the negation of a Σ_1-sentence, like $\phi \preccurlyeq \psi$, need not be Σ_1. The assertion $\psi \prec \phi$ is, however, a good candidate for a *strong* Σ_1-denial of $\phi \preccurlyeq \psi$. Our next Lemma isolates a few useful properties of this strong negation:

1.6. LEMMA. For Σ-sentences A, B,

i. $B \prec A$ is a Σ-sentence

ii. $\mathsf{CML} \vdash B \prec A \rightarrow \sim(A \preccurlyeq B)$

iii. $\mathsf{CML} \vdash A \vee B \rightarrow \big(B \prec A. \leftrightarrow .\sim(A \preccurlyeq B)\big)$

iv. $\mathsf{CML} + B \prec A \vdash \Box\sim(A \preccurlyeq B).$

Assertions i-iii are fairly obvious; iv follows from *A4* and *A6* of Definition 1.2.

We are now in position to state and prove the main Theorem of this section:

1.7. UBIQUITOUS FIXED POINT CALCULATION. Let CML^+ be CML augmented by simple consistency and super consistency. Then:

$$\mathsf{CML}^+ + \boxed{s}\big(p \leftrightarrow .(\Delta\sim p \vee \sigma) \preccurlyeq (\nabla p \vee \tau)\big) \vdash$$
$$\vdash (p \leftrightarrow \Box p \leftrightarrow \nabla p \leftrightarrow \sigma \preccurlyeq \tau) \wedge (\Box\sim p \leftrightarrow \Delta\sim p \leftrightarrow \tau \prec \sigma).$$

(The chains of equivalences are taken pairwise, e.g. $p \leftrightarrow \Box p$ and $\Box p \leftrightarrow \nabla p$, etc.)

Proof: Let

A: $\Delta\sim p \vee \sigma. \preccurlyeq .\nabla p \vee \tau$

B: $\nabla p \vee \tau . \prec . \Delta{\sim}p \vee \sigma$.

Then: A,B are strong negations of each other and $\boxed{s}(p \leftrightarrow A)$ is assumed.

Letting "$\vdash$" denote provability in $CML^{+} + \boxed{s}(p \leftrightarrow A)$, note

$A \vdash \Box A$, by $A4$ | $B \vdash \Box B \vdash \Box{\sim}A$, by $A4$, $A6$

$\vdash \Box p$ (a1) | $\vdash \Box{\sim}p$ (b1)

$\vdash {\sim}\Delta{\sim}p$ (a2) | $\vdash {\sim}\nabla p$ (b2)

$\vdash \sigma \preccurlyeq \tau$ (a3); | $\vdash \tau \prec \sigma$ (b3).

Consequences (a1) and (b1) follow from Lemma 1.5.i; (a2) and (b2) from (a1) and (b1) via super consistency; and (a3) and (b3) by some calculations involving $\preccurlyeq$ (specifically, the order axioms $A5$): For example,

$$A \wedge {\sim}\Delta{\sim}p \vdash (\Delta{\sim}p \vee \sigma . \preccurlyeq . \nabla p \vee \tau) \wedge {\sim}\Delta{\sim}p$$
$$\vdash \sigma \preccurlyeq . \nabla p \vee \tau$$
$$\vdash \sigma \preccurlyeq \tau.$$

The proof now consists of a number of trivial applications of (a1)-(a3) and (b1)-(b3) to establish the near infinity of equivalences of the conclusion.

$p \rightarrow \Box p$: For, $p \vdash A$

$\vdash \Box p$, by (a1).

$\Box p \rightarrow \nabla p$: This is just $A7$.

$\nabla p \rightarrow . \sigma \preccurlyeq \tau$: For, $\nabla p \vdash A \vee B$

$\vdash A$, by (b2)

$\vdash \sigma \preccurlyeq \tau$, by (a3).

$\sigma \preccurlyeq \tau . \rightarrow p$: For, $\sigma \preccurlyeq \tau \vdash A \vee B$

$\vdash A$, by (b3)

$\vdash p$.

$\Box{\sim}p \rightarrow \Delta{\sim}p$: This is just $A7$.

$\Delta{\sim}p \rightarrow . \tau \prec \sigma$: For, $\Delta{\sim}p \vdash A \vee B$

$\vdash B$, by (a2)

$\vdash \tau \prec \sigma$, by (b3).

$\tau \prec \sigma . \rightarrow \Box{\sim}p$: For, $\tau \prec \sigma \vdash A \vee B$

$\vdash B$, by (a3)

$$\vdash \Box{\sim}p, \quad \text{by (b2).} \qquad \text{QED}$$

Well, that's it. The Theorem and its proof are not nearly as impressive as the introduction to it might make one believe it should be. However, the unification it offers to many disparate applications of self-reference is (if I may say so about my own result) impressive and I suggest the reader look into a few of these applications in the next section to get some motivation for doing the following Exercises.

EXERCISES

1. i. Show:

   a. CML + *Extra Consistency* $\vdash$ *Simple Consistency*

   b. CML + *Super Consistency* $\vdash$ *Extra Consistency*

   c. CML + *Compatibility* $\vdash$ *Super Consistency*

   d CML + *A8* + *Extra Consistency* $\vdash$ *Super Consistency*

   ii. Even for $\Delta = \nabla$, extra consistency does not imply compatibility without a stronger form of *A8*. Use one of the interpretations of Chapter 4 to show this.

2. Prove Lemma 1.5. Show that 1.5.ii and 1.5.iii cannot be proven in CML by constructing suitable arithmetic interpretations of Δ and ∇.

3. Prove Lemma 1.6.

4. Let CML^+ be CML + *A8* + *Compatibility*.

   i. Show: $\mathrm{CML}^+ + \boxed{s}\big(p \leftrightarrow .C \vee D \wedge (\Delta{\sim}p \vee \sigma. \preccurlyeq .\nabla p \vee \tau)\big) \vdash$
   $\vdash \big(\nabla p \leftrightarrow .\nabla(C \vee D) \wedge (\sigma \preccurlyeq \tau) \vee \nabla C\big) \wedge$
   $\wedge \big(\Delta{\sim}p \leftrightarrow .\Delta({\sim}C \wedge {\sim}D) \vee \Delta{\sim}C \wedge (\tau \prec \sigma)\big)$

   ii. Show: $\mathrm{CML}^+ + \boxed{s}\big(p \leftrightarrow \bigvee_I (C_i \wedge (\Delta{\sim}p \vee \sigma_i. \preccurlyeq .\nabla p \vee \tau_i))\big) \vdash$
   $\vdash \big(\nabla p \leftrightarrow \bigvee_{\emptyset \neq J \subseteq I} (\nabla \bigvee_J C_j \wedge \bigwedge_J (\sigma_j \preccurlyeq \tau_j))\big) \wedge$
   $\wedge \big(\Delta{\sim}p \leftrightarrow \bigwedge_{\emptyset \neq J \subseteq I} (\Delta \bigwedge_J {\sim}C_j \vee \bigvee_J (\tau_j \prec \sigma_j))\big).$

   iii. Show: $\mathrm{CML}^+ + \boxed{s}\big(p \leftrightarrow \bigwedge_I (C_i \vee (\Delta{\sim}p \vee \sigma_i. \preccurlyeq .\nabla p \vee \tau_i))\big) \vdash$
   $\vdash \big(\nabla p \leftrightarrow \bigwedge_{\emptyset \neq J \subseteq I} (\nabla \bigwedge_J C_j \vee \bigvee_J (\sigma_j \preccurlyeq \tau_j))\big) \wedge$
   $\wedge \big(\Delta{\sim}p \leftrightarrow \bigvee_{\emptyset \neq J \subseteq I} (\Delta \bigvee_J {\sim}C_j \wedge \bigwedge_J (\tau_j \prec \sigma_j))\big).$

5. Under the assumptions of Exercise 4,

   i. Show: $\mathrm{CML}^+ + \boxed{s}\big(p \leftrightarrow \bigvee_I (C_i \wedge (\Box{\sim}p \vee \sigma_i. \preccurlyeq .\Box p \vee \tau_i))\big) \vdash$

$$\vdash \Box\sim p \leftrightarrow \bigwedge_I (\Box\sim C_i \vee (\tau_i \prec \sigma_i))$$

ii. Show: $\mathsf{CML}^+ + \boxed{s}\,(p \leftrightarrow \bigvee_I (\Delta\sim p \vee \sigma_i . \preccurlyeq . \nabla p \vee \tau_i)) \vdash$

$$\vdash (p \leftrightarrow \Box p \leftrightarrow \nabla p \leftrightarrow \bigvee_I (\sigma_i \preccurlyeq \tau_i)) \wedge$$

$$\wedge (\Box\sim p \leftrightarrow \Delta\sim p \leftrightarrow \bigwedge_I (\tau_i \prec \sigma_i)).$$

6. What does the theory CML^+ of Theorem 1.7 say about p,q if we assume

$$\boxed{s}\,(p \leftrightarrow . (\Delta q \vee \sigma) \preccurlyeq (\nabla p \vee \tau))$$

$$\boxed{s}\,(q \leftrightarrow . (\nabla p \vee \tau) \prec (\Delta q \vee \sigma))?$$

2. APPLICATIONS

Applications of the Ubiquitous Fixed Point Calculation 1.7 assume the following form: We choose Σ_1-formulae ρ_1,ρ_2 to interpret Δ,∇ respectively; we choose some theory T like PRA or Q, the provability predicate of which is to interpret $\Box$; and we choose Σ_1-sentences ψ,χ to interpret σ,τ. Then, for ϕ satisfying

$$T \vdash \phi \leftrightarrow . (\rho_1(\ulcorner\sim\phi\urcorner) \vee \psi) \preccurlyeq (\rho_2(\ulcorner\phi\urcorner) \vee \chi),$$

we conclude the *truth* of

$$\phi \leftrightarrow Pr_T(\ulcorner\phi\urcorner) \qquad\qquad Pr_T(\ulcorner\sim\phi\urcorner) \leftrightarrow \rho_1(\ulcorner\sim\phi\urcorner)$$

$$\leftrightarrow \rho_2(\ulcorner\phi\urcorner) \qquad\qquad \leftrightarrow . \chi \prec \psi.$$

$$\leftrightarrow . \psi \preccurlyeq \chi;$$

Because of the consistency assumptions of 1.7, we cannot conclude the provability of any of these equivalences in PRA, but rather in PRA plus some consistency assumption (which may at first sight appear weaker than those of 1.7) *and*, if *A7* turns out *not* to be provable, the assumption *A7*.

2.1. APPLICATION. (Rosser's Theorem). Let T be a consistent *RE* extension of Q. Suppose

$$Q \vdash \phi \leftrightarrow . Pr_T(\ulcorner\sim\phi\urcorner) \preccurlyeq Pr_T(\ulcorner\phi\urcorner). \qquad (*)$$

Then: i. $T \nvdash \phi,\sim\phi$

ii. $\mathsf{PRA} \vdash Con_T \rightarrow \sim Pr_T(\ulcorner\phi\urcorner)$

iii. $\mathsf{PRA} \vdash Con_T \rightarrow \sim Pr_T(\ulcorner\sim\phi\urcorner)$.

(I remark that the Diagonalisation Theorem depends only on the Σ_1-Completeness of a theory and is provable in Q, whence (*) is not a vacuous assumption.)

Proof: Interpret both Δ and ∇ by Pr_T and $\Box$ by Pr_Q. The Rosser sentence (*) is not quite exactly the form of an interpretation of the fixed point of 1.7 because the disjuncts corresponding to σ,τ are missing. This is no problem: Missing σ's and τ's can be taken to be interpreted by the false, refutable $\exists v(v \neq v)$ by default. Observe: From (*) and the order properties of $\preccurlyeq$, $\prec$,

$$Q \vdash \phi \leftrightarrow . (Pr_T(\ulcorner \sim\phi \urcorner) \vee \exists v(v \neq v)) \preccurlyeq (Pr_T(\ulcorner \phi \urcorner) \vee \exists v(v \neq v)).$$

To see that Theorem 1.7 now applies, notice that all the axioms of CML are provable and, for the given interpretation of Δ,∇, Con_T and $A7$ entail super consistency, i.e. the schema,

$$Pr_Q(\ulcorner \theta \urcorner) \rightarrow \sim Pr_T(\ulcorner \sim\theta \urcorner).$$

i. By 1.7, we conclude the truth of

$$Pr_T(\ulcorner \phi \urcorner) \leftrightarrow . \exists v(v \neq v) \preccurlyeq \exists v(v \neq v)$$

$$Pr_T(\ulcorner \sim\phi \urcorner) \leftrightarrow . \exists v(v \neq v) \prec \exists v(v \neq v),$$

i.e. the falsity of $Pr_T(\ulcorner \phi \urcorner)$, $Pr_T(\ulcorner \sim\phi \urcorner)$.

ii-iii. Since PRA proves all schemata of CML and since the only assumption needed to guarantee the additional superconsistency of CML^+ is Con_T, we get

$$PRA + Con_T \vdash Pr_T(\ulcorner \phi \urcorner) \leftrightarrow . \exists v(v \neq v) \preccurlyeq \exists v(v \neq v)$$

$$\vdash \sim Pr_T(\ulcorner \phi \urcorner)$$

$$PRA + Con_T \vdash Pr_T(\ulcorner \sim\phi \urcorner) \leftrightarrow . \exists v(v \neq v) \prec \exists v(v \neq v)$$

$$\vdash \sim Pr_T(\ulcorner \sim\phi \urcorner).$$

QED

It may be worth remarking that, in this example, we also have

$$PRA + Con_T \vdash \phi \leftrightarrow Pr_T(\ulcorner \phi \urcorner)$$

$$\vdash \sim\phi, \text{ by part iii.}$$

The converse, $\sim\phi \rightarrow Con_T$, is not provable in PRA (if T is sufficiently strong), as follows from the Second Incompleteness Theorem:

$$PRA + Con_T \nvdash Con(T + Con_T)$$

$$\nvdash \sim Pr_T(\ulcorner \sim Con_T \urcorner)$$

$$\nvdash \sim Pr_T(\ulcorner \phi \urcorner), \text{ if } PRA \vdash \sim\phi \leftrightarrow Con_T.$$

The year 1960 saw two generalisations of Rosser's Theorem of major importance. Both have been widely ignored. The first of these that we shall discuss is Mostowski's Theorem and was the source of inspiration for the research behind Chapter 4.

2.2. APPLICATION. (Mostowski's Theorem). Let $T_0, T_1, \ldots$ be an *RE* sequence of consistent extensions of Q. Define ϕ by

$$Q \vdash \phi \leftrightarrow . \bigvee Pr_{T_i}(\ulcorner \sim\phi \urcorner) \preccurlyeq \bigvee Pr_{T_i}(\ulcorner \phi \urcorner).$$

Then: i. $\forall i(T_i \nvdash \phi, \sim\phi)$

ii. $PRA + \forall i Con(T_i) + \forall x i(Pr_Q(x) \rightarrow Pr_{T_i}(x)) \vdash$

$$\vdash \sim \bigvee Pr_{T_i}(\ulcorner \phi \urcorner) \wedge \sim \bigvee Pr_{T_i}(\ulcorner \sim\phi \urcorner).$$

Proof: Again, interpret Δ and ∇ by $\bigvee Pr_{T_i}$ (assumed written with only one unbounded existential quantifier), σ, τ by $\exists v(v \neq v)$, and p by ϕ. All axiom schemata of CML other than *A7* become provable under such an interpretation; *A7* is true by the assumption that all the T_i's extend Q; and, using *A7*, super consistency follows readily from $\forall i Con(T_i)$. Thus,

$$\bigvee Pr_{T_i}(\ulcorner \phi \urcorner) \leftrightarrow . \exists v(v \neq v) \preccurlyeq \exists v(v \neq v) \quad (1)$$

$$\bigvee Pr_{T_i}(\ulcorner \sim\phi \urcorner) \leftrightarrow . \exists v(v \neq v) \prec \exists v(v \neq v) \quad (2)$$

are true, whence

$$\bigvee Pr_{T_i}(\ulcorner \phi \urcorner), \quad \bigvee Pr_{T_i}(\ulcorner \sim\phi \urcorner)$$

are false and ϕ is not decided by any T_i.

Assertion ii follows directly: Equivalences (1) and (2) are provable once one adds the assumptions $\forall i Con(T_i)$ and $\forall x i(Pr_Q(x) \rightarrow Pr_{T_i}(x))$ to PRA. QED

It was Shepherdson who added the side formulae σ, τ and thereby raised the level of application of the self-referential formulae from mere independence to a higher plateau, as we shall shortly see. First, the simplest version of his Theorem:

2.3. APPLICATION. (Shepherdson's Theorem, I). Let T be a consistent *RE* extension of Q and let ψ, χ be Σ_1-sentences. Define ϕ by

$$Q \vdash \phi \leftrightarrow . (Pr_T(\ulcorner \sim\phi \urcorner) \vee \psi) \preccurlyeq (Pr_T(\ulcorner \phi \urcorner) \vee \chi).$$

Then: i. $T \vdash \phi$ iff $Q \vdash \phi$

iff $\psi \preccurlyeq \chi$ is true

iff ϕ is true

ii. $T \vdash \sim\phi$ iff $Q \vdash \sim\phi$

iff $\chi \prec \psi$ is true

iii. $PRA + Con_T \vdash Pr_T(\ulcorner \phi \urcorner) \leftrightarrow . \psi \preccurlyeq \chi$

$$\vdash Pr_T(\ulcorner \sim\phi \urcorner) \leftrightarrow .\chi \prec \psi$$

iv. $\mathrm{PRA} + Con_T \vdash \phi \leftrightarrow .\psi \preccurlyeq \chi$.

The only point to the proof worth mentioning is that, by interpreting both Δ and ∇ as Pr_T and assuming T to extend Q, super consistency follows from the simple consistency of T.

Shepherdson's Theorem is a generalisation of Rosser's Theorem and instantly yields the latter as an application. Somewhat more interesting is the following corollary due, originally under a different proof, to Peter Päppinghaus:

<u>2.4. COROLLARY</u>. Let T_0, T_1 be consistent *RE* extensions of Q. Let

$$Q \vdash \phi \leftrightarrow .(Pr_{T_0}(\ulcorner \sim\phi \urcorner) \vee Pr_{T_1}(\ulcorner \sim Pr_{T_0}(\ulcorner \phi \urcorner)\urcorner)) \preccurlyeq$$
$$\preccurlyeq (Pr_{T_0}(\ulcorner \phi \urcorner) \vee Pr_{T_1}(\ulcorner \sim Pr_{T_0}(\ulcorner \sim\phi \urcorner)\urcorner)).$$

Then: i. $T_0 \nvdash \phi, \sim\phi$

ii. $T_1 \nvdash \sim Pr_{T_0}(\ulcorner \phi \urcorner), \quad \sim Pr_{T_0}(\ulcorner \sim\phi \urcorner)$.

Proof: By Shepherdson's Theorem. Here,

ψ: $Pr_{T_1}(\ulcorner \sim Pr_{T_0}(\ulcorner \phi \urcorner)\urcorner)$

χ: $Pr_{T_1}(\ulcorner \sim Pr_{T_0}(\ulcorner \sim\phi \urcorner)\urcorner)$,

and 2.3.i yields

$$T_0 \vdash \phi \quad \Rightarrow \quad \psi \preccurlyeq \chi \text{ is true}$$
$$\Rightarrow \quad \psi \text{ is true} \qquad (1)$$
$$\Rightarrow \quad T_1 \vdash \sim Pr_{T_0}(\ulcorner \phi \urcorner) \qquad (2)$$

where (1) follows by an order axiom. But T_1 contains Q, whence it proves all true Σ_1-sentences and

$$T_0 \vdash \phi \quad \Rightarrow \quad T_1 \vdash Pr_{T_0}(\ulcorner \phi \urcorner),$$

contrary to (2) and the consistency of T_1. Thus, $T_0 \nvdash \phi$. A similar argument shows $T_1 \nvdash \sim\phi$.

From $T_1 \vdash \sim Pr_{T_0}(\ulcorner \phi \urcorner)$ or $T_1 \vdash \sim Pr_{T_0}(\ulcorner \sim\phi \urcorner)$, we get the truth of $\psi \vee \chi$, whence that of $\psi \preccurlyeq \chi. \vee .\chi \prec \psi$. But the truth of this disjunction entails the truth of one of the disjuncts, which, by 2.3.i-ii, entails the decidability of ϕ in T_0:

$$T_0 \vdash \phi \quad \text{or} \quad T_1 \vdash \sim\phi,$$

contrary to what has already been proven. Hence,

$$T_1 \nvdash \sim Pr_{T_0}(\ulcorner\phi\urcorner) \quad \text{and} \quad T_1 \nvdash \sim Pr_{T_0}(\ulcorner\sim\phi\urcorner). \qquad \text{QED}$$

I note, with respect to this Corollary, that it is a result we would expect from looking at Kripke models for PRL_1-- interpreting $\Box$ as Pr_{T_0} and Δ as Pr_{T_1} (assuming $T_0 \subseteq T_1$). However, in the arithmetic completeness result obtained we could fix T_0, but had to vary T_1. The exception was for T_1 sufficiently strong relative to T_0, so that PRL_{ZF} axiomatised completely the schemata provable in T_0 (or PRA) about $\Box$ and Δ and the Kripke models sufficed for the proof (cf. Exercise 4 of Chapter 4, section 4).

Corollary 2.4 is interesting, but it is still an independence result and I promised to ascend to a higher plateau:

<u>2.5. COROLLARY</u>. Let T be a consistent *RE* extension of PRA. The following are equivalent:

i. (Σ_1-Disjunction Property). For all Σ_1-sentences θ_1, θ_2,

$$T \vdash \theta_1 \vee \theta_2 \quad \Rightarrow \quad T \vdash \theta_1 \quad \text{or} \quad T \vdash \theta_2$$

ii. (Σ_1-Soundness). For all Σ_1-sentences θ,

$$T \vdash \theta \quad \Rightarrow \quad \theta \text{ is true.}$$

Proof: ii $\Rightarrow$ i is an immediate corollary of the provability in T of all true Σ_1-sentences.

i $\Rightarrow$ ii. We argue by contraposition. Suppose ψ is a false, but provable Σ_1-sentence. Choose ϕ such that

$$Q \vdash \phi \leftrightarrow .(Pr_T(\ulcorner\sim\phi\urcorner) \vee \psi) \preccurlyeq Pr_T(\ulcorner\phi\urcorner).$$

Let θ_1, θ_2 be

$$\theta_1\colon \ (Pr_T(\ulcorner\sim\phi\urcorner) \vee \psi) \preccurlyeq Pr_T(\ulcorner\phi\urcorner)$$

$$\theta_2\colon \ Pr_T(\ulcorner\phi\urcorner) \prec (Pr_T(\ulcorner\sim\phi\urcorner) \vee \psi).$$

Obviously,

$$PRA \vdash \psi \rightarrow \theta_1 \vee \theta_2, \qquad (*)$$

whence $\quad T \vdash \theta_1 \vee \theta_2$.

But $T \nvdash \theta_1$ since ψ is false and $T \nvdash \theta_2$ since χ is vacant, whence equivalent to the false $\exists v(v \neq v)$. QED

Note the use (*) of induction, whence the restriction to *RE* theories T extending PRA rather than Q.

Corollary 2.5 is due independently to Harvey Friedman, Guaspari, and Don Jensen and Andrzej Ehrenfeucht. Our next corollary is due to Leo Harrington and Friedman, and uses a bit more of Shepherdson's Theorem. Till now we have only used 2.3.i-ii; the Harrington-Friedman result makes use of 2.3.iii.

2.6. COROLLARY. Let T be a consistent *RE* extension of PRA. Let $\theta \in \Pi_1$. There are $\theta_1 \in \Sigma_1$ and $\theta_2 \in \Pi_1$ such that

i. $PRA + Con_T \vdash \theta \leftrightarrow Con(T + \theta_1)$

ii. $PRA + Con_T \vdash \theta \leftrightarrow Con(T + \theta_2)$.

Proof: i. Choose $\theta_1 = \phi$ such that

$$PRA \vdash \phi \leftrightarrow .Pr_T(\ulcorner \sim\phi \urcorner) \preccurlyeq (Pr_T(\ulcorner \phi \urcorner) \vee \sim\theta),$$

$\sim\theta$ assumed written in Σ_1-form. By 2.3.iii,

$$PRA + Con_T \vdash Pr_T(\ulcorner \sim\phi \urcorner) \leftrightarrow \sim\theta$$

$$\vdash Con(T + \phi) \leftrightarrow \theta.$$

ii. Choose $\theta_2 = \sim\phi$, where

$$PRA \vdash \phi \leftrightarrow .(Pr_T(\ulcorner \sim\phi \urcorner) \vee \sim\theta) \preccurlyeq Pr_T(\ulcorner \phi \urcorner).$$

QED

By this Corollary, every Π_1-sentence is, modulo Con_T, provably equivalent to a consistency statement.

For an application that uses 2.3.iv, we first need a definition:

2.7. DEFINITION. Let T be a given theory, Γ a set of sentences, and θ a sentence. θ is Γ-*conservative over* T (or: Γ-*con over* T) if, for all $\gamma \in \Gamma$,

$$T + \theta \vdash \gamma \Rightarrow T \vdash \gamma.$$

2.8. EXAMPLE. (Kreisel). If T is a consistent *RE* extension of PRA, $\sim Con_T$ is Π_1-con over T.

Proof: We have come close to proving this before with Corollary 0.6.24 (or: Exercise 2.1.i of Chapter 3, section 3): $T \vdash \pi \Rightarrow PRA + Con_T \vdash \pi$, for any Π_1-sentence π. We need only go one step farther: Let π be a Π_1-sentence.

$$T + \sim Con_T \vdash \pi \Rightarrow T + Con(T + \sim Con_T) \vdash \pi, \quad \text{by 0.6.24}$$

$$\Rightarrow \quad T + Con_T \vdash \pi, \qquad (*)$$

by the Formalised Second Incompleteness Theorem: $Con_T \rightarrow Con(T + \sim Con_T)$. But, with (*), we get

$$T + \sim Con_T \vdash \pi \quad \Rightarrow \quad T + \sim Con_T \vee Con_T \vdash \pi$$
$$\Rightarrow \quad T \vdash \pi.$$

QED

As the attentive reader has noticed, Example 2.8 did not depend on Shepherdson's Theorem. The next example, due partly to Guaspari and partly to the author, is the application I had in mind.

2.9. COROLLARY. Let T be a consistent *RE* extension of PRA. The following are equivalent:

i. Con_T is Σ_1-con over T

ii. every consistent Π_1-sentence is Σ_1-con over T

iii. T is Σ_1-sound, i.e. $T \vdash \sim\pi \;\Rightarrow\; \sim\pi$ is true, for $\pi \in \Pi_1$.

Proof: iii $\Rightarrow$ ii $\Rightarrow$ i are easy exercises for the reader.

i $\Rightarrow$ iii. By contraposition. Let ψ be a false Σ_1-sentence provable in T and let ϕ satisfy

$$\mathrm{PRA} \vdash \phi \leftrightarrow .(Pr_T(\ulcorner \sim\phi \urcorner) \vee \psi) \preccurlyeq Pr_T(\ulcorner \phi \urcorner)$$

I claim that

a. $T \nvdash \phi$

b. $T + Con_T \vdash \phi$,

whence Con_T is not Σ_1-con over T. Assertion a follows directly by 2.3.i and the falsity of ψ. But 2.3.iv yields b:

$$T + Con_T \vdash \phi \leftrightarrow \psi$$
$$\vdash \phi, \quad \text{since } T \vdash \psi.$$

QED

These are enough applications of the simple Shepherdson Theorem for now. Let us look at the more sophisticated version of Shepherdson's Theorem:

2.10. APPLICATION. Let T be a consistent *RE* extension of Q and let $\psi v, \chi v$ be Σ_1-formulae with only v free. Choose ϕv so that

$$Q \vdash \phi v \leftrightarrow .(Pr_T(\ulcorner \sim\phi\dot{v} \urcorner) \vee \psi v) \preccurlyeq (Pr_T(\ulcorner \phi\dot{v} \urcorner) \vee \chi v).$$

Then: For any $x \in \omega$,

i. $T \vdash \phi\bar{x}$ iff $Q \vdash \phi\bar{x}$

iff $\psi\bar{x} \preccurlyeq \chi\bar{x}$ is true

iff $\phi\bar{x}$ is true

ii. $T \vdash \sim\phi\bar{x}$ iff $Q \vdash \sim\phi\bar{x}$

iff $\chi\bar{x} \prec \psi\bar{x}$ is true

iii. $PRA + Con_T \vdash Pr_T(\ulcorner\phi\bar{x}\urcorner) \leftrightarrow .\psi\bar{x} \preccurlyeq \chi\bar{x}$

$\vdash Pr_T(\ulcorner\sim\phi\bar{x}\urcorner) \leftrightarrow .\chi\bar{x} \prec \psi\bar{x}$

iv. $PRA + Con_T \vdash \phi\bar{x} \leftrightarrow .\psi\bar{x} \preccurlyeq \chi\bar{x}.$

This follows immediately from the simpler version 2.3 by observing that, for each $x \in \omega$, $\phi\bar{x},\psi\bar{x},\chi\bar{x}$ can replace ϕ,ψ,χ, respectively, in 2.3. To explain why 2.10 is so interesting, we need a few definitions.

2.11. DEFINITION. A set $X \subseteq \omega$ is an *RE set* if there is a Σ_1-formula ϕv such that

$$X = \{x \in \omega: \phi\bar{x} \text{ is true}\}.$$

2.12. DEFINITIONS. Let T be a theory in the language of arithmetic and let $X \subseteq \omega$. A formula ϕv *semi-represents* X *in* T, if, for all $x \in \omega$,

$x \in X$ iff $T \vdash \phi\bar{x}$;

ϕ *correctly semi-represents* X *in* T if one also has, for all $x \in \omega$,

$T \vdash \phi\bar{x}$ iff $\phi\bar{x}$ is true.

X is *semi-representable in* T if a semi-representation ϕv of X in T exists; X is *correctly semi-representable in* T if a correct semi-representation for X in T exists.

2.13. DEFINITION. Let T be a theory in the language of arithmetic and let X,Y be disjoint sets of natural numbers. A formula ϕv *dually semi-represents* X,Y *in* T if, for all $x \in \omega$,

$x \in X$ iff $T \vdash \phi\bar{x}$

$x \in Y$ iff $T \vdash \sim\phi\bar{x}$.

2.14. COROLLARY. Let T be a consistent *RE* extension of Q. Let X,Y be disjoint *RE* sets of natural numbers. There is a Σ_1-formula ϕv such that

i. ϕv dually semi-represents X,Y in T

ii. ϕv correctly semi-represents X in T.

This follows directly from 2.10.i-ii by letting $\psi v,\chi v$ be Σ_1-formulae defining

X,Y, respectively, and choosing ϕv as in the statement of Application 2.10.

A few remarks about Corollary 2.14 are in order:

2.15. REMARKS. i. It was 2.14.i that led Shepherdson to derive his fixed point.

ii. If T is Σ_1-sound, the semi-representability of any *RE* set in T is unremarkable: Simply use the Σ_1-formula exhibiting the set as *RE* as the semi-representation. For theories proving false Σ_1-formulae, an *RE*-formula defining a set X can semi-represent a larger set.

iii. The existence of semi-representations of *RE* sets in non-Σ_1-sound *RE* theories and of dual semi-representations of disjoint pairs of such sets in such theories was originally proven by more recursion theoretic means. Such proofs did not yield the correctness of the semi-representation given by the positive Σ_1-formula ϕv, which correctness comes for free using the self-reference. The author showed how to obtain 2.14 by recursion theory, but this recursion theory went beyond the traditional means.

iv. One can only have correctness of both semi-representations of X and Y by ϕv and $\sim\phi v$ if X and Y are complements. In particular, this means that X is recursive. Correct dual semi-representations of recursive sets and their complements are easily constructed and don't require any fancy self-reference or recursion theory.

EXERCISES

1. Let ϕ be as in Application 2.3.

   i. Show: $\mathsf{Q} \vdash \phi$ iff $\mathsf{PRA} \vdash \phi$.

   ii. How should one modify the definition of ϕ to guarantee additionally,

   $\mathsf{Q} \vdash \sim\phi$ iff $\mathsf{PRA} \vdash \sim\phi$?

2. Define a sentence θ to be T-*provably* Δ_1 if there are Σ_1-sentences ψ,χ such that

   $$\mathsf{T} \vdash \theta \leftrightarrow \psi, \qquad \mathsf{T} \vdash \theta \leftrightarrow \sim\chi.$$

   Show: T is Σ_1-sound iff, for any T-provably Δ_1-sentence θ, $\mathsf{T} \vdash \theta$ or $\mathsf{T} \vdash \sim\theta$.

3. Apply Mostowski's Theorem to construct an infinite set $\phi_0,\phi_1,\ldots$ of Σ_1-sentences which are independent over PRA, i.e. for any disjoint sets X,Y of natural numbers, $\mathsf{PRA} + \bigwedge_{i \in X} \phi_i + \bigwedge_{j \in Y} \sim\phi_j$ is consistent. (Compare with Exercise 5.i of Chapter 3, section 3.)

4. Let $T_0, T_1, \ldots$ be an *RE* sequence of consistent extensions of Q. Show: The following are equivalent:

    i. Each T_i is Σ_1-sound

    ii. For any Σ_1-sentences θ_1, θ_2,

    $$\exists i(T_i \vdash \theta_1 \vee \theta_2) \Rightarrow \exists j(T_j \vdash \theta_1 \text{ or } T_j \vdash \theta_2).$$

5. Let $T_0, T_1, \ldots$ be an *RE* sequence of consistent extensions of Q. Let X, Y be disjoint *RE* sets. Show: There is a Σ_1-formula ϕv such that

    i. ϕv dually semi-represents X, Y in each T_i

    ii. ϕv correctly semi-represents X in each T_i.

    Why is ii no longer remarkable? (Hint: Why can one assume T_0 is Σ_1-sound?)

6. Let T_0, T_1 be consistent *RE* extensions of Q.

    i. Suppose $T_0 \subseteq T_1$, $T_1 \vdash \theta$, $T_0 \nvdash \theta$, and $X_0 \subseteq X_1$ are *RE* sets. Show: There is a formula ϕv that semi-represents X_i in T_i.

    ii. Suppose $T_i \vdash \theta_i$, $T_i \nvdash \theta_{1-i}$ for $i = 0, 1$ and X_0, X_1 are *RE* sets. Show: There is a formula ϕv that semi-represents X_i in T_i.

    iii. Suppose $T_0 \vdash \theta$, $T_1 \vdash \sim\theta$ and (X_0, Y_0), (X_1, Y_1) are pairs of disjoint *RE* sets. Show: There is a formula ϕv that dually semi-represents X_i, Y_i in T_i.

    (Hint: i. Choose $\psi_0 v$ to uniformly semi-represent X_0 in T_0 and $T_0 + \sim\theta$, choose $\psi_1 v$ to uniformly semi-represent X_1 in T_0 and T_1, and let ϕv be $(\sim\theta \rightarrow \psi_0 v) \wedge \psi_1 v$. ii-iii are treated similarly.) (Remark: The original proofs of these results used the fixed point of Exercise 4.i of the previous section (along with an appeal to the present Exercise 5 in part i). Lindström and Visser pointed out that the use of the stronger fixed point is unnecessary. The reader might wish, nonetheless, to try his had at deriving these results by appeal to that Exercise. (Hint: Don't use the modal sentence *D* of 4.i in proving part i of this Exercise.))

7. Construct a sentence ϕ such that neither GB + ϕ nor GB + $\sim\phi$ is interpretable in GB, i.e. neither ϕ nor $\sim\phi$ is relatively interpretable in GB.

8. By Exercise 7, it is clear that there are sentences θ_1, θ_2 such that $\theta_1 \vee \theta_2$ is relatively interpretable in GB, but neither θ_1 nor θ_2 is. Show that one can take θ_1, θ_2 to be Σ_1-sentences. (Hint: GB + $\sim Con_{GB}$ is interpetable in GB.)

9. Show that the *RE* set

$\{ \ulcorner\phi\urcorner : \text{GB} + \phi \text{ is interpretable in GB} \}$

does not have an *RE* complement. (Hint: Prove an analogue for $Relint_{GB}$ to Application 2.10.)

10. Construct a sentence ϕ such that neither $\text{ZF} + \phi$ nor $\text{ZF} + \sim\phi$ is provably consistent relative to ZF, i.e.

i. $\text{PRA} \nvdash Con_{ZF} \rightarrow Con_{ZF + \phi}$

ii. $\text{PRA} \nvdash Con_{ZF} \rightarrow Con_{ZF + \sim\phi}$.

11. Find Σ_1-sentences θ_1, θ_2 such that

i. $\text{PRA} \vdash Con_{ZF} \rightarrow Con(\text{ZF} + \theta_1 \vee \theta_2)$

ii. $\text{PRA} \nvdash Con_{ZF} \rightarrow Con(\text{ZF} + \theta_1)$

iii. $\text{PRA} \nvdash Con_{ZF} \rightarrow Con(\text{ZF} + \theta_2)$.

(Remark: Both Exercises 10 and 11 can be proven by appeal to Carlson's Arithmetic Completeness Theorem of Chapter 4, section 4. They can also be proven by appeal to the Ubiquitous Fixed Point Calculation of the present Chapter. If one replaces ZF by an arbitrary consistent *RE* extension T of PRA, one is stuck with this latter Calculation.)

12. (For the expert.) Recalling the Selection Theorem of Chapter 0 (0.6.9), let $\psi v_0 v_1$ be a Σ_1-formula defining the graph of a partial recursive function F, let T be a consistent *RE* extension of PRA, and define $\phi v_0 v_1$ by

$$\text{PRA} \vdash \phi v_0 v_1 \leftrightarrow Sel\left(Pr_T(\ulcorner \sim\phi \dot{v}_0 \dot{v}_1 \urcorner) \vee \psi v_0 v_1\right).$$

Show: For any $x, y \in \omega$,

i. $\text{T} \vdash \phi \overline{x} \overline{y}$ iff $Fx = y$

ii. $\text{T} \vdash \sim\phi \overline{x} \overline{y}$ iff $\exists z \neq y (Fx = z)$

iii. $\text{T} \vdash \phi v_0 v_1 \wedge \phi v_0 v_2 \rightarrow v_1 = v_2$.

(Exercise 12 establishes a fairly strong semi-representability result for partial recursive functions in any consistent *RE* extension of PRA. A slightly weaker result was originally proven by recursion theoretic means by R.W. Ritchie and P.R. Young; W.E. Ritter gave a proof by appeal to a fixed point slightly different from the one given above. The full result can be obtained recursion theoretically. I remark also that the result uniformises to an *RE* sequence of consistent extensions of PRA.)

3. RELATIVISATION TO A PARTIAL TRUTH DEFINITION

Hierarchical generalisations of results are generally routine, but occasionally some new behaviour occurs and the only way to find out if this is the case is to check. Back in Chapter 3, section 3 (3.3.6-3.3.9, to be exact), we saw that there was no trouble in generalising Solovay's First Completeness Theorem to a truth-relativised provability predicate. In the present context, say that of the fixed point

$$p \leftrightarrow .(\Delta{\sim}p \vee \sigma) \preccurlyeq (\nabla p \vee \tau),$$

some difficulties occur. We shall take a brief look at this.

There is another reason for considering these generalisations: A quick application of one such yields a result supplying a non-Σ_1-sentence that cannot, as Guaspari demonstrated, be produced by the self-reference available in the theory R of Chapter 6. We shall look at this.

When dealing with Σ_n- and Π_n-formulae, it is convenient not to keep careful track of how much induction is needed by assuming full induction and working in extensions of PA.

Let us begin with a definition:

3.1. DEFINITION. Let Γ be a class of formulae and let Tr_Γ be a formula in Γ defining truth for sentences of Γ in PA. Let T be a consistent *RE* extension of PA and define

$$Pr_{T,\Gamma}(v):\quad \exists v_0 \exists v_1 v_2 \le v_0 \big(Tr_\Gamma(v_1) \wedge Prov_T(v_2, v_1 \dot{\to} v)\big).$$

The exact form of $Pr_{T,\Gamma}$ is not the most natural. More natural would be

$$\exists v_0 \big(Tr_\Gamma(v_0) \wedge Pr_T(v_0 \dot{\to} v)\big).$$

However, we will need to bound the *proof* that $v_0 \dot{\to} v$ and not v_0 or any witness to it. If Γ is Σ_n, this still leaves an unbounded existential quantifier in Tr_Γ, with the unpleasant consequence that

$$Pr_{T,\Gamma}(\ulcorner{\sim}\phi\urcorner) \preccurlyeq Pr_{T,\Gamma}(\ulcorner\phi\urcorner)$$

will be Σ_{n+1} rather than Σ_n. If Γ is Π_n, this formula will also be Σ_{n+1}, but this won't matter as the following Lemma underlining the (not very apparent) asymmetry between Σ_n and Π_n shows:

3.2. LEMMA. Let T be a consistent *RE* extension of PA, let $\Gamma = \Pi_n$, and let ϕv be a Σ_{n+1}-formula. Then:
$$\mathsf{PA} \vdash \phi v \to Pr_{T,\Gamma}(\ulcorner \phi \dot{v} \urcorner).$$

Proof: Write $\phi v \;=\; \exists v_0 \psi v_0 v$, with $\psi \in \Pi_n$, and observe:

$$\mathsf{PA} \vdash \phi v \to \exists v_0 Tr_\Gamma(\ulcorner \psi \dot{v}_0 \dot{v} \urcorner)$$
$$\vdash \phi v \to \exists v_0 \big(Tr_\Gamma(\ulcorner \psi \dot{v}_0 \dot{v} \urcorner) \wedge Pr_T(\ulcorner \psi \dot{v}_0 \dot{v} \to \exists v_0 \psi v_0 \dot{v} \urcorner)$$
$$\vdash \phi v \to \exists v_1 \big(Tr_\Gamma(v_1) \wedge Pr_T(v_1 \dot{\to} \ulcorner \phi \dot{v} \urcorner)\big)$$
$$\vdash \phi v \to Pr_{T,\Gamma}(\ulcorner \phi \dot{v} \urcorner).$$

QED

Thus, for $\Gamma = \Pi_n$, interpreting $\Box$, Δ, and ∇ all by $Pr_{T,\Gamma}$ yields a valid interpretation of CML. However, for application, we need also to interpret super consistency. What is super consistency in this case? Well, it is the schema,

$$Pr_{T,\Gamma}(\ulcorner \theta \urcorner) \to {\sim} Pr_{T,\Gamma}(\ulcorner {\sim}\theta \urcorner),$$

which clearly reduces to the single instance

$${\sim} Pr_{T,\Gamma}(\ulcorner \overline{0} = \overline{1} \urcorner).$$

("Clearly"-- because the Derivability Conditions hold: We verified this in 3.3.7 for $\Gamma = \Sigma_n$, but the proof given there works for $\Gamma = \Pi_n$ as well.) Now, this last formula can be written out in full:

$${\sim} \exists v_1 \big(Tr_\Gamma(v_1) \wedge Pr_T(v_1 \dot{\to} \ulcorner \overline{0} = \overline{1} \urcorner)\big),$$

which is equivalent to

$$\forall v_0 \big(Pr_T(v_0) \to Tr_{\sim\Gamma}(v_0)\big),$$

i.e. Uniform Σ_n-Reflexion for T, $RFN_{\Sigma_n}(\mathsf{T})$ (cf. 0.6.20 and following). Similarly, an assertion ${\sim} Pr_{T,\Gamma}(\ulcorner \phi \urcorner)$ is equivalent to

$$\forall v_0 \big(Pr_{T + \sim\phi}(v_0) \to Tr_{\sim\Gamma}(v_0)\big),$$

i.e. $RFN_{\Sigma_n}(\mathsf{T} + {\sim}\phi)$, asserting $\mathsf{T} + {\sim}\phi$ to be Σ_n-sound. Thus, we have:

3.3. THEOREM. Let T be a Σ_n-sound *RE* extension of PA. Let $\Gamma = \Pi_n$ and choose ϕ such that

$$\mathsf{PA} \vdash \phi \leftrightarrow .Pr_{T,\Gamma}(\ulcorner {\sim}\phi \urcorner) \preccurlyeq Pr_{T,\Gamma}(\ulcorner \phi \urcorner).$$

Then:
i. $\mathsf{T} + \phi$ is Σ_n-sound
ii. $\mathsf{T} + {\sim}\phi$ is Σ_n-sound.

For, by the Calculation,

$$\mathsf{PA} + RFN_{\Sigma_n}(\mathsf{T}) \vdash {\sim} Pr_{T,\Gamma}(\ulcorner \phi \urcorner)$$

$$\vdash RFN_{\Sigma_n}(\mathsf{T} + {\sim}\phi),$$

and
$$\mathsf{PA} + RFN_{\Sigma_n}(\mathsf{T}) \vdash {\sim}Pr_{T,\Gamma}(\ulcorner{\sim}\phi\urcorner)$$
$$\vdash RFN_{\Sigma_n}(\mathsf{T} + \phi).$$

Now, Theorem 3.3 is not bad. It seems to be what the Fixed Point Calculation 1.7 yields in the present context. But... much more can be said about the sentence ϕ of the Theorem, i.e. in the present context, the modal analysis comes up short. Of course, this suggests a new modal analysis. The result in need of analysis is the following Theorem of Guaspari:

3.4. THEOREM. Let Γ_1, Γ_2 be some Σ_m or Π_n. Let T be a consistent *RE* extension of PA, and let ψ, χ be Σ_1-sentences. Choose ϕ so that

$$\mathsf{PA} \vdash \phi \leftrightarrow .(Pr_{T,\Gamma_1}(\ulcorner{\sim}\phi\urcorner) \vee \psi) \preccurlyeq (Pr_{T,\Gamma_2}(\ulcorner\phi\urcorner) \vee \chi).$$

Then:
i. $\mathsf{T} \vdash \phi$ iff $\psi \preccurlyeq \chi$ is true

ii. $\mathsf{T} \vdash {\sim}\phi$ iff $\chi \prec \psi$ is true

iii. if ψ, χ are false and $\gamma_1 \in \Gamma_1$, $\gamma_2 \in \Gamma_2$ are sentences consistent with T,

then
a. $\mathsf{T} + \gamma_1 \nvdash {\sim}\phi$

b. $\mathsf{T} + \gamma_2 \nvdash \phi$.

In terms of the notion of Γ-conservatism of the last section (Definition 2.7), we can restate iii.a-b as follows:

iii.a'. ϕ is ${\sim}\Gamma_1$-con over T

iii.b'. ${\sim}\phi$ is ${\sim}\Gamma_2$-con over T.

As for the complexity of ϕ, the most interesting case is when $\Gamma_1 = \Sigma_n$ and $\Gamma_2 = \Pi_n$. Then ϕ is Σ_n and, assuming ψ, χ false, it is Π_n-con over T; and ${\sim}\phi$ is a Π_n-sentence that is Σ_n-con over T.

Instead of proving Theorem 3.4, which I leave as an Exercise to the reader, I shall here prove the following simpler, related result also due to Guaspari:

3.5. THEOREM. Let T be a consistent *RE* extension of PA and let

$$\mathsf{PA} \vdash \phi \leftrightarrow .Pr_{T,\Sigma_1}(\ulcorner{\sim}\phi\urcorner) \preccurlyeq Pr_T(\ulcorner\phi\urcorner).$$

Then: i. $\mathsf{T} \nvdash \phi$

ii. ϕ is Π_1-con over T' for any T' satisfying $\mathsf{PA} \subseteq \mathsf{T}' \subseteq \mathsf{T}$.

Proof: i. Suppose $\mathsf{T} \vdash \phi$. Let x be minimum so that $Prov_T(\overline{x}, \ulcorner\phi\urcorner)$ is true and observe

$$PA \vdash \phi \leftrightarrow \exists v \leq \bar{x} \exists v_1 v_2 \leq v\left(Tr_{\Sigma_1}(v_1) \wedge Prov_T(v_2, v_1 \dot{\rightarrow} \ulcorner\sim\phi\urcorner)\right)$$

$$\vdash \phi \leftrightarrow \bigvee Tr_{\Sigma_1}(\ulcorner\psi_i\urcorner)$$

$$\vdash \phi \leftrightarrow \bigvee \psi_i,$$

where the ψ_i's range over those Σ_1-sentences ψ for which $Prov_T(\bar{y}, \ulcorner\psi \rightarrow \sim\phi\urcorner)$ is true for some $y \leq x$. (Such must exist or else $PA \vdash \sim\phi$, whence $T \vdash \sim\phi$, contrary to the consistency of T.) But then we have

$$PA \vdash \bigvee \psi_i, \qquad T \vdash \bigwedge (\psi_i \rightarrow \sim\phi)$$

whence $\quad T \vdash \bigvee \psi_i \rightarrow \sim\phi$

$$\vdash \sim\phi,$$

a contradiction.

ii. Suppose ϕ is not Π_1-con over some $T' \subseteq T$. Then, there is a Π_1-sentence unprovable in T' such that

$$T' + \phi \vdash \pi.$$

In particular,

$$T + \phi \vdash \pi, \qquad PA \nvdash \pi.$$

From the former of these two assertions, we see that

$$T \vdash \sim\pi \rightarrow \sim\phi$$

i.e. $\quad PA \vdash Pr_T(\ulcorner\sim\pi \rightarrow \sim\phi\urcorner)$

$$PA + \sim\pi \vdash Tr_{\Sigma_1}(\ulcorner\sim\pi\urcorner) \wedge Pr_T(\ulcorner\sim\pi \rightarrow \sim\phi\urcorner). \qquad (1)$$

Now, choosing x large enough to exceed $\ulcorner\sim\pi\urcorner$ and a proof of $\sim\pi \rightarrow \sim\phi$ in T, we have

$$PA + \sim\pi \vdash \phi \leftrightarrow \exists v_1 v_2 \leq \bar{x}\left(Tr_{\Sigma_1}(v_1) \wedge Prov_T(v_2, v_1 \dot{\rightarrow} \ulcorner\sim\phi\urcorner) \wedge \right.$$

$$\left. \wedge \; \forall v_3 \leq \bar{x} \sim Prov_T(v_3, \ulcorner\phi\urcorner)\right) \qquad (2)$$

$$\vdash \phi \leftrightarrow \exists v_1 v_2 \leq \bar{x}\left(Tr_{\Sigma 1}(v_1) \wedge Prov_T(v_2, v_1 \dot{\rightarrow} \ulcorner\sim\phi\urcorner)\right) \qquad (3)$$

$$\vdash \phi, \qquad (4)$$

contrary to the assumptions that $PA + \sim\pi \vdash \sim\phi$ and $PA \nvdash \pi$. (Here, (2) follows by the definition of ϕ, (3) by the unprovability of ϕ in T established in part i, and (4) by (1).) QED

The hereditary property ii of the Π_1-conservatism of ϕ appears to be special. It is not known to what extent Theorems 3.4 and 3.5 can uniformise to several (even: two) theories T_i.

Theorem 3.5 gives us what we need to return to modal questions. Throughout this monograph, we have referred to the schema,

$$A \to \Box A,$$

as holding for Σ-sentences A and referred to Demonstrable Σ_1-Completeness in explaining it. However,

$$\phi \to Pr_{PA}(\ulcorner \phi \urcorner)$$

holds for many non-Σ_1-sentences ϕ:

3.6. COROLLARY. There is a sentence ψ such that

$$PA \vdash \psi \to Pr_{PA}(\ulcorner \psi \urcorner),$$

yet ψ is not provably equivalent to any Σ_1-sentence in PA.

This Corollary was originally proven by Clark Kent. The following proof actually yields a refinement and is due to Guaspari:

Proof: Let ϕ be as in Theorem 3.5 for $T = PA + \sim Con_{PA}$ and choose

$$\psi = \sim\phi \wedge Pr_{PA}(\ulcorner \sim\phi \urcorner).$$

It is clear that $PA \vdash \psi \to Pr_{PA}(\ulcorner \psi \urcorner)$. Suppose, however, that

$$PA \vdash \psi \leftrightarrow \sigma \qquad (1)$$

for some Σ_1-sentence σ. Observe,

$$PA \vdash \sigma \to \sim\phi \wedge Pr_{PA}(\ulcorner \sim\phi \urcorner)$$

$$\vdash \phi \to \sim\sigma$$

$\vdash \sim\sigma$, since ϕ is Π_1-con over PA

$\vdash \sim\psi$, by (1)

$$\vdash \phi \vee \sim Pr_{PA}(\ulcorner \sim\phi \urcorner), \qquad (2)$$

by definition of ψ. But then

$$PA + \sim Con_{PA} \vdash Pr_{PA}(\ulcorner \sim\phi \urcorner)$$

$$\vdash \phi,$$

but this contradicts the independence of ϕ over $PA + \sim Con_{PA}$. QED

(Remark: We did not use the fact that ϕ is Π_1-con over $PA + \sim Con_{PA}$, but merely that it is independent over this theory.)

Where does this bring us? Let us express this as a problem about R.

3.7. DEFINITION. Let A be a formula in the language of R. We say that A produces a *non-*Σ_1*-self prover* if

i. $\mathsf{R} \vdash A \rightarrow \Box A$

ii. for every arithmetic interpretation $*$, A^* is *not* equivalent to a Σ_1-sentence.

Non-modally, we have just seen that there are non-Σ_1-self-provers. Modally, we might also expect such to exist, e.g. if A is $\sim\Box B \wedge \Box\sim\Box B$, for some B. Guaspari proved this expectation false:

3.8. THEOREM. There are no non-Σ_1-self-provers in R.

Proof: Let $\underline{K}$ be any linear A-sound model of R^- in which A is false, say

$$K = \{\beta_n < \beta_{n-1} < \ldots < \beta_0\}.$$

If A is nowhere forced, $\Box\sim A$ is forced at β_n and $\underline{K}$ is $\Box\sim A$-sound, whence $(\Box\sim A)^*$ is true, i.e. A^* is refutable in PA, whence equivalent to $\exists v(v \neq v)$, under the interpretation $*$ constructed from $\underline{K}$ in Chapter 6.

If A is forced somewhere, let m be maximum so that $\beta_m \Vdash A$. Observe that A is forced at exactly the nodes at which $\Box^{m+1}f$ is forced, whence $\underline{K}$ is a $\Box(\Box^{m+1}f \leftrightarrow A)$-sound model of $\Box(\Box^{m+1}f \leftrightarrow A)$. But then A^* is provably equivalent to the Σ_1-sentence $(\Box^{m+1}f)^*$. QED

EXERCISES

1. Prove Theorem 3.4.
2. Show arithmetically that the Σ_1-Rosser sentence is not Π_1-con over T.
3. (Guaspari). Theorem 3.8 is a little too weak. Prove that the Theorem still holds if we add finitely many propositional constants c_i and axioms $c_i \leftrightarrow A_i(c_i)$, where p is boxed in $A_i(p)$. (Remark: Guaspari's full result is still a bit more general than this.)

4. ŠVEJDAR'S SELF-REFERENTIAL FORMULAE

There is yet one more type of self-referential statement that demands discussion. It was originally introduced by Petr Hájek, but was raised to the level of high art by his student Vítezslav Švejdar, whence the title of the present section. The interest of this type of self-reference is two-fold: For one thing, in attempting to understand this type of self-reference, Švejdar expanded the modal analysis of section 1 and a further expansion thereof by Hájek. But also, this type of self-reference is based directly on an important property that PA and ZF share, but GB and PRA do not-- namely, reflexiveness: PA and ZF prove the consistencies of each of their respective finite subtheories. In past sections, we have occasionally made PA rather than PRA our base theory simply to get a little extra induction because we were dealing with complex formulae. Now, we switch to PA for a far more fundamental reason: We are dealing with a property that essentially demands full induction.

We shall first concentrate on arithmetic matters and later take a look at the modal logic. Our starting point is the following Theorem of Mostowski:

4.1. THEOREM. Let T be an extension of PA in the language of PA. Then: For any finitely axiomatised subtheory T_0 of T, $T \vdash Con_{T_0}$.

Here, the formula Pr_{T_0} on which Con_{T_0} is based is assumed to be the one given by the method of Chapter 0 (0.5.11-0.5.12) for the trivial description of the non-logical axioms of T_0:

$$NonLogAx_{T_0}(v): \quad v = \ulcorner \phi_0 \urcorner \vee \ \ldots \vee \ v = \ulcorner \phi_{m-1} \urcorner,$$

where $\phi_0,\ldots,\phi_{m-1}$ are the axioms of T_0.

I shall not prove Theorem 4.1. The details of the proof are far removed from the application we wish to make of the result. This application can be partially explained by the observation that a theory T proves a sentence ϕ iff $\sim\phi$ is not consistent with T, iff $\sim\phi$ is not consistent with some finitely axiomatised subtheory of T. Thus, $Pr_T(\ulcorner \phi \urcorner)$ is equivalent to

"for some n, $\sim\phi$ is inconsistent with the first n axioms of T".

Thus, we can use such an n as a witness to the provability of ϕ in place of the witness referred to in Pr_T. Thus, the following:

4.2. DEFINITION. Let T be an *RE* theory with proof predicate Pr_T based on a formula $NonLogAx_T(v_0)$ defining the non-logical axioms of T. By $Con(T \restriction v)$ we mean the formula $\sim Pr_{T^*(v)}(\ulcorner \overline{0} = \overline{1} \urcorner)$ based on the formula

$$NonLogAx_{T^*}(v,v_0): \quad NonLogAx_T(v_0) \vee v_0 \leq v.$$

In words, $Con(T\restriction v)$ asserts the consistency of those axioms of T with code at most v. (Cf. 0.5.11-0.5.12 for details.)

4.3. DEFINITION. For T as in 4.2 and any sentence ϕ, by $Con(T\restriction v + \phi)$ we mean the consistency assertion for the provability predicate based on

$$NonLogAx_T(v) \wedge v_0 \le v. \vee v_0 = \ulcorner\phi\urcorner.$$

4.4. LEMMA. Let T be a consistent *RE* extension of PA. Let ϕ be any sentence. Then: $\quad PA \vdash Pr_T(\ulcorner\phi\urcorner) \leftrightarrow \exists v \sim Con(T\restriction v + \sim\phi).$

I omit the proof.

The formula $\exists v \sim Con(T\restriction v + \sim\phi)$ is Σ_1, but it is not written in the best form as it has two leading unbounded existential quantifiers-- that displayed and the one occurring in $\sim Con$-- whence the formula

$$\exists v \sim Con(T\restriction v + \phi) \preccurlyeq \exists v \sim Con(T\restriction v + \sim\phi)$$

is not Σ_1, but is Σ_2. However, the formula is still of some interest:

4.5. THEOREM. Let T be a consistent *RE* extension of PA in the language of PA and choose ϕ such that

$$PA \vdash \phi \leftrightarrow . \exists v \sim Con(T\restriction v + \phi) \preccurlyeq \exists v \sim Con(T\restriction v + \sim\phi).$$

Then: $\quad T \nvdash \phi, \sim\phi.$

I don't need to prove this because the diligent reader has already done so! Back in Exercise 8 of Chapter 6, section 1, we looked at the following modal system:

4.6. DEFINITION. The modal system Z^- is the system of modal logic in the language of R^- (Definition 6.1.2) with the axiom schema *A5* of that system replaced by

$$A5'. \quad \Box A \to \Box(\sim B \to (\Box A \prec \Box B)).$$

In the Exercise cited, the reader proved

$$Z^- \vdash \Box(p \leftrightarrow .\Box\sim p \preccurlyeq \Box p) \wedge \sim\Box f. \to .\sim\Box p \wedge \sim\Box\sim p.$$

Interpreting $\Box$ by $\exists v \sim Con(T\restriction v + \cdot)$, Theorem 4.5 will follow directly from the modal theorem-- once we've shown that Z^- is sound with respect to the interpretation, which amounts to verifying axiom schema *A5'*:

4.7. LEMMA. *A5'* translates to a provable schema of PA under the interpretation cited.

Proof: We argue informally: Suppose ϕ, ψ are sentences and ϕ is provable, so

$\sim\phi$ is not consistent with some finite subtheory of T. Choose x so that $\sim Con(T\upharpoonright\overline{x} + \sim\phi)$. Assume $\sim\psi$. Applying Theorem 4.1 to $T + \sim\psi$, we see that

$$T + \sim\psi \vdash Con(T\upharpoonright\overline{x} + \sim\psi),$$

whence

$$\sim\psi \rightarrow Con(T\upharpoonright\overline{x} + \sim\psi) \wedge \sim Con(T\upharpoonright\overline{x} + \sim\phi)$$
$$\rightarrow . \exists v \sim Con(T\upharpoonright v + \sim\phi) \prec \exists v \sim Con(T\upharpoonright v + \sim\psi).$$

Formalising this, we get

$$Pr_T(\ulcorner\phi\urcorner) \rightarrow Pr_T(\ulcorner\sim\psi \rightarrow . \exists v \sim Con(T\upharpoonright v + \sim\phi) \prec \exists v \sim Con(T\upharpoonright v + \sim\psi)\urcorner),$$

which, by Lemma 4.4, is equivalent to the formula desired. QED

With this Lemma, Theorem 4.5 follows immediately. Unfortunately, Theorem 4.5 isn't very interesting. One really wants the following conclusions:

i. for all $x \in \omega$, $T \vdash Con(T\upharpoonright\overline{x} + \phi)$

ii. for all $x \in \omega$, $T \vdash Con(T\upharpoonright\overline{x} + \sim\phi)$.

I don't see any way of obtaining these modally using Z^-. We can prove this arithmetically (cf. Exercise 4, below), but, in this monograph, we want a modal treatment. Švejdar accomplishes this-- or, rather, gets what we want i,ii for-- by adding a new infix binary modal operator $\lhd$.

Before describing $\lhd$, let me state what i-ii are good for. We have already remarked that T proves ϕ iff T can prove $\sim\phi$ *not* to be consistent with some finite truncation of T. But what does it mean if T proves ϕ consistent with each such truncation? By Theorem 4.1, this will happen if T proves ϕ; by i-ii, this is not strong enough to imply that T proves ϕ. It turns out that condition i has a couple of nice equivalents. One such can be found in Exercise 1. Another is the following:

4.8. THEOREM. Let T be a consistent *RE* extension of PA in the language of PA, and let ϕ be a sentence. The following are equivalent:

i. for all $x \in \omega$, $T \vdash Con(T\upharpoonright\overline{x} + \phi)$

ii. $T + \phi$ is interpretable in T.

Again, I will not prove Theorem 4.8. Its proof, due to Steven Orey, consists of formalising the completeness theorem for the predicate calculus in PA and lies beyond the scope of this monograph. As in Theorem 4.1, we wish only to cite this result.

Švejdar's method of getting what he wants from our fixed point consists of extending Z^- by adding to it a modal operator $\lhd$, with arithmetic interpretation $(A \lhd B)^*$ being

"$T + A^*$ is interpretable in $T + B^*$".

This formula, let us denote it by $Intp_T(v_0, v_1)$, is Π_2.

The first step is to extend Z^-:

4.9. DEFINITION. Z is the modal system extending Z^- by the addition of the rule of inference,

$$R. \quad A \,/\, \sim B \rightarrow (\Box A \prec \Box B).$$

Note that this rule of inference is just an informal version of $A5'$, whence valid under the present interpretation. (Note too that it yields the usual $R2$: Let B be f.)

The next step is to add the new operator:

4.10. DEFINITION. The modal system ZI is given by extending the language, axiom schemata, and rules of inference of Z to include the new binary modal operator $\lhd$ and adding the following axiom schemata:

$I1.\quad \Box(p \rightarrow q) \rightarrow (q \lhd p)$

$I2.\quad (p \lhd q) \wedge (q \lhd r) \rightarrow (p \lhd r)$

$I3.\quad (p \lhd q) \wedge (p \lhd r) \rightarrow (p \lhd q \vee r)$

$I4.\quad (p \lhd q) \rightarrow .\Box \sim p \rightarrow \Box \sim q$

$I5.\quad (\sim q \lhd r) \rightarrow . \sim p \lhd (r \wedge \sim(\Box p \prec \Box q)).$

4.11. LEMMA. Interpreting $\Box$ as before and $\lhd$ by

$$(A \lhd B)^*: \quad Intp_T(\ulcorner A^* \urcorner, \ulcorner B^* \urcorner),$$

every theorem of ZI is schematically provable in PA.

Proof: The only non-obvious thing is the validity of the schema $I5$. We must show $\quad Intp_T(\ulcorner \sim\psi \urcorner, \ulcorner \chi \urcorner) \rightarrow Intp_T(\ulcorner \sim\phi \urcorner, \ulcorner \chi \wedge (\Box\phi \prec \Box\psi) \urcorner),$

where I write $\Box$ to save space. Let us argue informally: We are assuming $T + \sim\psi$ is interpretable in $T + \chi$, whence

$$T + \chi \vdash Con(T \restriction \overline{x} + \sim\psi) \qquad (1)$$

for all x. Assuming

$$\exists v \sim Con(T \upharpoonright v + \sim\phi) \prec \exists v \sim Con(T \upharpoonright v + \sim\psi), \qquad (2)$$

we have

$$Con(T \upharpoonright \bar{x} + \sim\psi) \rightarrow Con(T \upharpoonright \bar{x} + \sim\phi) \qquad (3)$$

for any ϕ. Thus, (1), (2), and (3) yield

$$T + \chi + (2) \vdash Con(T \upharpoonright \bar{x} + \sim\phi),$$

for all x, i.e. $\sim\phi$ is interpretable in $T + \chi + (2)$, as desired. QED

Now for some modal work:

4.12. THEOREM. $ZI \vdash \Box(p \leftrightarrow (\Box\sim p \preccurlyeq \Box p)) \rightarrow (p \lhd t) \wedge (\sim p \lhd t)$.

Proof: Let A abbreviate $\Box(p \leftrightarrow (\Box\sim p \preccurlyeq \Box p))$. Observe

$$ZI \vdash (p \lhd p) \rightarrow (\sim p \lhd (p \wedge \sim(\Box p \prec \Box\sim p)), \quad \text{by } I5$$

$$\vdash \sim p \lhd (p \wedge \sim(\Box p \prec \Box\sim p)), \quad \text{by } I1.$$

Thus, $\qquad ZI + A \vdash \sim p \lhd p, \qquad (*)$

by $I2$, since

$$ZI + A \vdash \Box(p \rightarrow .\Box\sim p \preccurlyeq \Box p)$$

$$\vdash \Box(p \rightarrow \sim(\Box p \prec \Box\sim p))$$

$$\vdash p \wedge \sim(\Box p \prec \Box\sim p). \lhd p, \quad \text{by } I1.$$

Now apply $I3$ to (*) and the obvious $\sim p \lhd \sim p$ to get

$$ZI + A \vdash (\sim p \lhd p) \wedge (\sim p \lhd \sim p)$$

$$\vdash \sim p \lhd (p \vee \sim p), \quad \text{by } I3$$

$$\vdash \sim p \lhd t, \quad \text{by } I1.$$

Similarly, one shows $ZI + A \vdash p \lhd t$. QED

4.13. COROLLARY. Let ϕ be as in Theorem 4.5. Then: $T + \phi$ and $T + \sim\phi$ are interpretable in T.

4.14. COROLLARY. Let ϕ be as in Theorem 4.5. Then: For all $x \in \omega$,

i. $T \vdash Con(T \upharpoonright \bar{x} + \phi)$

ii. $T \vdash Con(T \upharpoonright \bar{x} + \sim\phi)$.

Let me close with the remark that Švejdar's system ZI can also be used to study other things, e.g. interpretability in GB. Also, we could have obtained 4.14 without reference to interpretability by directly interpreting $(A \lhd B)^*$ as

$$\forall x(T + B^* \vdash Con(T\restriction \overline{x} + A^*)).$$

I hope, however, the interpretative digression was illuminating.

EXERCISES

1. Let T extend PA in the language of PA. Show: ϕ is Π_1-con over T iff $T \vdash Con(T_0 + \phi)$ for all finite $T_0 \subseteq T$.

2. Prove Lemma 4.4.

3. Do Exercise 8 of Chapter 6, section 1.

4. Let T be as usual, let ψ, χ be Σ_1-sentences, and let

$$PA \vdash \phi \leftrightarrow .(\exists v \sim Con(T\restriction v + \phi) \vee \psi) \preceq (\exists v \sim Con(T\restriction v + \sim\phi) \vee \chi).$$

Write $\psi = \exists v \psi' v$, $\chi = \exists v \chi' v$. Choose x minimum (assuming such exists) such that $\psi' \overline{x}$ is true, $\chi' \overline{x}$ is true, $PA \vdash \sim Con(T\restriction \overline{x} + \phi)$, or $PA \vdash \sim Con(T\restriction \overline{x} + \sim\phi)$. Observe that

$$PA \vdash \phi \leftrightarrow \exists v \le \overline{x}((\sim Con(T\restriction v + \phi) \vee \psi' v) \wedge$$
$$\wedge \forall v_1 < v(Con(T\restriction v_1 + \sim\phi) \wedge \sim \chi' v_1)).$$

i. Show: $T \vdash \phi \Rightarrow \psi \preceq \chi$ is true

ii. Show: $T \vdash \sim\phi \Rightarrow \chi \prec \psi$ is true

iii. Show: $\psi \preceq \chi$ is true $\Rightarrow T \vdash \phi$

iv. Show: $\chi \prec \psi$ is true $\Rightarrow T \vdash \sim\phi$

v. Assume ψ, χ are false. Show: For all $x \in \omega$,

a. $T \vdash Con(T\restriction \overline{x} + \phi)$

b. $T \vdash Con(T\restriction \overline{x} + \sim\phi)$.

(Hints: iii. $T + \sim\phi \vdash \forall v < \overline{x} Con(T\restriction v + \sim\phi)$. Show: $T + \sim\phi \vdash \phi$. iv. Ditto. v. By 4.1, $T \vdash Con(T\restriction \overline{x} + \phi) \vee Con(T\restriction \overline{x} + \sim\phi)$. Show: $T + \sim Con(T\restriction \overline{x} + \phi) \vdash \phi$. What does this prove?)

5. Show: No unprovable Π_1-sentence is interpretable in PA. (Hint: Exercise 1.)

6. Show directly that ZI is valid when one interprets $(A \lhd B)^*$ by

$$\forall x(T + B^* \vdash Con(T\restriction \overline{x} + A^*)).$$

Bibliography

S.N. Artyomov

1980 Arithmetically complete modal theories, (Russian), Semiotics and Information Science, no. 14, (Russian), pp. 115-133, Akad. Nauk SSSR, Vsesojuz. Inst. Naucn. i. Tehn. Informacii, Moscow.

A. Avron

1984 On modal systems having arithmetical interpretations, J. Symbolic Logic 49 pp. 935-942.

C. Bernardi

1975 The fixed-point theorem for diagonalizable algebras, Studia Logica 34, pp. 239-251.

1976 The uniqueness of the fixed-point in every diagonalizable algebra, Studia Logica 35, pp. 335-343.

G. Boolos

1979 The Unprovability of Consistency, Cambridge University Press.

1982 Extremely undecidable sentences, J. Symbolic Logic 47, pp. 191-196.

R. Bull & K. Segerberg

1984 Basic modal logic, in: D. Gabbay & F. Guenthner, eds., Handbook of Philosophical Logic, II, Reidel, Dordrecht.

T. Carlson

A Modal logics with several operators and provability interpretations, to appear.

H. Friedman

1975 The disjunction property implies the numerical existence property, Proc. Nat. Acad. Sci. 62, no. 8, pp. 2877-2878.

K. Gödel

1931 Über formal unentscheidbare Sätze der Principia Mathematica und verwandter Systeme I, Monatsh. Math. Phys. 38, pp. 173-198.

D. Guaspari

1979 Partially conservative extensions of arithmetic, Trans. AMS 254, pp. 47-68.

D. Guaspari (continued)

1983 Sentences implying their own provability, J. Symbolic Logic 48, pp. 777-789.

D. Guaspari & R.M. Solovay

1979 Rosser sentences, Annals Math. Logic 16, pp. 81-99.

P. Hájek

1971/1972 On interpretability in set theories I, II, Comment. Math. Univ. Carol. 12, pp. 73-79; 13, pp. 445-455.

1981 On interpretability in theories containing arithmetic II, Comment. Math. Univ. Carol. 22, pp. 667-688.

D. Hobby

A Finite fixed point algebras are subdiagonalisable, to appear.

D. Jensen & A. Ehrenfeucht

1976 Some problem in elementary arithmetics, Fund. Math. 92, pp. 223-245.

C. Kent

1973 The relation of A to Prov ⌜A⌝ in the Lindenbaum sentence algebra, J. Symbolic Logic 38, pp. 295-298.

G. Kreisel & G. Takeuti

1974 Formally self-referential propositions in cut-free classical analysis and related systems, Diss. Math. 118, pp. 1-50.

P. Lindström

1979 Some results on interpretability, in: F.V. Jensen, B.H. Mayoh, & K.K. Møller, Proceedings of the 5th Scandinavian Logic Symposium, Aalborg Univ. Press.

M.H. Löb

1955 Solution of a problem of Leon Henkin, J. Symbolic Logic 20, pp. 115-118.

A. Macintyre & H. Simmons

1973 Gödel's diagonalization technique and related properties of theories, Colloq. Math. 28, pp. 165-180.

F. Montagna

1979 On the diagonalizable algebra of Peano arithmetic, Bolletino U.M.I. (5) 16-B, pp. 795-812.

1984 The predicate modal logic of provability, Notre Dame J. Formal Logic 25, pp. 179-189.

R. Montague

1963 Syntactical treatments of modality, with corollaries on reflexion principles and finite axiomatizability, Acta Phil. Fennica 16, pp. 153-167.

A. Mostowski

1961 A generalization of the incompleteness theorem, Fund. Math. 49, pp. 205-232.

J. B. Rosser

1936 Extensions of some theorems of Gödel and Church, J. Symbolic Logic, pp. 87-91.

G. Sambin

1974 Un estensione del theorema di Löb, Rendiconti del Seminario Matematico dell'Universita di Padova 52, pp. 193-199.

1976 An effective fixed-point theorem in intuitionistic diagonalizable algebras, Studia Logica 35, pp. 345-361.

J. Shepherdson

1960 Representability of recursively enumerable sets in formal theories, Arch. f. math. Logik 5, pp. 119-127.

C. Smoryński

1978 Beth's theorem and self-referential sentences, in: A. Macintyre, L. Pacholski, and J. Paris, eds., Logic Colloquium '77, North-Holland, Amsterdam.

1979 Calculating self-referential statements I: explicit calculations, Studia Logica 38, pp. 17-36.

1980 Calculating self-referential statements, Fund. Math. 109, pp. 189-210.

1981A Calculating self-referential statements: Guaspari sentences of the first kind, J. Symbolic Logic 46, pp. 329-344.

1981B Fifty years of self-reference in arithmetic, Notre Dame J. Formal Logic 22, pp. 357- 374.

1982A Fixed point algebras, Bull. AMS (NS) 6, pp. 317-356; corrigendum: Bull. AMS (NS) 8 (1983), pp. 407-408.

1982B Commutativity and self-reference, Notre Dame J. Formal Logic 23, pp. 443-452.

1984 Modal logic and self-reference, in: D. Gabbay & F. Guenthner, eds., Handbook of Philosophical Logic II, Reidel, Dordrecht.

A Quantified modal logic and self-reference, to appear.

R.M. Solovay

1976 Provability interpretations of modal logic, Israel J. Math. 25, pp. 287-304.

A Infinite fixed point algebras, to appear.

V. Švejdar

1983 Modal analysis of generalized Rosser sentences, J. Symbolic Logic 48, pp. 986-999.

A. Tarski, A. Mostowski, & R. Robinson

1953 Undecidable Theories, North-Holland, Amsterdam.

A. Visser

1981 Aspects of diagonalization and provability, dissertation, Utrecht.

A The provability logics of recursively enumerable theories extending Peano arithmetic at arbitrary theories extending Peano arithmetic, to appear.